pytest 框架与自动化测试应用

房荔枝　梁丽丽 ◎ 编著

清华大学出版社
北京

内 容 简 介

pytest 是 Python 的第三方测试框架，可以实现软件测试各层次自动化。本书系统讲述 pytest 的主要技术及其在各层次自动化测试的应用方法。

第 1 章介绍 pytest 框架。第 2 章讲解框架运行管理及如何对测试用例、断言进行管理，以及运行管理中所包括的各种形式和层次的执行，展示框架的灵活性和全面性。第 3 章详细介绍框架的核心技术 fixture、fixture 使用依赖注入技术完成方法、类、文件级、会话级关联、数据关联和共享，可完美解决各种情况耦合和共享。第 4 章结合测试中最重要的数据驱动技术所产生的参数化技术，并配合 fixture 技术实现一组代码测试多组数据的数据与代码分离技术。第 5 章系统讲解各种实用插件。第 6~10 章是实践，从单元自动化测试、接口自动化测试、Web 自动化测试及 App 自动化测试计划开始讲解设计、实现及执行的全流程。第 11 章介绍 BDD，让非技术人员也能参与测试的全过程。第 12 章介绍 pytest 的一些相关配置。

本书可作为从事软件自动化测试的技术人员的参考书，也可作为高等院校软件测试专业师生的参考书。

本书封面贴有清华大学出版社防伪标签，无标签者不得销售。
版权所有，侵权必究。举报：010-62782989，beiqinquan@tup.tsinghua.edu.cn。

图书在版编目(CIP)数据

pytest 框架与自动化测试应用/房荔枝，梁丽丽编著. —北京：清华大学出版社，2021.9(2022.11 重印)
ISBN 978-7-302-58715-6

Ⅰ. ①p… Ⅱ. ①房… ②梁… Ⅲ. ①软件工具－程序设计 Ⅳ. ①TP311.561

中国版本图书馆 CIP 数据核字(2021)第 141321 号

责任编辑：赵佳霓
封面设计：郭 媛
责任校对：时翠兰
责任印制：宋 林

出版发行：清华大学出版社
网　　址：http://www.tup.com.cn，http://www.wqbook.com
地　　址：北京清华大学学研大厦 A 座　　邮　编：100084
社 总 机：010-83470000　　　　　　　　邮　购：010-62786544
投稿与读者服务：010-62776969，c-service@tup.tsinghua.edu.cn
质量反馈：010-62772015，zhiliang@tup.tsinghua.edu.cn
课件下载：http://www.tup.com.cn,010-83470236

印 装 者：三河市龙大印装有限公司
经　　销：全国新华书店
开　　本：186mm×240mm　　印　张：26.5　　字　数：594 千字
版　　次：2021 年 11 月第 1 版　　　　　　印　次：2022 年 11 月第 2 次印刷
印　　数：1501～2500
定　　价：100.00 元

产品编号：089124-01

前言
PREFACE

在整个开发过程中,由于市场变化频繁导致需求变化频繁,公司层面需要尽快知道做出的各种决策的结果和反馈,也就是希望在整个产品开发的生命周期中各个环节都能快速跟上这种变化,迅速做出正确响应。

以前环境的自动化和测试的自动化是整个过程的难点,现在有了 Docker 和 k8s,可以快速实现环境的部署。基于质量的原因,编写测试自动化脚本时需分层(单元层测试、服务层测试、UI 层测试)实现,能否在不同层使用同一框架测试,决定自动化脚本的开发效率。pytest 框架可以实现各分层和各质量特性的测试。

Python 语言在数据采集、人工智能等技术中被广泛使用,因此基于这些技术实现的产品中的测试使用 Python 语言的占比会越来越大。本书介绍的 pytest 框架是基于 Python 的,满足上述编程语言和提升开发效率的要求。相信 pytest 框架在未来一定会火起来,无论是开发、测试、运维人员,还是运用 DevOps 模型的公司或是希望降低开发成本的公司,都必然会使用 pytest 这个框架。

本书主要介绍 pytest 框架的核心知识,从框架技术开始,对每项实用技术进行详细阐述,并有练习和示例,之后系统讲解实用插件,尤其是 Allure 框架。Allure 框架可以让测试报告内容更丰富。本书的重点是教读者在各层次测试自动化中如何运用 pytest 框架进行测试自动化设计。实践章节的思路和技术是编者多年测试经验的缩影。本书有细节、有深度、有实践,按技术成长路线为读者量身定做案例,帮助读者快速掌握自动化测试。通过学习本书,读者可以设计并开发自动化测试框架和脚本。本书也可以作为工具书,从三级目录查找某些功能的实现方法。

建议读者先将前几章重点技术掌握熟练,再通过后几章的实践来理解思路并自己编写代码,在示例中实践。在工作中可首先进行接口测试自动化,再结合 Jenkins 部署执行,实现公司层面持续集成(CI)的落地,以及根据工作需要进行 UI 层测试自动化实践及持续部署交付(CD)的落地,最后可加入单元层面测试自动化,达成整个 DevOps 的落地。

本书所涉及的非测试专业技术,例如接口层测试所涉及的网络方面的知识、Web 测试中所涉及的前端技术、App测试中所涉及的Android相关技术知识等，读者可自行学习。

由于编者水平有限，书中难免存在不妥之处，请读者见谅，并提宝贵意见。最后感谢在本书写作过程中帮助我们的每个人。

编　者

2021 年 8 月

本书源代码下载

目录
CONTENTS

第 1 章　pytest 框架介绍 ··· 1

1.1　pytest 框架引入 ··· 1
1.1.1　测试框架能为我们解决什么问题 ··· 1
1.1.2　测试框架的分类 ··· 2
1.1.3　什么是 pytest ··· 2
1.2　技术前提要求 ··· 3
1.2.1　技术前提 ··· 3
1.2.2　适合人群 ··· 3
1.3　环境准备及资料准备 ··· 3
1.3.1　Python 的环境验证 ··· 3
1.3.2　安装 pytest ··· 4
1.4　pytest 初体验 ··· 6
1.4.1　在终端建立测试方法及执行过程 ··· 6
1.4.2　在 PyCharm 建立测试方法及执行过程 ··· 7
1.4.3　pytest 的框架结构 ··· 9
1.4.4　在 PyCharm 中配置运行工具 ··· 15
1.4.5　右击以 pytest 方式执行代码 ··· 17
1.4.6　去掉 main 方法执行测试 ··· 17
1.4.7　PyCharm 中执行某个测试方法 ··· 18
1.4.8　运行窗口的工具栏含义 ··· 19
1.5　执行的查找原则和测试类及测试方法的命名 ··· 20
1.6　本章小结 ··· 21

第 2 章　pytest 的测试用例管理及运行管理 ··· 22

2.1　测试用例的命名管理 ··· 22
2.2　用例执行顺序 ··· 22
2.3　测试用例的断言管理 ··· 23

- 2.3.1 什么是断言 …… 23
- 2.3.2 断言的时机 …… 23
- 2.3.3 断言的分类与使用 …… 24
- 2.3.4 触发一个指定异常的断言 …… 31
- 2.3.5 为失败断言添加自定义的说明 …… 33
- 2.3.6 Assert 各种类型断言 …… 36
- 2.4 测试用例的运行管理 …… 37
  - 2.4.1 获取帮助信息 …… 38
  - 2.4.2 最常用运行测试用例方式 …… 39
  - 2.4.3 通过 python -m pytest 调用 pytest …… 46
  - 2.4.4 在 Python 代码中调用 pytest …… 47
  - 2.4.5 pytest 执行结束时返回的状态码 …… 52
  - 2.4.6 输出代码中的控制台信息 …… 52
  - 2.4.7 显示详细信息 …… 52
  - 2.4.8 不显示详细信息 …… 53
  - 2.4.9 显示简单总结结果 …… 53
  - 2.4.10 执行指定的测试用例 …… 55
  - 2.4.11 执行指定目录下所有的测试用例 …… 56
  - 2.4.12 -k 参数执行包含特定关键字的测试用例 …… 56
  - 2.4.13 执行指定 nodeid 的测试用例 …… 58
  - 2.4.14 -m 参数执行指定标记的用例 …… 60
  - 2.4.15 执行指定包中的测试用例 …… 63
  - 2.4.16 修改回溯信息的输出模式 …… 65
- 2.5 运行的失败管理 …… 67
  - 2.5.1 最多允许失败的测试用例数 …… 67
  - 2.5.2 失败运行管理的原理 …… 67
- 2.6 跳过 skip 测试用例的执行 …… 76
  - 2.6.1 @pytest.mark.skip 装饰器 …… 77
  - 2.6.2 pytest.skip 方法 …… 77
  - 2.6.3 @pytest.mark.skipif 装饰器 …… 81
  - 2.6.4 pytest.importorskip 方法 …… 83
  - 2.6.5 跳过测试类 …… 85
  - 2.6.6 跳过测试模块 …… 86
  - 2.6.7 跳过指定文件或目录 …… 87
  - 2.6.8 各种跳过小结 …… 90
- 2.7 标记用例为预期失败 …… 90

2.7.1　@pytest.mark.xfail 标记用例 ··· 91
　　2.7.2　使用 pytest.xfail 标记用例 ·· 98
　　2.7.3　xfail 标记如何失效 ··· 100
2.8　中断调试及错误处理 ··· 102
　　2.8.1　失败时加载 PDB 环境 ··· 102
　　2.8.2　开始执行时就加载 PDB 环境 ·· 104
　　2.8.3　设置断点 ·· 104
　　2.8.4　使用内置的中断函数 ··· 106
　　2.8.5　错误句柄 ·· 106
2.9　结果分析及报告 ·· 108
　　2.9.1　分析测试执行时间 ·· 108
　　2.9.2　创建及定制 JUnitXML 格式的测试报告 ······································ 108
2.10　不稳定测试用例处理 ··· 112
　　2.10.1　为什么不稳定测试是个问题 ·· 112
　　2.10.2　潜在的根本原因是什么 ·· 112
　　2.10.3　pytest 为我们提供的解决策略 ·· 113
　　2.10.4　pytest_CURRENT_TEST ··· 113
　　2.10.5　可以重新运行的插件 ··· 113
　　2.10.6　测试人员采用的解决策略 ··· 113
2.11　本章小结 ·· 114

第 3 章　pytest 中最闪亮的 fixture 功能 ··· 115

3.1　fixture 介绍 ·· 115
3.2　fixture 目标 ·· 115
3.3　fixture 基本的依赖注入功能 ··· 116
3.4　fixture 应用在初始化设置 ·· 117
3.5　fixture 应用在配置销毁 ··· 119
　　3.5.1　使用 yield 代替 return ·· 119
　　3.5.2　使用 with 写法 ··· 120
　　3.5.3　使用 addfinalizer 方法 ·· 120
　　3.5.4　yield 与 addfinalizer 的区别 ··· 121
3.6　fixture 方法源码详细讲解 ·· 123
3.7　不同层级 scope 使用 fixture 实例 ·· 125
　　3.7.1　模块(module)级别使用 fixture 实例 ·· 125
　　3.7.2　类(class)级别使用 fixture 实例 ·· 127
　　3.7.3　会话(session)级别使用 fixture 与 conftest.py 配合 ·························· 127

3.7.4 session 级别实例 ································ 129
3.8 使用 params 传递不同数据 ································ 132
 3.8.1 测试方法使用两个简单测试数据 ································ 132
 3.8.2 二(多)个测试方法共用两个简单测试数据 ································ 133
 3.8.3 有效测试数据与预期失败 xfail 的测试数据 ································ 133
 3.8.4 params 与 ids 的应用 ································ 135
 3.8.5 params 综合实例 ································ 137
3.9 自动调用 fixture ································ 138
 3.9.1 使用 fixture 中参数 autouse＝True 实现 ································ 138
 3.9.2 使用@pytest.mark.usefixtures ································ 140
 3.9.3 数据库自动应用的实例 ································ 141
3.10 第三方插件通过文件夹共享测试数据 ································ 142
3.11 fixture 的并列与嵌套调用 ································ 142
 3.11.1 并列使用 fixture ································ 142
 3.11.2 嵌套调用 fixture ································ 143
 3.11.3 多个 fixture 的实例化顺序 ································ 144
 3.11.4 fixture 返回工厂函数 ································ 146
 3.11.5 高效地利用 fixture 实例 ································ 147
3.12 在不同的层级上重写 fixture ································ 149
 3.12.1 在文件夹(conftest.py)层级重写 fixture ································ 149
 3.12.2 在模块层级重写 fixture ································ 151
 3.12.3 在用例参数中重写 fixture ································ 152
 3.12.4 参数化的 fixture 可重写非参数化的 fixture,反之亦然 ································ 152
3.13 本章小结 ································ 154

第 4 章 pytest 的数据驱动和参数传递 ································ 155

4.1 参数化介绍 ································ 155
4.2 参数化的应用 ································ 155
 4.2.1 单一参数化应用 ································ 155
 4.2.2 多参数应用 ································ 156
 4.2.3 多个参数化 ································ 158
 4.2.4 参数化与 fixture 的结合 ································ 158
 4.2.5 pytestmark 实现参数化 ································ 158
4.3 parametrize 源码详细讲解 ································ 159
4.4 argnames 参数 ································ 160
 4.4.1 argnames 与测试方法中的参数关系 ································ 160

- 4.4.2 argnames 调用覆盖同名的 fixture ············· 161
- 4.5 argvalues 参数 ············· 162
  - 4.5.1 argvalues 来源于 Excel 文件 ············· 162
  - 4.5.2 使用 pytest.param 为 argvalues 赋值 ············· 163
- 4.6 indirect 参数 ············· 166
- 4.7 ids 参数 ············· 167
  - 4.7.1 ids 的长度 ············· 167
  - 4.7.2 ids 相同 ············· 168
  - 4.7.3 ids 中使用中文 ············· 168
  - 4.7.4 通过函数生成 ids ············· 169
  - 4.7.5 ids 的覆盖 ············· 170
  - 4.7.6 ids 的作用 ············· 171
- 4.8 scope 参数 ············· 171
  - 4.8.1 module 级别 ············· 172
  - 4.8.2 未指定 scope ············· 172
- 4.9 pytest_generate_tests 钩子方法 ············· 173
- 4.10 本章小结 ············· 175

第 5 章 pytest 的相关插件及插件管理 ············· 176

- 5.1 pytest 的插件安装 ············· 176
- 5.2 常见插件介绍 ············· 176
- 5.3 常用插件的使用 ············· 177
  - 5.3.1 pytest-assume 断言报错后依然执行 ············· 177
  - 5.3.2 pytest-cov 测试覆盖率 ············· 182
  - 5.3.3 pytest-freezegun 冰冻时间 ············· 187
  - 5.3.4 pytest-flakes 静态代码检查 ············· 188
  - 5.3.5 pytest-html 生成 HTML 报告 ············· 189
  - 5.3.6 pytest-httpserver 模拟 HTTP 服务 ············· 189
  - 5.3.7 pytest-instafail 用于用例失败时立刻显示错误信息 ············· 192
  - 5.3.8 pytest-mock 模拟未实现的部分 ············· 192
  - 5.3.9 pytest-ordering 调整执行顺序 ············· 196
  - 5.3.10 pytest-pep8 自动检测代码规范 ············· 197
  - 5.3.11 pytest-picked 运行未提交 git 的用例 ············· 198
  - 5.3.12 pytest-rerunfailures 失败重试 ············· 203
  - 5.3.13 pytest-repeat 重复运行测试 ············· 203
  - 5.3.14 pytest-random-order 随机顺序执行 ············· 205

5.3.15　pytest-sugar 显示彩色进度条 ……………………………………………… 209
　　　5.3.16　pytest-selenium 浏览器兼容性测试 …………………………………… 210
　　　5.3.17　pytest-timeout 设置超时时间 …………………………………………… 211
　　　5.3.18　pytest-xdist 测试并发执行 ……………………………………………… 211
　5.4　插件管理 ……………………………………………………………………………… 212
　　　5.4.1　在测试模块或 conftest 文件中加载插件 ………………………………… 212
　　　5.4.2　找出哪些插件处于活动状态 ……………………………………………… 212
　　　5.4.3　通过名称停用/注销插件 …………………………………………………… 212
　5.5　本章小结 ……………………………………………………………………………… 213

第 6 章　与 Allure 框架结合定制测试报告 …………………………………………… 214

　6.1　Allure 框架介绍 ……………………………………………………………………… 214
　6.2　Allure 如何生成测试报告 …………………………………………………………… 214
　6.3　Allure 报告组成 ……………………………………………………………………… 214
　　　6.3.1　总览 …………………………………………………………………………… 215
　　　6.3.2　类别 …………………………………………………………………………… 215
　　　6.3.3　测试套件 ……………………………………………………………………… 215
　　　6.3.4　功能 …………………………………………………………………………… 216
　　　6.3.5　图形 …………………………………………………………………………… 217
　　　6.3.6　时间轴 ………………………………………………………………………… 217
　　　6.3.7　包 ……………………………………………………………………………… 218
　6.4　Allure 的初体验 ……………………………………………………………………… 218
　　　6.4.1　Allure 在 Windows 系统下安装 …………………………………………… 218
　　　6.4.2　Allure 在 Linux 环境下安装 ………………………………………………… 219
　　　6.4.3　Allure 在 Mac OS 系统下安装 ……………………………………………… 220
　　　6.4.4　Allure 的简单用法 …………………………………………………………… 221
　　　6.4.5　Allure 的帮助说明 …………………………………………………………… 222
　6.5　定制测试报告 ………………………………………………………………………… 224
　　　6.5.1　定制详细的步骤说明 ………………………………………………………… 224
　　　6.5.2　不同类型的附件补充测试说明 ……………………………………………… 228
　　　6.5.3　定制各种类型内容描述 ……………………………………………………… 230
　　　6.5.4　定制测试标题 ………………………………………………………………… 233
　　　6.5.5　各种链接 ……………………………………………………………………… 234
　　　6.5.6　自定义各种标签 ……………………………………………………………… 236
　　　6.5.7　严重性标记 …………………………………………………………………… 238
　　　6.5.8　重试信息展示 ………………………………………………………………… 241
　6.6　本章小结 ……………………………………………………………………………… 242

第 7 章　单元自动化测试实践 243

- 7.1　什么是单元测试 243
- 7.2　pytest 测试框架是单元测试的框架 243
- 7.3　单元测试与质量 243
- 7.4　单元测试一个函数 243
- 7.5　单元测试一个类 246
  - 7.5.1　类的说明 246
  - 7.5.2　开发的调用 247
  - 7.5.3　类持续开发：功能的增加及修改 248
  - 7.5.4　类的单元测试 248
- 7.6　本章小结 251

第 8 章　API 自动化测试实践 252

- 8.1　测试微信公众号接口 252
  - 8.1.1　熟悉接口文档以便获取信息 252
  - 8.1.2　接口测试用例设计 257
- 8.2　执行测试 257
  - 8.2.1　使用 get()、post() 方法发送请求，返回响应 258
  - 8.2.2　使用 conftest 共享数据 259
  - 8.2.3　读取 yaml 数据文件进行 parametrize 260
  - 8.2.4　关联接口数据传递及更新删除接口测试 260
  - 8.2.5　fixture 的依赖接口需要测试，也需要参数化 262
- 8.3　使用 Allure 定制报告 267
- 8.4　使用 pytest 进行各种执行 269
- 8.5　本章小结 269

第 9 章　Web 自动化测试持续集成实践 270

- 9.1　Web 自动化测试及持续集成源起 270
- 9.2　被测试系统的安装和介绍 270
  - 9.2.1　人力资源管理系统安装 270
  - 9.2.2　人力资源管理系统介绍 272
- 9.3　Web 项目自动化原理及 Web 测试框架 273
  - 9.3.1　自动化测试要达到的目标和涉及的技术 273
  - 9.3.2　Web 自动化测试框架 Selenium 介绍 273
  - 9.3.3　Selenium 框架技术简述 274

9.4 整合 Web 自动化测试框架 ······ 274
9.4.1 自动化测试准备 ······ 274
9.4.2 创建工程目录 ······ 278
9.4.3 页面元素定位 ······ 278
9.4.4 页面元素操作 ······ 281
9.4.5 提高代码的复用性和灵活性——封装 ······ 285
9.4.6 编写测试用例 ······ 286
9.4.7 测试执行 ······ 289
9.4.8 生成 Allure 报告 ······ 290
9.5 Web 自动化测试本地环境持续集成 ······ 291
9.5.1 Jenkins 2 实现自动化执行测试及持续集成流程 ······ 291
9.5.2 使用自由风格配置 Python 自动化测试 ······ 291
9.5.3 使用 pipeline 配置 Python 自动化测试 ······ 296
9.5.4 使用 BlueOcean 配置 Python 自动化测试 ······ 300
9.6 本章小结 ······ 305

第 10 章 App 自动化测试项目实践 ······ 306

10.1 App 自动化测试框架选择 ······ 306
10.2 App 自动化测试环境的搭建 ······ 306
10.2.1 安装和验证 Java JDK——Windows 系统 ······ 306
10.2.2 安装和验证 Node.js ······ 309
10.2.3 安装 Android SDK ······ 309
10.2.4 安装模拟器或连接真机 ······ 311
10.2.5 安装 appium-desktop ······ 311
10.2.6 安装 appium-client ······ 311
10.2.7 appium-doctor 环境检查 ······ 314
10.3 使用 pytest 和 Allure 建立 App 自动化混合框架 ······ 315
10.3.1 安装所需要的包和插件 ······ 315
10.3.2 建立目录结构 ······ 315
10.3.3 连接 App 的配置及启动 App ······ 315
10.3.4 使用各种工具进行元素定位 ······ 316
10.3.5 使用 PO 方式建立元素定位 locators 类 ······ 319
10.3.6 使用 PO 方式建立元素操作方法基类 ······ 320
10.3.7 使用 PO 方式建立每个页面或功能的元素操作方法类 ······ 321
10.3.8 结合 pytest 的特性建立公共数据共享文件 conftest.py ······ 322
10.3.9 使用 PO 的方式建立测试类 ······ 323

    10.3.10 使用 yaml 文件及 pytest 中的 parametrize 作为数据驱动程序……… 324
    10.3.11 使用 Allure 标签定制报告 ……………………………………… 324
    10.3.12 封装的一些公共的方法 …………………………………………… 325
    10.3.13 在文件中读取配置文件数据 ……………………………………… 326
    10.3.14 在测试用例中添加 log 日志 ……………………………………… 327
    10.3.15 组织测试用例(添加运行标记) …………………………………… 328
    10.3.16 实现持续集成——在 Jenkins 运行测试代码 ……………………… 329
  附:本次运行的部分 appium 日志翻译 ……………………………………………… 333

第 11 章 行为驱动开发(BDD)实现自动化测试 …………………………………… 335

  11.1 什么是 BDD ……………………………………………………………………… 335
  11.2 BDD 开发过程 …………………………………………………………………… 335
  11.3 BDD 的功能和作用 ……………………………………………………………… 336
  11.4 中国 BDD 现状 …………………………………………………………………… 337
  11.5 pytest-bdd 实现 BDD 开发 ……………………………………………………… 337
    11.5.1 pytest-bdd 安装 ……………………………………………………… 337
    11.5.2 pytest-bdd 的项目结构 ……………………………………………… 338
    11.5.3 BDD 的标准语法 ……………………………………………………… 338
    11.5.4 BDD 实现步骤 ………………………………………………………… 338
    11.5.5 BDD 单元测试实践:添加功能(单一数据) ………………………… 340
    11.5.6 BDD 单元测试实践:添加和删除功能(数据通过参数传递) …… 343
    11.5.7 BDD 单元测试实践:数据参数化 …………………………………… 345
    11.5.8 BDD 接口测试实践:requests 和 pytest-bdd 实现 bing 的
           搜索接口 ……………………………………………………………… 346
    11.5.9 BDD UI 自动化测试实践:selenium 和 pytest-bdd 实现
           搜索功能 ……………………………………………………………… 348
    11.5.10 BDD UI 自动化测试实践:selenium 和 pytest-bdd 实现
           搜索功能参数化 ……………………………………………………… 349
  11.6 本章小结 ………………………………………………………………………… 352

第 12 章 pytest.ini 配置及其他配置 ………………………………………………… 353

  12.1 pytest 中的各种配置 …………………………………………………………… 353
    12.1.1 @pytest.marker 标记用例 …………………………………………… 356
    12.1.2 添加测试用例路径 …………………………………………………… 357

- 12.1.3 指定 pytest 忽略哪些搜索目录 …… 357
- 12.1.4 usefixtures 的默认配置 …… 358
- 12.1.5 修改测试用例的搜索匹配规则 …… 358
- 12.1.6 ids 中解决中文显示乱码问题 …… 360
- 12.1.7 console_output_style 输出样式配置 …… 360
- 12.1.8 xfail_strict 不是预期的失败显示 FAILED …… 361
- 12.1.9 cache_dir 缓存目录设置 …… 361
- 12.1.10 filterwarnings 警告过滤 …… 363
- 12.1.11 log 相关配置 …… 363
- 12.1.12 添加 pytest 执行默认参数选项 …… 364
- 12.1.13 minversion 的设置及限制 …… 366
- 12.1.14 required_plugins 需要的插件 …… 366

12.2 警告相关配置 …… 367
- 12.2.1 警告信息的默认捕获行为 …… 367
- 12.2.2 命令行配置警告是否捕获 …… 368
- 12.2.3 将警告转换成异常失败 …… 369
- 12.2.4 通过 pytest.ini 设置 filterwarnings 实现 …… 369
- 12.2.5 使用@pytest.mark.filterwarnings 装饰器实现警告忽略 …… 371
- 12.2.6 设置 pytestmark 变量实现添加警告 …… 372
- 12.2.7 命令行选项参数去掉警告信息 …… 373
- 12.2.8 通过触发警告自定义失败时的提示消息 …… 373

12.3 内置 fixture 之临时目录 …… 374
- 12.3.1 tmp_path …… 374
- 12.3.2 tmp_path_factory …… 376
- 12.3.3 tmpdir …… 377
- 12.3.4 tmpdir_factory …… 378

12.4 输出及捕获级别配置 …… 380
- 12.4.1 标准输出/标准错误输出/标准输入的默认捕获行为 …… 380
- 12.4.2 修改和去掉捕获行为 …… 380
- 12.4.3 在测试用例中访问捕获到的信息 …… 381

12.5 Mock …… 382
- 12.5.1 使用 Mock 对象模拟测试情景 …… 383
- 12.5.2 Mock 类的原型 …… 384
- 12.5.3 MonkeyPatching 返回的对象：临时修改全局配置 …… 385
- 12.5.4 修改 MonkeyPatching 函数功能或者类属性 …… 385
- 12.5.5 修改 MonkeyPatching 环境变量 …… 388
- 12.5.6 修改 MonkeyPatching 字典 …… 389

12.6	钩子——Hook 方法的作用	391
	12.6.1　pytest_runtest_makereport 修改测试报告内容	391
	12.6.2　pytest_collection_modifyitems 改变用例执行顺序	396
	12.6.3　pytest_terminal_summary	399
12.7	本章小结	404

第 1 章

pytest 框架介绍

1.1 pytest 框架引入

框架是什么？其实就是别人写好的一些代码帮助我们做某些事情。一般会把重复工作通过代码的方式封装好，这样我们就可以调用，这些封装好的代码就是框架。

1.1.1 测试框架能为我们解决什么问题

在整个测试的过程当中，通常流程化的东西包括编写测试计划、编写测试用例、执行测试用例、提交 Bug、编写测试报告，还有沟通和开会等。

作为测试人员，平时做的最多的工作是什么？如图 1-1 所示。

图 1-1　测试人员平时重复的工作

执行测试和提交 Bug 这个过程是重复出现的。那么我们可以把整个执行的过程封装成一个框架，每次我们只需把测试用例提交，之后让它帮助我们进行一些用例的执行及管理。提交 Bug 这个活动也是重复的，在提交报告的过程中，我们需要一些运行结果的依据，最好能够生成一个非常漂亮的报告供我们参考，那么这部分需要有相应的框架来帮助我们实现。

总结来讲，所有的重复工作，我们都可用框架的方式实现，所以需要自己亲自设计和编写脚本代码。执行测试及提交精美测试报告这些任务可以由 pytest 和它的插件帮助我们完成。

1.1.2 测试框架的分类

- 进行测试执行操作：Selenium、Appium、Requests；
- 测试用例管理及执行：pytest、Unittest、Nose；
- 测试报告：Allure；
- 数据驱动：DDT；
- 环境部署及持续集成：Docker、Jenkins。

1.1.3 什么是 pytest

pytest 是一个基于 Python 的测试框架，用于编写和执行测试代码。pytest 应用在自动化测试场ంతో、单元自动化测试、API 自动化测试、Web/App 自动化测试等领域。我们可以使用 pytest 编写从简单到复杂的测试，即可以编写代码来测试 API、数据库、UI 等。

我们可以设想一下企业中实现 pytest 自动化测试执行之后的场景：开发人员在编写代码之后提交代码，然后我们便可以触发整个测试脚本的运行，运行之后，我们很快便可以得到测试的结果，那么可能当你还没有下班之前，测试结果便发到了你的邮箱，或者以短信的形式通知你，你需要把 Bug 改了才可以下班。

这种快速自动化测试出结果的方式最适合目前所需求的快速迭代，更解决了领导希望通过快速的反馈做下一步决策的问题。

1. pytest 的特点和优点

pytest 自动化测试框架会随着 DevOps 的落地实施而变得会越来越流行。它之所以流行，也是因为有很多的优点。

- pytest 是免费和开源的；
- pytest 有活跃的社区和维护组织；
- pytest 的语法简单灵活，容易上手；
- 支持参数化，也就是支持数据驱动；
- 支持测试用例的 skip 和 xfail 处理；
- pytest 可以自动检测测试文件和测试功能；
- pytest 允许我们运行整个测试套件的一部分；
- 能够支持简单的单元测试和复杂的功能测试；

注：可以与 Selenium/Appium 等一起进行 UI 自动化测试，也可以与 Requests 一起进行接口自动化测试。

- 可以很好地和 Jenkins 集成；
- pytest 具有很多第三方插件，并且可以自定义扩展；
- pytest-allure 可以生成完美的 HTML 测试报告；
- pytest 可以并行运行多个测试，从而减少测试套件的执行时间 pytest-xdist；

- 与以前的测试框架兼容,可执行由 Unittest、Nose 所写的测试脚本。

2. pytest 的官网及资料地址

pytest 官网及帮助文档网址 https://docs.pytest.org/en/latest/。

pypi 网址 https://pypi.org/project/pytest/。

GitHub 网址 https://github.com/pytest-dev/pytest/。

1.2 技术前提要求

1.2.1 技术前提

- 懂点 Python 语言;
- 做过测试或了解测试。

1.2.2 适合人群

- 黑盒测试人员转测试开发;
- 高校计算机相关专业学生为进入名企测试岗位或测试开发岗位;
- 开发人员想要提高自己的代码质量。

1.3 环境准备及资料准备

pytest 是一个基于 Python 的框架,需要安装在 Python 上,所以我们首先需要验证 Python 是否成功安装,然后安装 pytest。我们的编码是在 Python 的集成开发环境 PyCharm 上进行。

1.3.1 Python 的环境验证

Python 的命令在 Windows 和 Mac 系统中的命令基本一致,如有不同,则会分别说明。在终端(cmd)中执行:

```
python -- version
```

在 Windows 系统的 cmd 中执行结果如下:

```
C:\Users\lindaw> python -- version
Python 3.6.8
```

在 Mac 系统的终端中执行结果如下:

```
lindafang@bogon ~ % python -- version
Python 3.6.8
```

1.3.2 安装pytest

接下来开始安装pytest，我们可以安装任何版本的pytest，在本书中pytest 5.4.1是我们要安装的版本。需要执行以下命令：

```
pip install pytest == 5.4.1
```

安装显示过程如下：

```
Collecting pytest
  Downloading https://files.Pythonhosted.org/packages/c7/e2/c19c667f42f72716a7d03e8dd4d6f
63f47d39feadd44cc1ee7ca3089862c
/pytest-5.4.1-py3-none-any.whl (246kB)
     100% |████████████████████████████████| 256kB 10kB/s
Collecting wcwidth (from pytest)
  Downloading https://files.Pythonhosted.org/packages/f6/d5/1ecdac957e3ea12c1b319fcdee8b6
917ffaff8b4644d673c4d72d2f20b49
/wcwidth-0.1.9-py2.py3-none-any.whl
Requirement already satisfied, skipping upgrade: attrs>=17.4.0 in c:\programs\Python\
Python36\lib\site-packages (from pytest) (19.1.0)
Collecting importlib-metadata>=0.12; Python_version <"3.8" (from pytest)
  Downloading https://files.Pythonhosted.org/packages/ad/e4/891bfcaf868ccabc619942f27940c
77a8a4b45fd8367098955bb7e152fb1
/importlib_metadata-1.6.0-py2.py3-none-any.whl
Collecting pluggy<1.0,>=0.12 (from pytest)
  Downloading https://files.Pythonhosted.org/packages/a0/28/85c7aa31b80d150b772fbe4a22948
7bc6644da9ccb7e427dd8cc60cb8a62
/pluggy-0.13.1-py2.py3-none-any.whl
```

如果要安装最新版本的pytest，则执行的命令如下：

```
pip install pytest
```

如果已经安装一个版本了，则可以更新成最新版本，执行的命令如下：

```
pip install -U pytest
```

执行以下命令可以显示pytest的帮助部分：

```
pytest -h
```

帮助显示如下，此处未显示完整，我们会在后续的课程中介绍这些参数的使用。

```
usage: pytest [options] [file_or_dir] [file_or_dir] [...]

positional arguments:
  file_or_dir

general:
  -k EXPRESSION            only run tests which match the given substring
                           expression. An expression is a Python evaluatable
                           expression where all names are substring-matched
                           against test names and their parent classes. Example:
                           -k 'test_method or test_other' matches all test
                           functions and classes whose name contains
                           'test_method' or 'test_other', while -k 'not
                           test_method' matches those that don't contain
                           'test_method' in their names. -k 'not test_method and
                           not test_other' will eliminate the matches.
                           Additionally keywords are matched to classes and
                           functions containing extra names in their
                           'extra_keyword_matches' set, as well as functions
                           which have names assigned directly to them.
  -m MARKEXPR              only run tests matching given mark expression.
                           example: -m 'mark1 and not mark2'.
  --markers                show markers (builtin, plugin and per-project ones).
  -x, --exitfirst          exit instantly on first error or failed test.
  --maxfail=num            exit after first num failures or errors.
  --fixtures, --funcargs
                           show available fixtures, sorted by plugin appearance
                           (fixtures with leading '_' are only shown with '-v')
  --capture=method         per-test capturing method: one of fd|sys|no.
  -s                       shortcut for --capture=no.
  --runxfail               report the results of xfail tests as if they were not
                           marked

pytest-warnings:
  -W PythonWARNINGS,       --Pythonwarnings=PythonWARNINGS
                           set which warnings to report, see -W option of Python
                           itself.

logging:
  --no-print-logs          disable printing caught logs on failed tests.
  --log-level=LOG_LEVEL
                           logging level used by the logging module

reporting:
  --alluredir=DIR          Generate Allure report in the specified directory (may
                           not exist)
  --clean-alluredir        Clean alluredir folder if it exists
```

```
                        --allure-no-capture    Do not attach pytest captured logging/stdout/stderr to
...

to see available markers type: pytest --markers
to see available fixtures type: pytest --fixtures
(shown according to specified file_or_dir or current dir if not specified; fixtures with
leading '_' are only shown with the '-v' option
...)
```

如果要查看pytest的详细介绍,则可以执行的命令如下:

```
pip show pytest
```

显示版本信息、主页信息、作者信息、版权信息、安装位置信息、关联包信息等。

```
Name: pytest
Version: 5.4.1
Summary: pytest: simple powerful testing with Python
Home-page: https://docs.pytest.org/en/latest/
Author: Holger Krekel, Bruno Oliveira, Ronny Pfannschmidt, Floris Bruynooghe, Brianna Laugher,
Florian Bruhin and others
Author-email: None
License: MIT license
Location: /Library/Frameworks/Python.framework/Versions/3.6/lib/Python3.6/site-packages
Requires: attrs, py, pluggy, more-itertools, importlib-metadata, packaging, wcwidth
Required-by: pytest-xdist, pytest-sugar, pytest-rerunfailures, pytest-ordering, pytest
-metadata, pytest-forked, pytest-bdd, pytest-assume
```

1.4　pytest初体验

1.4.1　在终端建立测试方法及执行过程

(1) 在计算机磁盘上创建test_one.py文件。

(2) 可以先使用记事本打开该文件,输入如下信息后保存并关闭文件。此文件定义了方法,以及断言1是否等于1。

```
def test_one():
    assert 1 == 1
```

(3) 在终端输入命令。

```
pytest test_one.py
```

（4）或输入 pytest 加空格后把文件拖曳过去。

（5）按回车键执行。

执行结果如图 1-2 所示，它们的含义分别对应如下：

平台 win32、Python 的版本、pytest 的版本、自带插件及版本。

根目录。

第三方插件及版本。

收集到 1 个测试。

文件名后面的点表示通过。

最下面一行是 1，表示通过，所花费的时间为 0.03s。

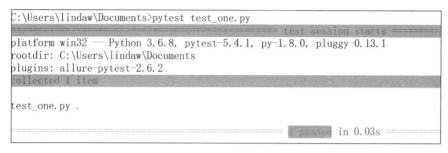

图 1-2　Windows 下 pytest 执行测试脚本的结果

注意：通常 pytest 执行测试用例的执行方式都是这种命令行的方式。

1.4.2　在 PyCharm 建立测试方法及执行过程

如果需要进行大量 Python 脚本的编写，我们则需要 Python 的 IDE 工具。下面的操作是在 PyCharm 中完成的。已建立项目 pytest_book。

（1）新建 test_first.py 文件。

（2）输入 import pytest ♯这是将 pytest 框架导入。

（3）建立一个测试方法 def test_1()：。

（4）缩进后输入 print("这是第一个测试方法")。

（5）在下面输入 if __name__ == "__main__"：

注：name 和 main 前后都是两个小短横。

（6）缩进后输入 pytest.main(["-s", "test_first.py"])。

（7）单击 if 左边的 ▶ 按钮，执行代码，如图 1-3 所示。

完整代码如下：

```
import pytest    ♯导入 pytest 包


def test_1():
```

```python
        print("这是第一个测试方法")

if __name__ == '__main__':
    # -s 是参数,输入控制台信息,后面是文件名
    pytest.main("-s", "test_first.py")
```

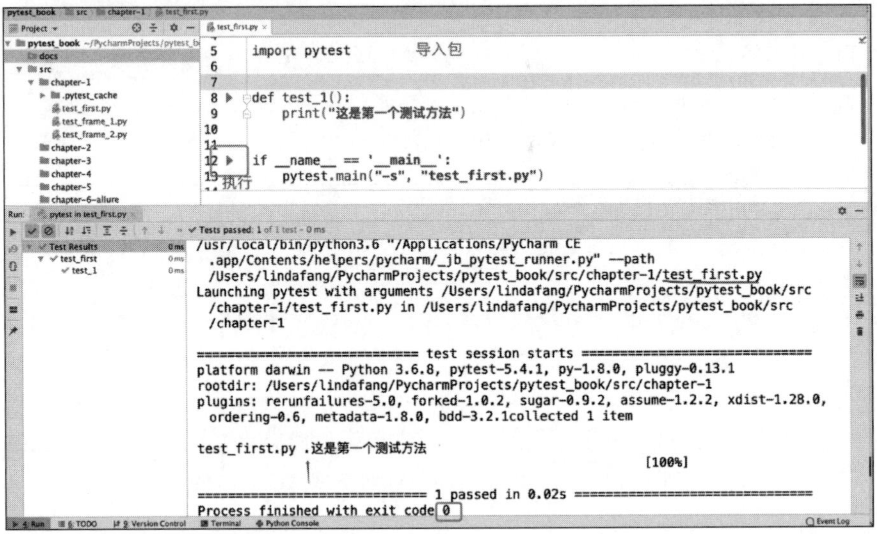

图 1-3　在 PyCharm 上执行测试方法

执行结果如下：最后一句中的 code 0 表示没有执行错误。如果执行失败,则显示为 code 1。

```
Testing started at 20:08 ...
/usr/local/bin/Python3.6 "/Applications/PyCharm
CE.app/Contents/helpers/PyCharm/_jb_pytest_runner.py" -- path
/Users/lindafang/PyChARMProjects/pytest_book/src/chapter-1/test_first.py
Launching pytest with arguments /Users/lindafang/PyCharmProjects/pytest_book/src/chapter-
1/test_first.py in /Users/lindafang/PyCharmProjects/pytest_book/src/chapter-1

=============== test session starts =======================
platform darwin -- Python 3.6.8, pytest-5.4.1, py-1.8.0, pluggy-0.13.1
rootdir: /Users/lindafang/PyCharmProjects/pytest_book/src/chapter-1
plugins: rerunfailures-5.0, forked-1.0.2, sugar-0.9.2, assume-1.2.2, xdist-1.28.0,
ordering-0.6, metadata-1.8.0, bdd-3.2.1collected 1 item

test_first.py .这是第一个测试方法
                                                                    [100%]

==================== 1 passed in 0.02s ====================
Process finished with exit code 0
```

注意：最后一句的快捷方式：直接输入 main，按回车键后会自动建立主函数调用。import pytest 如果有红色下波浪线，则表示没有导入成功。通常是由于计算机中除默认的环境以外，还有其他 Python 环境。更换到已经安装完 pytest 的 Python 环境便可以解决此问题，或把每个环境都安装上。

1.4.3　pytest 的框架结构

在执行测试的过程中，我们经常会做些准备工作，再进行测试，测试完成后通常也会将现场恢复原状，因此我们希望执行测试的框架能提供各层次的测试前和测试后的方法。

绝大多数的测试框架只提供 setup 与 teardown。在 setup 的方法中进行准备代码，在 teardown 方法中进行销毁代码。

pytest 提供了相对自由和层次丰富的 setup 与 teardown 框架结构。pytest 支持 5 个层次的 setup 和 teardown，包括：session 会话级、module 模块级、function 函数级、class 类级、method 方法级。

（1）模块级（setup_module/teardown_module）：开始于模块始末（不在类中）。
（2）函数级（setup_function/teardown_function）：对函数用例生效（不在类中）。
（3）方法级（setup_method/teardown_method）：开始于方法始末（在类中）。
（4）类级（setup_class/teardown_class）：只在类中前后运行一次（在类中）。
（5）还有兼容的类里面的层级（setup/teardown）：运行在调用方法的前后。

这些级别用在配置初始化 setup 和销毁 teardown 中，同样可以配合 pytest 中 fixture 功能而变得更强大。将在 3.3 节具体讲解。

下面给大家逐步演示常用级别所影响的范围层次。先演示模块级和函数级，之后演示方法级、类级及兼容的 setup 和 teardown，最后将整体 5 个常用级别放在一起演示执行效果。

模块和函数级实践步骤如下：
（1）新建 Python 文件 test_frame_1.py。
（2）在文件中输入如下代码。
（3）单击 ▶ 执行代码。
（4）查看运行结果是否与方法中输入的文字含义相同。

完整代码如下：

```python
# Author: lindafang
# File: test_frame_1.py

import pytest


# 模块中的方法
def setup_module():
```

```python
    print("\nsetup_module: 整个test_module.py模块只执行一次")

def teardown_module():
    print("teardown_module: 整个test_module.py模块只执行一次")


def setup_function():
    print("setup_function: 每个用例开始前都会执行")


def teardown_function():
    print("teardown_function: 每个用例结束后都会执行")


# 测试模块中的用例1
def test_one():
    print("正在执行测试模块----test_one")


# 测试模块中的用例2
def test_two():
    print("正在执行测试模块----test_two")

if __name__ == "__main__":
    pytest.main(["-s", "test_frame_1.py"])
```

执行结果如下：module 级在 py 文件中只执行一次，function 则在非类的测试方法前后执行，每个测试方法都会调用一次。

```
Testing started at 19:42 ...
/usr/local/bin/Python3.6 "/Applications/PyCharm
CE.app/Contents/helpers/PyCharm/_jb_pytest_runner.py" -- path
/Users/lindafang/PyCharmProjects/pytest_book/src/chapter-1/test_frame_1.py
Launching pytest with arguments /Users/lindafang/PyCharmProjects/pytest_book/src/chapter-
1/test_frame.py in /Users/lindafang/PyCharmProjects/pytest_book/src/chapter-1

============ test session starts =======================
platform darwin -- Python 3.6.8, pytest-5.4.1, py-1.8.0, pluggy-0.13.1
rootdir: /Users/lindafang/PyCharmProjects/pytest_book/src/chapter-1
plugins: rerunfailures-5.0, forked-1.0.2, sugar-0.9.2, assume-1.2.2, xdist-1.28.0,
ordering-0.6, metadata-1.8.0, bdd-3.2.1collected 2 items

test_frame_1.py
setup_module: 整个test_module.py模块只执行一次
setup_function: 每个用例开始前都会执行
```

```
.正在执行测试模块----test_one
teardown_function：每个用例结束后都会执行
setup_function：每个用例开始前都会执行
.正在执行测试模块----test_two
teardown_function：每个用例结束后都会执行
teardown_module：整个test_module.py模块只执行一次
                                                    [100%]

================== 2 passed in 0.03s ==================
Process finished with exit code 0
```

整体文件结构、代码及运行结果如图1-4所示。

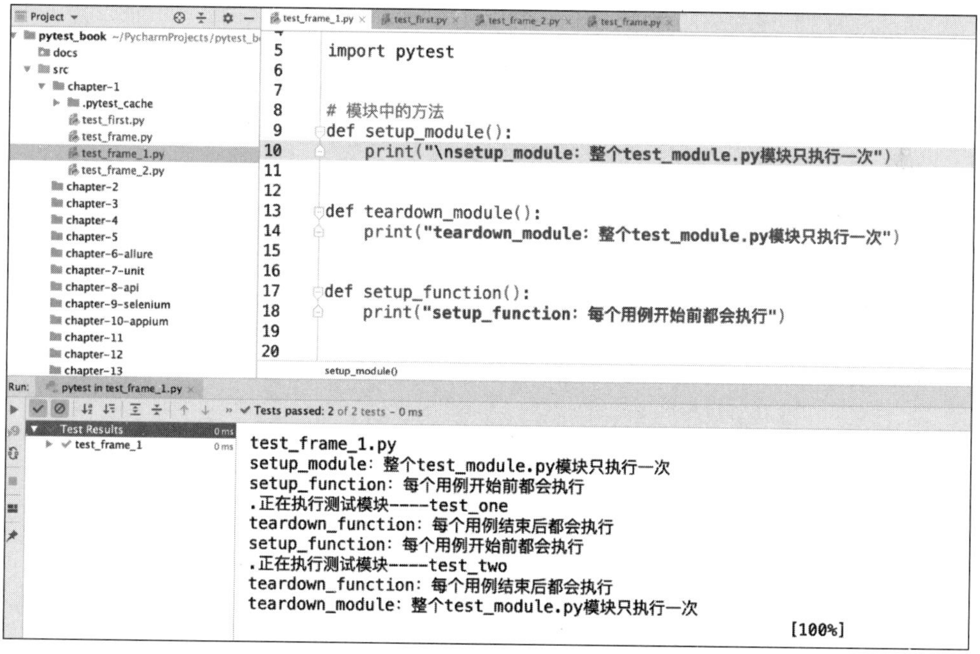

图1-4　test_frame_1.py文件执行结果

类和方法级实践如下：

（1）新建Python文件test_frame_2.py。

（2）在文件中输入如下代码。

（3）单击 ▶ 执行代码。

（4）查看运行结果是否与方法中输入的文字含义相同。

完整代码如下：

```
# Author: lindafang
# File: test_frame_2.py
```

```python
import pytest


# 测试类
class TestCase(object):
    def setup_class(self):
        print("\nsetup_class: 所有用例执行之前")

    def teardown_class(self):
        print("teardown_class: 所有用例执行之后")

    def setup_method(self):
        print("setup_method: 每个用例开始前执行")

    def teardown_method(self):
        print("teardown_method: 每个用例结束后执行")

    def setup(self):
        print("setup: 每个用例开始前都会执行")

    def teardown(self):
        print("teardown: 每个用例结束后都会执行")

    def test_three(self):
        print("正在执行测试类----test_three")

    def test_four(self):
        print("正在执行测试类----test_four")


if __name__ == "__main__":
    pytest.main(["-s", "test_frame_2.py"])
```

执行结果如下:

```
Testing started at 21:29 ...
/usr/local/bin/Python3.6 "/Applications/PyCharm
CE.app/Contents/helpers/PyCharm/_jb_pytest_runner.py" --path
/Users/lindafang/PyCharmProjects/pytest_book/src/chapter-1/test_frame_2.py
Launching pytest with arguments /Users/lindafang/PyCharmProjects/pytest_book/src/chapter-
1/test_frame_2.py in /Users/lindafang/PyCharmProjects/pytest_book/src/chapter-1

================== test session starts =========================
platform darwin -- Python 3.6.8, pytest-5.2.1, py-1.8.0, pluggy-0.13.1
rootdir: /Users/lindafang/PyCharmProjects/pytest_book/src/chapter-1
```

```
plugins: rerunfailures-5.0, forked-1.0.2, sugar-0.9.2, assume-1.2.2, xdist-1.28.0,
ordering-0.6, metadata-1.8.0, bdd-3.2.1collected 2 items

test_frame_2.py
setup_class：所有用例执行之前
setup_method：每个用例开始前执行
setup：每个用例开始前都会执行
.正在执行测试类----test_three
teardown：每个用例结束后都会执行
teardown_method：每个用例结束后执行
setup_method：每个用例开始前执行
setup：每个用例开始前都会执行
.正在执行测试类----test_four
teardown：每个用例结束后都会执行
teardown_method：每个用例结束后执行
teardown_class：所有用例执行之后
                                                [100%]
==================== 2 passed in 0.03s ================================
Process finished with exit code 0
```

将类中方法和非在类中的方法等所有级放在一起执行，代码如下：

```
# Author: lindafang
# File: test_frame.py
import pytest


# 模块中的方法
def setup_module():
    print("\nsetup_module: 整个 test_module.py 模块只执行一次")


def teardown_module():
    print("teardown_module: 整个 test_module.py 模块只执行一次")


def setup_function():
    print("setup_function: 每个非类函数测试用例开始前都会执行")


def teardown_function():
    print("teardown_function: 每个非类函数测试用例结束后都会执行")


# 测试模块中的用例1
```

```python
def test_one():
    print("正在执行测试模块----test_one")


# 测试模块中的用例2
def test_two():
    print("正在执行测试模块----test_two")


# 测试类
class TestCase(object):
    def setup_class(self):
        print("\nsetup_class: 在类中所有测试用例执行之前")

    def teardown_class(self):
        print("teardown_class: 在类中所有测试用例执行之后")

    def setup_method(self):
        print("setup_method: 每个类中测试方法用例开始前执行")

    def teardown_method(self):
        print("teardown_method: 每个类中测试方法用例结束后执行")

    def setup(self):
        print("setup: 每个类中测试方法用例开始前都会执行")

    def teardown(self):
        print("teardown: 每个类中测试方法用例结束后都会执行")

    def test_three(self):
        print("正在执行测试类----test_three")

    def test_four(self):
        print("正在执行测试类----test_four")


if __name__ == "__main__":
    pytest.main(["-s", "test_frame.py"])
```

执行结果如下：

```
Testing started at 22:02 ...
/usr/local/bin/Python3.6 "/Applications/PyCharm
CE.app/Contents/helpers/PyCharm/_jb_pytest_runner.py" --path
/Users/lindafang/PyCharmProjects/pytest_book/src/chapter-1/test_frame.py
```

```
Launching pytest with arguments /Users/lindafang/PyCharmProjects/pytest_book/src/chapter-
1/test_frame.py in /Users/lindafang/PyCharmProjects/pytest_book/src/chapter-1

================== test session starts ==============================
platform darwin -- Python 3.6.8, pytest-5.2.1, py-1.8.0, pluggy-0.13.1
rootdir: /Users/lindafang/PyCharmProjects/pytest_book/src/chapter-1
plugins: rerunfailures-5.0, forked-1.0.2, sugar-0.9.2, assume-1.2.2, xdist-1.28.0,
ordering-0.6, metadata-1.8.0, bdd-3.2.1collected 4 items

test_frame.py
setup_module: 整个 test_module.py 模块只执行一次
setup_function: 每个非类函数测试用例开始前都会执行
.正在执行测试模块----test_one
teardown_function: 每个非类函数测试用例结束后都会执行
setup_function: 每个非类函数测试用例开始前都会执行
.正在执行测试模块----test_two
teardown_function: 每个非类函数测试用例结束后都会执行

setup_class: 在类中所有测试用例执行之前
setup_method: 每个类中测试方法用例开始前执行
setup: 每个类中测试方法用例开始前都会执行
.正在执行测试类----test_three
teardown: 每个类中测试方法用例结束后都会执行
teardown_method: 每个类中测试方法用例结束后执行
setup_method: 每个类中测试方法用例开始前执行
setup: 每个类中测试方法用例开始前都会执行
.正在执行测试类----test_four
teardown: 每个类中测试方法用例结束后都会执行
teardown_method: 每个类中测试方法用例结束后执行
teardown_class: 在类中所有测试用例执行之后
teardown_module: 整个 test_module.py 模块只执行一次
                                                        [100%]

========================= 4 passed in 0.04s ==========================
Process finished with exit code 0
```

1.4.4 在 PyCharm 中配置运行工具

刚才我们执行脚本是通过 main 方法执行的,也就是通过 Python 调用 pytest 的 main 方法执行。PyCharm 支持 pytest 工具的执行方式,接下来讲解如何配置。

Mac 系统中的操作如下:

单击左上角 PyCharm→Preferences-> Tools-> Python Integrated Tools,在 Testing 的 Default test runner 中选择 pytest。

具体配置如图 1-5 所示,不同版本可能稍有差别。

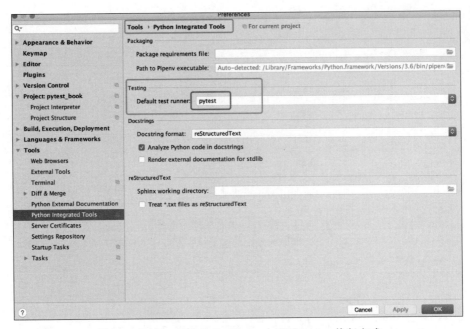

图 1-5　在 Mac 系统的 PyCharm 中配置 pytest 执行方式

注意：配置完成后有些功能需要重启计算机才能生效。

Windows 系统下的配置如图 1-6 所示，单击 File→Setting→Tools Python Intergrated Tools，在 Default test runner 中选择 py.test，单击 OK 或 Apply 按钮保存。

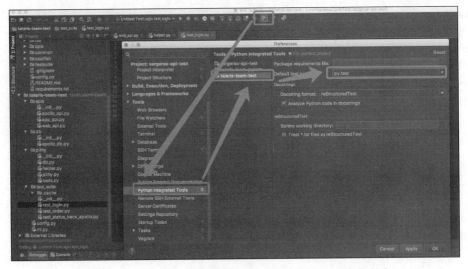

图 1-6　在 Windows 系统的 PyCharm 中配置 pytest 执行方式

配置完成后，执行方式可以很灵活。既可以通过 main 方法执行，还可以通过 pytest 直接执行。

1.4.5 右击以 pytest 方式执行代码

我们可以通过右击选择运行 test_first.py，如图 1-7 所示。

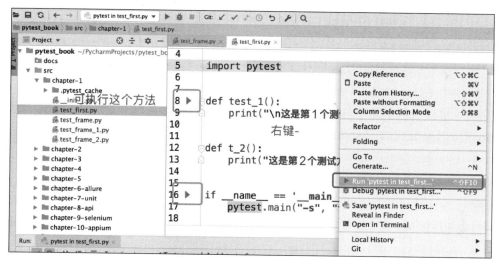

图 1-7 以右击选择 pytest 的执行方式执行测试

1.4.6 去掉 main 方法执行测试

真正编写代码时我们很少通过 main 方法执行，所以写代码时便不加 main 方法了。修改 test_first 的代码如下：

```
def test_1():
    print("\n 这是第 1 个测试方法")

def test_3():
    print("\n 这是第 3 个测试方法")

def t_2():
    print("这是第 2 个测试方法")
```

在不是方法名的任何行用鼠标右击选择 pytest in test_first 并执行。
执行结果如下：

```
test_first.py .
这是第 1 个测试方法
.
这是第 3 个测试方法
                                    [100%]
```

```
=================== 2 passed in 0.02s ===============================
Process finished with exit code 0
```

1.4.7 PyCharm 中执行某个测试方法

通常有以下两种方式：

（1）在某个测试方法所在行单击鼠标右键，只执行这个测试方法。

（2）使用左边 ▶ 按钮，只执行这个测试方法。

大家在实践过程中如果出现按上面方式操作仍执行全部测试方法，或单击鼠标右键执行时出现下面的结果，如图 1-8 所示，则需要修改一下执行配置。

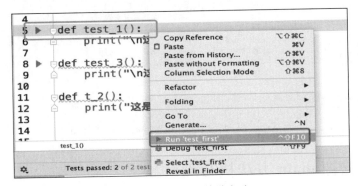

图 1-8　右击运行单种方法

执行配置修改步骤如下：

（1）选择工具栏中的配置：单击 ▼。

（2）单击 Edit Configurations，如图 1-9 所示。

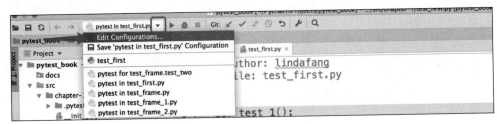

图 1-9　工具栏中的配置

（3）在弹出的运行调试配置窗口中单击"＋"。

（4）选择 Python tests。

（5）在出现的窗口右侧的 Configuration 中选择 Module name。

（6）在下面的文本框中输入 test_first.test_1（前面是文件名，点后面是方法名）。

（7）如果此时 Name 没有自动填上，则可输入如图 1-10 所示的名字。

（8）保存。

第1章　pytest框架介绍　19

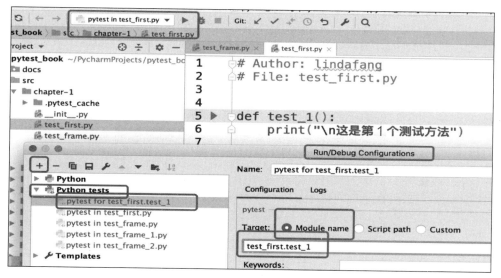

图 1-10　运行调试配置说明

修改完配置后,可以验证通过上述两种方式执行某一个测试方法是否成功。

注意：有些版本配置需要先关闭,然后重新打开 PyCharm 才配置成功。

1.4.8　运行窗口的工具栏含义

在 pytest 初体验过程中,我们了解了如何编写测试方法。如果需要执行并查看执行的结果,那么运行窗口中的工具都有什么含义呢?

运行窗口如图 1-11 所示,具体含义依次是:

- 对号表示可以查看所有通过的测试用例;
- 禁止号表示可以查看所有忽略的测试用例;
- 按 a～z 的顺序排序;
- 按执行的时间排序;
- 展开所有结果;
- 收起所有结果;
- 向上箭头表示执行前一个失败的用例;
- 向下箭头表示执行下一个失败的用例;
- 导出测试结果;
- 钟表表示导入执行过的测试结果可选;
- 最后一个是工具选项。

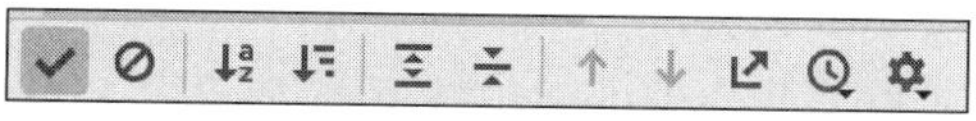

图 1-11　执行结果窗口工具栏

左侧框中工具如图 1-12 所示,其含义从上到下依次是:

- 绿色 ▶ 为运行测试方法;
- 重新运行失败的测试用例;

- 切换自动测试；
- 停止测试；
- 默认布局；
- 钉住窗口。

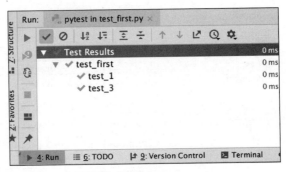

图 1-12　执行结果窗口左侧工具

1.5　执行的查找原则和测试类及测试方法的命名

pytest 根据一定的规则搜索并运行测试。标准的测试搜索规则如下：

（1）从一个或多个目录开始查找，可以在命令行指定文件名或目录名。如果未指定，则使用当前目录。

（2）在该目录和所有子目录下递归查找测试模块。

（3）测试模块指文件名为 test_*.py 或 *_test.py 的文件。

（4）在测试模块中查找以 test_开头的函数名。

（5）查找名字以 Test 开头的类，其中，首先筛选掉包含 __init__ 函数的类，再查找以 test_ 开头类中的方法。

以上是标准的测试搜索规则，也可以更改它们，我们将在第 12 章讲解如何更改测试搜索的规则。

注意：所有包都要有 __init__.py 文件。pytest 可以执行由 Unittest 框架所写的用例和方法。

通常测试模块（文件）、测试类、测试函数和方法的命名都遵循这个原则。如果不遵守这个原则就不会被搜索到，便不会有预期的效果。

我们举例说明，如图 1-13 所示。

（1）在 test_first.py 文件中新建一个测试方法。

（2）命名为 t_2，输入打印语句。

（3）右击选择以 pytest 方式执行。

（4）执行结果如图 1-13 所示，只执行了一种方法。

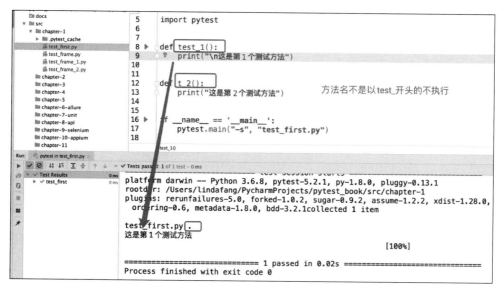

图 1-13　不以 test_开头的测试方法不执行

1.6　本章小结

本章主要讲解 pytest 的入门内容，了解 pytest 可以帮助我们执行测试方法，可以配合各种插件帮助我们更完美执行测试。以下是本章的要点：

（1）pytest 的特点和各种优点。

（2）pytest 提供的框架结构灵活而富有层次。

（3）pytest 的命令行执行与 PyCharm 中代码的执行方式。

（4）运行窗口及运行结果的含义。

（5）pytest 搜索用例的方式和原则。

第 2 章 pytest 的测试用例管理及运行管理

通过第 1 章的讲解，我们对 pytest 有了一个初步了解，明白了 pytest 用例组织的框架结构，并建立简单的测试方法。本章重点内容包括用例管理中的断言管理和运行管理。

2.1 测试用例的命名管理

第 1 章讲解了 pytest 搜索的原则，如果自动化的脚本是从用例的开始进行的，建议先建立命名规范以便统一用例的命名。命名时不能使用关键字命名，而应以_间隔有实际意义的单词为主，不要担心字符串过长，标记清晰即可。同时也要考虑测试用例的执行可通过用例的名称组合进行，同一类的用例可以有相同的部分名称。2.3 节会详细说明如何通过用例或文件名执行不同用例。

2.2 用例执行顺序

用例的执行顺序通常遵循一定的基本原则，但执行顺序也可通过插件改变。其基本原则是：根据名称的字母逐一进行 ASCII 比较，其值越小越先执行。当含有多个测试模块（.py 文件）时，根据基本原则执行。在一个测试模块（.py 文件）中，先执行测试函数，后执行测试类。如果含有多个测试类，则遵循基本原则，类中的测试方法遵循方法输入顺序。

如果想要改变执行顺序，则可通过插件实现，也可修改用例的位置。

具体示例如图 2-1 所示。文件夹和文件的执行顺序按字母顺序、数字顺序执行，测试方法按行号从小到大顺序执行，参数数据按参数的先后执行。

```
19 failed
  - chapter-2/test_assert_1.py:11 test_f
  - chapter-2/test_assert_2.py:5 test_long_str_comparison
  - chapter-2/test_assert_2.py:11 test_eq_list
  - chapter-2/test_assert_2.py:15 test_dict_comparison
  - chapter-2/test_assert_2.py:27 test_set_comparison
  - chapter-2/test_assert_3.py:15 test_true
  - chapter-2/test_assert_3.py:21 test_in_dict
  - chapter-2/test_assert_3.py:33 test_long_list
  - chapter-2/test_assert_3.py:39 test_long_copy
  - chapter-2/test_assert_except.py:13 test_mytest
  - chapter-2/test_assert_except.py:26 test_match
  - chapter-2/test_assert_sample.py:29 test_approx_simple_fail
  - chapter-2/test_assert_sample.py:45 test_warrior_long_description
  - chapter-2/test_assert_sample.py:65 test_get_starting_equiment
  - chapter-2/test_fail_cus.py:16 test_foo_compare
  - chapter-2/test_fail_cus.py:22 test_num
  - chapter-2/test_nodeid.py:20 TestNodeId.test_two[3-4]
  - chapter-2/test_raise.py:19 test_mytest_error
  - chapter-2/test_windows_skip_1.py:39 test_cakan
```

图 2-1　测试文件及用例执行顺序

2.3　测试用例的断言管理

2.3.1　什么是断言

测试的本质是验证预期结果与实际结果是否一致,那么在语言或框架中如何验证呢？通过断言的方式,即通过断言比对两个结果是否一致。

pytest 的断言使用 Python 中的 assert 函数。不同于其他框架单独设置一套 assertEqual 等函数,pytest 使用的就是 Python 自带的 assert 函数,使应用更加灵活。也就是编程语言可以判断什么,测试验证就可以判断什么。

2.3.2　断言的时机

到底什么时候使用断言比较好呢？可以从开发和测试两个方面为大家简单介绍断言的时机及不使用断言的情况。

开发人员　通常在下面时机点添加断言(本书以测试为主,对开发只做简单介绍)：

- 防御性地编程,也就是不满足条件时直接显示失败。示例代码如下：

```
assert target in(x, y, z)
if   target == x:
    run_x_code()
elif  target == y:
    run_y_code()
else:
    assert target == z
    run_z_code()
```

上述代码,当 target 为 x 时运行 run_x_code();当 target 为 y 时运行 run_y_code();当 target 为其他情况时运行 else 中的代码,也就是 target 如果为 z,则运行 run_z_code();如果不是上述内容,会直接引起一个简单而又直接的失败。

- 运行时对程序逻辑检测;
- 合约性检查(例如前置条件、后置条件);

例如:"如果你传给笔者一个非空字符串,笔者保证返回首字母转换成大写的字符串。"

- 程序中的常量;
- 检查文档。

测试人员 通常对所有需要比对的需求及功能进行验证时使用断言,下面分析一下断言的主要时机点:

- 验证页面是否跳转到正确的页面;
- 验证计算结果与正确结果是否一致;
- 验证类型是否一致;
- 验证添加功能是否成功添加;
- 验证接口返回的 JSON 数据是否正确;
- 验证接口返回的状态码是否正确;
- 验证接口性能是否在范围内;
- 验证元素是否已被选择;
- 验证元素是否为不可操作;
- 验证返回值是否与预期一致。

不使用断言的几种情况:

- 不要用于测试用户提供的数据,或者那些在所有情况下需要改变检查的地方;
- 不要用于检查你认为在通常使用中可能失败的地方;
- 用户绝看不到一个 AssertionError,如果看到了,那就是必须修复的缺陷;
- 特别地不要因为断言只比一个明确的测试加一个触发异常简单而使用它。断言不是懒惰的代码编写者的捷径;
- 不要将断言用于公共函数库输入参数的检查,因为用户不能控制调用者,并且不能保证它不破坏函数的执行;
- 不要将断言用于用户期望修改的任何地方。换句话说,用户没有任何理由在产品代码中捕获一个 AssertionError 异常;
- 不要太多使用断言,它们将使代码变得晦涩难懂。

2.3.3 断言的分类与使用

pytest 使用 Python 的 assert 函数,支持显示常见的 Python 子表达式的值,包括:调用、属性、比较、二进制和一元运算符。也就是 Python 语言有多少种判断,assert 就有多少种断言,包括断言函数返回值是否相等、断言表达式执行后的结果是否正确、各种不同比较运算符的断言、比较各种数据类型（字符串、列表、字典、集合）不同的断言。

1. 验证函数返回值是否相等

断言函数返回了某个值。如果此断言失败，则将看到函数调用的返回值。
示例代码如下：

```
def f():
    return 3

# 断言 f()的返回值是否等于 4
def test_f():
    assert f() == 4
```

执行效果如图 2-2 所示。

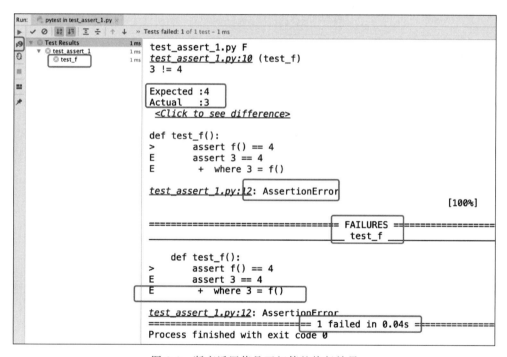

图 2-2　断言返回值是否相等的执行效果

图 2-2 中部分框的含义如下：

```
Excepted: 4    期望值是 4
Actual: 3      实际值是 3
Test_assert_1.py:12:AssertError  表示在文件的第 12 行出的错，是断言错误
+ Where 3 = f()    提供的建议，f()返回的值是 3
1 failed in 0.04s   表示总共有 1 个错误，整体运行时间为 0.04s
```

2. 表达式断言

通过表达式运算后的结果进行真假判断。

示例代码如下,如果值是偶数,不会报错。如果值是奇数,则表示断言失败,此时就会显示断言后面的提示信息:

```
a = 5
def test_exp():
    assert a % 2 == 0,"你的值是奇数,它应该是偶数"
```

执行结果如图 2-3 所示,a ％2==0 表达式后面的参数表示当出错时的提示信息:"你的值是奇数,它应该是偶数",F 表示失败。

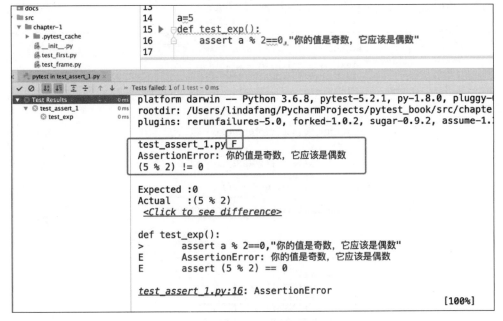

图 2-3　整除断言执行结果

3. 比较类型断言

比较类型断言包括相等、不等、包含、不包含、真、假等断言,以及长列表中的详细断言。下面通过代码分别演示。test_lessmore 测试方法用于比较小于或等于及大于或等于。test_true 测试方法用于比较真假,通过 1＜3 返回值再与真假比较。test_not_eq 测试方法用于比较不等于,以及判断是否不在其中。

示例代码如下:

```
♯比较
def test_lessmore():
    assert 1 <= 3
```

```
    assert 5 >= 1.3


#真假
def test_true():
    assert (1 < 3) is True
    assert (5 > 3) is False


#断言包含,当有'时用"号
def test_in_dict():
    x = 'linda1'
    assert x in "I'm linda"


#断言不等,以及判断是否不在其中
def test_not_eq():
    assert 1 != 2
    assert ['linda'] not in ['linda', 'tom', 'seven']


#断言长的列表中有哪些不同
def test_long_list():
    x = [str(i) for i in range(100)]
    y = [str(i) for i in range(0, 100, 2)]
    assert x == y

def test_long():
    assert [12] * 50 == [11] * 50
```

Python 的长列表比较智能,在 test_long_list 测试方法中:列表 x 的元素为 0~100 的所有自然数,列表 y 的元素为 0~100 的所有偶数,它们之间的不同会有 50 个,直接提示 '1' != '2'。test_long 测试方法将 50 个 12 和 50 个 11 进行比较,直接提示 12!=11。

代码执行结果如下,仔细体会。

```
test_assert_3.py .F
test_assert_3.py:12 (test_true)
def test_true():
        assert (1 < 3) is True
>       assert (5 > 3) is False
E       assert (5 > 3) is False

test_assert_3.py:15: AssertionError
F
```

```
test_assert_3.py:18 (test_in_dict)
def test_in_dict():
        x = 'linda1'
>       assert x in "I'm linda"
E       assert 'linda1' in "I'm linda"

test_assert_3.py:21: AssertionError
.F
test_assert_3.py:30 (test_long_list)
['0', '1', '2...'4', '5', ...] != ['0', '2', '4...8', '10', ...]

Expected :['0', '2', '4...8', '10', ...]
Actual   :['0', '1', '2...'4', '5', ...]
 <Click to see difference>

def test_long_list():
        x = [str(i) for i in range(100)]
        y = [str(i) for i in range(0, 100, 2)]
>       assert x == y
E       AssertionError: assert ['0', '1', '2...'4', '5', ...] == ['0', '2', '4...8', '10', ...]
E         At index 1 diff: '1' != '2'
E         Left contains 50 more items, first extra item: '50'
E         Use -v to get the full diff

test_assert_3.py:34: AssertionError
F
test_assert_3.py:35 (test_long)
[12, 12, 12, 12, 12, 12, ...] != [11, 11, 11, 11, 11, 11, ...]

Expected :[11, 11, 11, 11, 11, 11, ...]
Actual   :[12, 12, 12, 12, 12, 12, ...]
 <Click to see difference>

def test_long():
>       assert [12] * 50 == [11] * 50
E       assert [12, 12, 12, 12, 12, 12, ...] == [11, 11, 11, 11, 11, 11, ...]
E         At index 0 diff: 12 != 11
E         Use -v to get the full diff

test_assert_3.py:37: AssertionError                                    [100%]
```

4. 不同数据类型的比较断言

pytest 当遇到比较断言时，它提供了具有丰富细节且差异化的支持。同时，当断言失败时，pytest 也提供了非常人性化的失败说明，中间往往夹杂着相应变量的 introspection（自己反省）信息，这个被称为断言的自省。那么，pytest 是如何做到的呢？首先 pytest 发现测试模块,然后引入它们,与此同时,pytest 会复写断言语句,但是,不是测试模块的断言语句并不会被复写。

下面情况执行时有着丰富的细节比对:

(1) 比较字符串:显示上下文差异。
(2) 比较列表:显示第一个失败的索引。
(3) 比较字典:显示不同的条目。
(4) 比较集合:显示不同的元素。

```python
def test_long_str_comparison():
    str3 = 'abcdef'
    str4 = 'adcdef'
    assert str3 == str4


def test_eq_list():
    assert [0, 1, 2] == [0, 1, 3]


def test_dict_comparison():
    dict1 = {
        'name': 'linda',
        'age': 18,
    }
    dict2 = {
        'name': 'linda',
        'age': 88,
    }
    assert dict1 == dict2


def test_set_comparison():
    set1 = set("1308")
    set2 = set("8035")
    assert set1 == set2
```

执行结果如图 2-4~图 2-7 所示,分别比较的是字符串、列表、字典和集合,比对的细节很明确。

字符串的比较会具体到每个不同的字符,如图 2-4 所示。

列表中元素的比较会具体到每个索引,如图 2-5 所示。

字典类型的比较会具体到每个键值,相同的不显示,只显示不同的键,如图 2-6 所示。

集合类型会从左右两边显示不同,如图 2-7 所示。

```
test_assert_1.py F
test_assert_1.py:14 (test_long_str_comparison)
abcdef != adcdef

Expected :adcdef
Actual   :abcdef
 <Click to see difference>

def test_long_str_comparison():
        str3 = 'abcdef'
        str4 = 'adcdef'
>       assert str3 == str4
E       AssertionError: assert 'abcdef' == 'adcdef'
E         - abcdef
E         ?  ^
E         + adcdef
E         ?  ^

test_assert_1.py:18: AssertionError
```

图 2-4　长字符串比较执行结果

```
F
test_assert_1.py:20 (test_eq_list)
[0, 1, 2] != [0, 1, 3]

Expected :[0, 1, 3]
Actual   :[0, 1, 2]
 <Click to see difference>

def test_eq_list():
>       assert [0, 1, 2] == [0, 1, 3]
E       assert [0, 1, 2] == [0, 1, 3]
E         At index 2 diff: 2 != 3
E         Use -v to get the full diff

test_assert_1.py:22: AssertionError
```

图 2-5　列表类型比较执行结果

```
test_assert_1.py:24 (test_dict_comparison)
{'age': 18, 'name': 'linda'} != {'age': 88, 'name': 'linda'}

Expected :{'age': 88, 'name': 'linda'}
Actual   :{'age': 18, 'name': 'linda'}
 <Click to see difference>

def test_dict_comparison():
        dict1 = {
            'name': 'linda',
            'age': 18,
        }
        dict2 = {
            'name': 'linda',
            'age': 88,
        }
>       assert dict1 == dict2
E       AssertionError: assert {'age': 18, 'name': 'linda'} == {'age': 88, 'name': 'linda'}
E         Omitting 1 identical items, use -vv to show
E         Differing items:
E         {'age': 18} != {'age': 88}
E         Use -v to get the full diff

test_assert_1.py:34: AssertionError
```

图 2-6　字典类型比较执行结果

```
F
test_assert_1.py:36 (test_set_comparison)
{'0', '1', '3', '8'} != {'0', '3', '5', '8'}

Expected :{'0', '3', '5', '8'}
Actual   :{'0', '1', '3', '8'}
 <Click to see difference>

def test_set_comparison():
        set1 = set("1308")
        set2 = set("8035")
>       assert set1 == set2
E       AssertionError: assert {'0', '1', '3', '8'} == {'0', '3', '5', '8'}
E         Extra items in the left set:
E         '1'
E         Extra items in the right set:
E         '5'
E         Use -v to get the full diff

test_assert_1.py:40: AssertionError
```

图 2-7　集合类型比较执行结果

2.3.4 触发一个指定异常的断言

在进行异常测试时,会有这样的场景,程序希望在某时某地抛出一个指定的异常,如果的确抛出这个指定的异常,则程序是正确的。如果抛出的异常不是指定的那个异常或者不抛出异常,则表示程序是错误的。

使用 raises 引起一个指定的异常,再通过测试方法检查代码是否可抛出这个异常,如果抛出此异常,则表示程序是正确的,如果不抛出或者抛出的不正确,则表示程序是错误的。这样我们就可以检查代码是否抛出一个指定的异常。

引起一个解释器请求退出的异常,通过 test_mytest 测试方法实现断言并判断是否是指定的异常,示例代码如下:

```
import pytest

def f():
    #解释器请求退出
    raise SystemExit(1)

def test_mytest():
    #当调用 f()时出现 SystemExit 异常,则表示程序是正确的.出现其他异常表示程序是错误的
    with pytest.raises(SystemExit):
        f()
```

运行的结果是正常的,但如果把异常的类型修改就会出现执行测试不通过。大家可以动手试试,具体执行自己体会。

```
with pytest.raises(ValueError):
    f()
```

同时程序可以在抛出指定异常时,断言属性中的值是否正确。

其中,excinfo 是 ExceptionInfo 的一个实例,它封装了异常的信息。常用的属性包括:.type、.value 和 .traceback。

```
import pytest

def myfunc():
    #引起值错误
    raise ValueError("返回 40013 支付错误")


def test_match():
    #当调用 myfunc()时出现值错误,则表示程序是正确的
```

```
    #将值的信息保存到 excinfo 中,并且可能断言值中的属性 value 的内容
    with pytest.raises(ValueError) as excinfo:
        myfunc()
    assert '40013' in str(excinfo.value)
```

注意:在上下文管理器的作用域中,raises 代码必须是最后一行,否则其后面的代码将不会被执行。

所以,如果上述例子缩进到与函数调用为同一个层级,则测试将永远成功,代码如下:

```
def test_match():
    with pytest.raises(ValueError) as excinfo:
        myfunc()
        assert '456' in str(excinfo.value)    #缩进到与函数调用为同一个层级
```

这是因为 assert '456' in str(excinfo.value)并不会被执行。

也可以给 pytest.raises()传递一个关键字参数 match,来测试异常的字符串表示 str(excinfo.value)是否符合给定的正则表达式(和 Unittest 中的 TestCase.assertRaisesRegexp 方法类似),代码如下:

```
#异常类型可以写多个,后面可以跟正则表达式,与断言类似
with pytest.raises((ValueError, RunTimeError), match = r'.*40011.*'):
```

pytest 实际调用的是 re.search()方法,用来进行上述检查,match＝r'.*40011.*',r 表示原始,*表示所有,这个正则表达式的意思是字符串中包含 40011,同时 pytest.raises()也支持检查多个期望异常(以元组的形式传递参数),这时只需触发其中任意一个。

把要断言的属性内容写成 40011 而不是 40013,执行结果如下:

```
        def myfunc():
            #引起值错误
>           raise ValueError("返回 40013 支付错误")
E           ValueError: 返回 40013 支付错误

test_assert_except.py:23: ValueError

During handling of the above exception, another exception occurred:

        def test_match():
            #当调用 myfunc()时出现值错误,则表示程序是正确的
            #将值的信息保存到 excinfo 中,并且可能断言值中的属性 value 的内容
            with pytest.raises(ValueError) as excinfo:
            #异常类型可以写多个,后面可以跟正则表达式,与断言类似
            with pytest.raises((ValueError, RunTimeError), match = r'.*40011.*'):
>               myfunc()
```

```
E       AssertionError: Pattern '. * 40011. * 'not found in '返回 40013 支付错误'
test_assert_except.py:32: AssertionError
========================== 1 failed in 0.03s ==========================
Process finished with exit code 0
```

2.3.5　为失败断言添加自定义的说明

有时需要在断言时显示自定义的说明，可以直接在断言后面添加提示信息，代码如下：

```
def test_num():
    assert 1 == 2, '这两个值不相等'
```

执行的结果如下：

```
test_fail_cus.py F
AssertionError: 这两个值不相等
1 != 2

Expected :2
Actual   :1
 <Click to see difference>

def test_num():
>       assert 1 == 2, '这两个值不相等'
E       AssertionError: 这两个值不相等
E       assert 1 == 2

test_fail_cus.py:23: AssertionError
```

上述代码是比较两个基本数据，当比较两个对象实例时，可通过重写它的__eq__()方法比对实现。

代码如下：

```
# Author: lindafang
# Date: 2020 - 05 - 05 18:40
# File: test_fail_cus.py
class Foo:
    def __init__(self, val):
        self.val = val

    def __eq__(self, other):
```

```
            return self.val == other.val

def test_foo_compare():
    f1 = Foo(1)
    f2 = Foo(2)
    assert f1 == f2
```

但当执行这个用例时其结果并不能直观地从中看出来失败的原因 assert < test_fail_cus.Foo object at 0x109a7de10 > == < test_fail_cus.Foo object at 0x109a7dc88 >。

```
test_fail_cus.py F
test_fail_cus.py:11 (test_foo_compare)
< test_fail_cus.Foo object at 0x109a7de10 > != < test_fail_cus.Foo object at 0x109a7dc88 >

Expected :< test_fail_cus.Foo object at 0x109a7dc88 >
Actual   :< test_fail_cus.Foo object at 0x109a7de10 >
< Click to see difference >

def test_foo_compare():
    f1 = Foo(1)
    f2 = Foo(2)
>   assert f1 == f2
E   assert < test_fail_cus.Foo object at 0x109a7de10 > == < test_fail_cus.Foo object at 0x109a7dc88 >

test_fail_cus.py:15: AssertionError
                                                              [100%]
```

在这种情况下，如何比较两个实例呢？

通常有两种方式：第一种是重写 __repr__() 方法，第二种是用钩子的方法。下面将详细说明具体如何实现。

第一种方法：通过重写 Foo 的 __repr__() 方法，把详细信息显示出来。

在上述代码中添加下面的方法，代码如下：

```
def __repr__(self):
    return str(self.val)
```

当执行用例时，能看到失败的原因，其原因是因为 1==2 不成立。

代码如下：

```
test_fail_cus.py F
test_fail_cus.py:13 (test_foo_compare)
```

```
1 != 2

Expected :2
Actual   :1
<Click to see difference>

    def test_foo_compare():
            f1 = Foo(1)
            f2 = Foo(2)
>           assert f1 == f2
E           assert 1 == 2

test_fail_cus.py:17: AssertionError
1 failed in 0.05s
```

第二种方法：使用钩子的方法加断言的详细信息。

在当前目录下新建 conftest.py 文件，在 conftest.py 文件中使用 pytest _assertrepr_compare 这个钩子方法添加自定义的失败说明，在本目录下创建 conftest.py 文件（这个文件中的内容会供本目录中的所有文件共享，pytest 执行时先执行这个文件中的内容再执行测试用例，在 3.2 节和 5.4 节将分别详细讲解 conftest.py 和钩子的用法），在文件中输入下面代码。自定义失败说明后再执行 test_fail_cus.py。

代码如下：

```
# Author: lindafang
# File: conftest.py
from test_fail_cus import Foo

def pytest _assertrepr_compare(op, left, right):
    if isinstance(left, Foo) and isinstance(right, Foo) and op == " == ":
        return [
            "比较两个 Foo 实例：",        # 顶头写概要
            "值: {} != {}".format(left.val, right.val),    # 除了第 1 行,其余都可以缩进
        ]
```

执行后的结果如下，此时会看到一个更友好的失败说明，在出现错的代码行下面有具体的自定义的提示信息。

代码如下：

```
test_fail_cus.py F
test_fail_cus.py:15 (test_foo_compare)
    def test_foo_compare():
            f1 = Foo(1)
            f2 = Foo(2)
```

```
>       assert f1 == f2
E       assert 比较两个 Foo 实例:
E             值: 1 != 2

test_fail_cus.py:19: AssertionError
```

2.3.6 Assert 各种类型断言

为了能更好地理解 assert 的意义和使用，下面是实践部分，大家可以输入代码并执行，查看执行结果。test_mag 方法断言调用函数进行返回值的判断；test_simple_math 方法断言比较运算符，在计算机中小数的计算会有误差。我们当然希望像现实世界对小数进行计算那样，例如 0.1+0.2＝0.3，test_approx_simple 方法中的 approx 函数解决了上述问题。

test_warrior_long_description 方法进行长文本断言。test_get_starting_equiment 方法是防御性编程的例子。test_isinstance 方法用于测试返回类型是否一致。

代码如下：

```
# Author: lindafang
# File: test_assert_sample.py
import textwrap
from math import sqrt
from pytest import approx


# 勾股定理求斜边值, sqrt 表示求平方根
def magnitude(x, y):
    return sqrt(x * x + y * y)


def test_mag():
    assert magnitude(3, 4) == 5


# 添加求绝对值函数 abs, 这样当浮点小数的计算的误差小于 0.0001 时, 表示这个断言是正确的
def test_simple_math():
    assert abs(0.1 + 0.2) - 0.3 < 0.0001

# approx 函数是解决浮点小数问题的, 所以此断言也是正确的
def test_approx_simple():
    assert 0.1 + 0.2 == approx(0.3)


def test_approx_simple_fail():
    assert 0.1 + 0.2 == approx(0.35)
```

```python
# 比较一个长文本哪里不同
def get_long_class_description(class_name):
    assert class_name == "warrior"
    return textwrap.dedent(
        """\
A seasoned veteran of many battles. High Strength and Dexterity
allow to yield heavy armor and weapons, as well as carry
more equipment while keeping a light roll. Weak in magic.
"""
    )

# 比较一个长文本哪里不同
def test_warrior_long_description():
    desc = get_long_class_description("warrior")
    assert (
        desc
        == textwrap.dedent(
            """\
A seasoned veteran of many battles. Strength and Dexterity
allow to yield heavy armor and weapons, as well as carry
more equipment. Weak in magic.
"""
        )
    )

# 防御性编程,如果不是战士则直接退出,表示战士用来判断返回的值是否正确
def get_starting_equipment(class_name):
    assert class_name == "战士"
    return ["长剑", "战士装备", "盾"]


def test_get_starting_equiment():
    expected = ["长剑", "战士装备"]
    assert get_starting_equipment("战士") == expected, "装备不符"

# 测试返回类型是否一致
def test_isinstance():
    task_id = 10
    assert isinstance(task_id, int)
```

2.4 测试用例的运行管理

测试用例的运行管理基本通过 pytest 命令行选项来完成各种运行情况处理。先通过帮助来看一看有哪些命令行选项,下面详细讲解常用的几个选项。

注意：所有的命令执行都在终端下运行。

2.4.1 获取帮助信息

```
pytest --version          # 查看版本号和 pytest 的引入路径
pytest -h                 # 查看帮助信息
pytest --fixtures         # 可查看 fixtures 的相关功能帮助信息
pytest --markers          # 可查看 mark 的相关功能帮助信息
```

以下是 pytest --fixtures 的帮助返回信息。

代码如下：

```
lindafang@linda ~ % pytest --fixtures
Test session starts (platform: darwin, Python 3.6.8, pytest 5.2.1, pytest-sugar 0.9.2)
rootdir: /Users/lindafang
plugins: rerunfailures-5.0, forked-1.0.2, sugar-0.9.2, assume-1.2.2, xdist-1.28.0,
ordering-0.6, metadata-1.8.0, bdd-3.2.1
------------------- test session ---------------------
/Library/Frameworks/Python.framework/Versions/3.6/lib/python3.6/site-packages/py/_path/
common.py:377: in visit
    for x in Visitor(fil, rec, ignore, bf, sort).gen(self):
/Library/Frameworks/Python.framework/Versions/3.6/lib/python3.6/site-packages/py/_path/
common.py:429: in gen
    for p in self.gen(subdir):
    if self.fil is None or self.fil(p):
/Library/Frameworks/Python.framework/Versions/3.6/lib/python3.6/site-packages/_pytest/
main.py:612: in _visit_filter
    return f.check(file=1)
/Library/Frameworks/Python.framework/Versions/3.6/lib/python3.6/site-packages/py/_path/
local.py:383: in check
    return not kw["file"] ^ isfile(self.strpath)
/Library/Frameworks/Python.framework/Versions/3.6/lib/python3.6/genericpath.py:30:
in isfile
    st = os.stat(path)
cache
    Return a cache object that can persist state between testing sessions.

    cache.get(key, default)
    cache.set(key, value)

    Keys must be a ``/`` separated value, where the first part is usually the
    name of your plugin or application to avoid clashes with other cache users.

    Values can be any object handled by the json stdlib module.
```

```
capsys
    Enable text capturing of writes to ``sys.stdout`` and ``sys.stderr``.

    The captured output is made available via ``capsys.readouterr()`` method
    calls, which return a ``(out, err)`` namedtuple.
    ``out`` and ``err`` will be ``text`` objects.
... #以下代码太多,故省略,大家自行执行
```

2.4.2 最常用运行测试用例方式

```
pytest [options] [file_or_dir] [file_or_dir] [...]    #中括号是可选项
pytest [参数操作][文件或文件夹]                          #文件或文件夹
```

后面不加任何参数和文件时,pytest 自动收集当前路径下的所有符合要求的测试用例并执行。例如:pytest 执行的是第 2 章 chapter-2 文件夹所有测试用例的执行详细信息。

执行结果如下,这部分结果同时也包括 2.4.1 节实现的执行结果,可以比对是否与结果一致。

```
lindafang@linda chapter-2 % pytest
Test session starts (platform: darwin, Python 3.6.8, pytest 5.2.1, pytest-sugar 0.9.2)
rootdir: /Users/lindafang/PyCharmProjects/pytest_book/src/chapter-2
plugins: rerunfailures-5.0, forked-1.0.2, sugar-0.9.2, assume-1.2.2, xdist-1.28.0,
ordering-0.6, metadata-1.8.0, bdd-3.2.1
collecting ...

―――――――――――――――――― test_f ――――――――――――――――――

    def test_f():
>       assert f() == 4
E       assert 3 == 4
E        +  where 3 = f()

test_assert_1.py:12: AssertionError

 test_assert_1.py ✗
 4%  

―――――――――――― test_long_str_comparison ――――――――――――

    def test_long_str_comparison():
        str3 = 'abcdef'
        str4 = 'adcdef'
>       assert str3 == str4
```

```
E       AssertionError: assert 'abcdef' == 'adcdef'
E         - abcdef
E         ?   ^
E         + adcdef
E         ?   ^

test_assert_2.py:8: AssertionError

 test_assert_2.py ×
7% ▋

------------------------ test_eq_list ------------------------

    def test_eq_list():
>       assert [0, 1, 2] == [0, 1, 3]
E       assert [0, 1, 2] == [0, 1, 3]
E         At index 2 diff: 2 != 3
E         Use -v to get the full diff

test_assert_2.py:12: AssertionError

 test_assert_2.py ×
11% ▋

------------------ test_dict_comparison ------------------

    def test_dict_comparison():
        dict1 = {
            'name': 'linda',
            'age': 18,
        }
        dict2 = {
            'name': 'linda',
            'age': 88,
        }
>       assert dict1 == dict2
E       AssertionError: assert {'age': 18, 'name': 'linda'} == {'age': 88, 'name': 'linda'}
E         Omitting 1 identical items, use -vv to show
E         Differing items:
E         {'age': 18} != {'age': 88}
E         Use -v to get the full diff

test_assert_2.py:24: AssertionError

 test_assert_2.py ×
14% ▋▋
```

```
--------------------- test_set_comparison ----------------------

    def test_set_comparison():
        set1 = set("1308")
        set2 = set("8035")
>       assert set1 == set2
E       AssertionError: assert {'0', '1', '3', '8'} == {'0', '3', '5', '8'}
E         Extra items in the left set:
E         '1'
E         Extra items in the right set:
E         '5'
E         Use -v to get the full diff

test_assert_2.py:30: AssertionError

 test_assert_2.py ×
18%  ██
 test_assert_3.py √
21%  ███
```

```
--------------------------- test_true ---------------------------

    def test_true():
        assert (1 < 3) is True
>       assert (5 > 3) is False
E       assert (5 > 3) is False

test_assert_3.py:17: AssertionError

 test_assert_3.py ×
25%  ███
```

```
-------------------------- test_in_dict ------------------------

    def test_in_dict():
        x = 'linda1'
>       assert x in "I'm linda"
E       assert 'linda1' in "I'm linda"

test_assert_3.py:23: AssertionError

 test_assert_3.py ×√
32%  ████
```

```
------------------- test_long_list ------------------
```

```
        def test_long_list():
            x = [str(i) for i in range(100)]
            y = [str(i) for i in range(0, 100, 2)]
>           assert x == y
E           AssertionError: assert ['0', '1', '2...'4', '5', ...] == ['0', '2', '4...8', '10', ...]
E             At index 1 diff: '1' != '2'
E             Left contains 50 more items, first extra item: '50'
E             Use -v to get the full diff

test_assert_3.py:36: AssertionError

 test_assert_3.py ×
36 % ■■■☐


------------------------ test_long_copy ------------------------

        def test_long_copy():
            #50个12,与50个11比对,pytest会更好地提示它们本质的区别
>           assert [12] * 50 == [11] * 50
E           assert [12, 12, 12, 12, 12, 12, ...] == [11, 11, 11, 11, 11, 11, ...]
E             At index 0 diff: 12 != 11
E             Use -v to get the full diff

test_assert_3.py:41: AssertionError

 test_assert_3.py ×
39 % ■■■☐


------------------------ test_mytest ------------------------

        def test_mytest():
            #当调用f()时出现SystemExit的异常,表示程序是正确的。如果出现其他异常,
            #则表示程序是错误的
            with pytest.raises(SystemExit):
            with pytest.raises(ValueError):
>               f()

test_assert_except.py:17:
_____

        def f():
            #解释器请求退出
>           raise SystemExit(1)
E           SystemExit: 1
```

```
test_assert_except.py:10: SystemExit

  test_assert_except.py ×
43% ■■■■■

---------------- test_match ----------------------------------

    def test_match():
        #当调用 myfunc()时出现值错误,则表示程序是正确的
        #将值的信息保存到 excinfo 中,并且可能断言值中的属性 value 的内容
         with pytest.raises(ValueError) as excinfo:
        #异常类型可以写多个,后面可以跟正则表达式,与断言类似
        with pytest.raises((ValueError, RunTimeError), match = r'.*40011.*'):
>           myfunc()

test_assert_except.py:32:
_____

    def myfunc():
        #引起值错误
>       raise ValueError("返回 40013 支付错误")
E       ValueError: 返回 40013 支付错误

test_assert_except.py:23: ValueError

During handling of the above exception, another exception occurred:

    def test_match():
        #当调用 myfunc()时出现值错误,则表示程序是正确的
        #将值的信息保存到 excinfo 中,并且可能断言值中的属性 value 的内容
         with pytest.raises(ValueError) as excinfo:
        #异常类型可以写多个,后面可以跟正则表达式,与断言类似
        with pytest.raises((ValueError, RunTimeError), match = r'.*40011.*'):
>           myfunc()
E           AssertionError: Pattern '.*40011.*' not found in '返回 40013 支付错误'

test_assert_except.py:32: AssertionError

  test_assert_except.py ×
46% ■■■■■
  test_assert_sample.py √√√
57% ■■■■■■

---------------- test_approx_simple_fail ----------------

    def test_approx_simple_fail():
```

```
>       assert 0.1 + 0.2 == approx(0.35)
E       assert (0.1 + 0.2) == 0.35 ± 3.5e-07
E         + where 0.35 ± 3.5e-07 = approx(0.35)

test_assert_sample.py:30: AssertionError

 test_assert_sample.py ×
61% ■■■■■■■|

------------ test_warrior_long_description ------------------

    def test_warrior_long_description():
        desc = get_long_class_description("warrior")
>       assert (
            desc
            == textwrap.dedent(
        """\
    A seasoned veteran of many battles. Strength and Dexterity
    allow to yield heavy armor and weapons, as well as carry
    more equipment. Weak in magic.
    """
        )
    )
E       AssertionError: assert 'A seasoned v...              \n' == 'A seasoned v...            \n'
E         - A seasoned veteran of many battles. High Strength and Dexterity
E         ?                                     -----
E         + A seasoned veteran of many battles. Strength and Dexterity
E           allow to yield heavy armor and weapons, as well as carry
E         - more equipment while keeping a light roll. Weak in magic.
E         ?                ---------------------------
E         + more equipment. Weak in magic.

test_assert_sample.py:47: AssertionError

 test_assert_sample.py ×
64% ■■■■■■■|

-------------- test_get_starting_equiment ------------------

    def test_get_starting_equiment():
        expected = ["长剑", "战士装备"]
>       assert get_starting_equipment("战士") == expected, "装备不符"
E       AssertionError: 装备不符
E       assert ['长剑', '战士装备', '盾'] == ['长剑', '战士装备']
E         Left contains one more item: '盾'
```

```
E           Use -v to get the full diff

test_assert_sample.py:67: AssertionError

 test_assert_sample.py ×√
71% ■■■■■■■■
 test_assert_skip.py ss√×
86% ■■■■■■■■■

------------------------ test_foo_compare ------------------

    def test_foo_compare():
        f1 = Foo(1)
        f2 = Foo(2)
>       assert f1 == f2 , '第1个实例中的内容与第2个实例中的内容不相等'
E       assert 比较两个 Foo 实例:
E             值: 1 != 2

test_fail_cus.py:19: AssertionError

 test_fail_cus.py ×
89% ■■■■■■■■■

------------------- test_num ---------------------

    def test_num():
>       assert 1 == 2, '这两个值不相等'
E       AssertionError: 这两个值不相等
E       assert 1 == 2

test_fail_cus.py:23: AssertionError

 test_fail_cus.py ×
93% ■■■■■■■■■
 test_raise.py √
96% ■■■■■■■■■

---------------------- test_mytest_error --------------------

    def test_mytest_error():
        with pytest.raises(ImportError):
            print("如果不是解释器请求退出引起的异常,则表示测试不通过")
>           f()

test_raise.py:22:
-------------------------------------------------
```

```
        def f():
            #解释器请求退出
>           raise SystemExit(1)
E           SystemExit: 1

test_raise.py:10: SystemExit
------------------- Captured stdout call -------------------
```
如果不是解释器请求退出引起的异常,则表示测试不通过

```
 test_raise.py ×
100 % ██████████████

Results (0.17s):
        8 passed
       17 failed
         - test_assert_1.py:11 test_f
         - test_assert_2.py:5 test_long_str_comparison
         - test_assert_2.py:11 test_eq_list
         - test_assert_2.py:15 test_dict_comparison
         - test_assert_2.py:27 test_set_comparison
         - test_assert_3.py:15 test_true
         - test_assert_3.py:21 test_in_dict
         - test_assert_3.py:33 test_long_list
         - test_assert_3.py:39 test_long_copy
         - test_assert_except.py:13 test_mytest
         - test_assert_except.py:26 test_match
         - test_assert_sample.py:29 test_approx_simple_fail
         - test_assert_sample.py:45 test_warrior_long_description
         - test_assert_sample.py:65 test_get_starting_equiment
         - test_fail_cus.py:16 test_foo_compare
         - test_fail_cus.py:22 test_num
         - test_raise.py:19 test_mytest_error
        1 xfailed
        2 skipped
```

有些效果是安装了一些插件才有的,插件安装将在第 5 章详细介绍。

2.4.3　通过 python -m pytest 调用 pytest

可以通过 Python 的解释器来执行测试,命令如下:

```
python -m pytest [...]
```

但是,这和直接执行 pytest [...]命令的效果几乎一模一样。

执行结果如下:

```
Results (0.04s):
       1 failed
         - test_assert_1.py:11 test_f
lindafang@linda chapter-2 % python -m pytest test_assert_1.py
```

这是在终端下执行的，上段代码最后一行是执行命令，%号前面是计算机用户信息（lindafang@linda）和所在位置（chapter-2），以及后面的命令（python -m pytest test_assert_1.py）。

2.4.4　在 Python 代码中调用 pytest

可以直接在代码中调用 pytest，在 main()函数中调用 pytest.main()与在命令行中执行 pytest 的效果几乎是一样的，本书在第 1 章中就使用了这种方式调用并执行的。

pytest.main()方法中的选项和参数不同，这也将影响执行的结果。具体看下面的例子。

- 例一：传递选项和参数 pytest.main(['-q', __file__])。

pytest.main(['-q', __file__])中 q 是参数 quiet 的简写，表示静默执行，__file__是获得当前文件名，也是要执行的文件，但它不会触发 SystemExit，而是返回 exitcode。

执行结果如下：

```
# Author: lindafang
# Date: 2020-05-13 12:54
# File: test_invoke_via_main.py

import time


def test_one():
    time.sleep(10)


if __name__ == '__main__':
    import pytest
    ret = pytest.main(['-q', __file__])
    print("pytest.main() 返回 pytest.ExitCode.INTERRUPTED: ", ret == pytest.ExitCode.INTERRUPTED)
```

可以选择两种执行方式，不同的执行方式在中断后信息显示略有不同。

第一种，在 PyCharm 中使用 pytest 框架执行这个文件。

用例中有等待 10s 的操作，在这期间，打断停止执行，pytest.main()返回的是 INTERRUPTED 状态码：KeyboardInterrupt。

执行结果如下：

```
Testing started at 12:57 ...
/usr/local/bin/python3.6 "/Applications/PyCharm
CE.app/Contents/helpers/PyCharm/_jb_pytest_runner.py" -- path
/Users/lindafang/PyCharmProjects/pytest_book/src/chapter-2/test_invoke_via_main.py
Launching pytest with arguments /Users/lindafang/PyCharmProjects/pytest_book/src/chapter-
2/test_invoke_via_main.py in /Users/lindafang/PyCharmProjects/pytest_book/src/chapter-2

===================== test session starts =====================
......
collected 1 item

test_invoke_via_main.py

!!!!!!!!!!!!!!!!!!!!!!!!!!KeyboardInterrupt !!!!!!!!!!!!!!!!!!!!!!!!!!
/Users/lindafang/PyCharmProjects/pytest_book/src/chapter-2/test_invoke_via_main.py:
9: KeyboardInterrupt
(to show a full traceback on KeyboardInterrupt use --full-trace)
===================== no tests ran in 0.81s ==================
Process finished with exit code 0
```

- 第二种，选择终端执行 Python 文件名，它们之间有所不同，后者可以打印出输出信息。

```
lindafang@linda chapter-2 % python test_invoke_via_main.py
Test session starts (platform: darwin, Python 3.6.8, pytest 5.2.1, pytest-sugar 0.9.2)
rootdir: /Users/lindafang/PyCharmProjects/pytest_book/src/chapter-2, inifile: pytest.ini
plugins: rerunfailures-5.0, forked-1.0.2, sugar-0.9.2, assume-1.2.2, xdist-1.28.0,
ordering-0.6, metadata-1.8.0, bdd-3.2.1
^C
!!!!!!!!!!!!!!!!!!!!!!!!!!KeyboardInterrupt !!!!!!!!!!!!!!!!!!!!!!!!!!
/Users/lindafang/PyCharmProjects/pytest_book/src/chapter-2/test_invoke_via_main.py:
9: KeyboardInterrupt
(to show a full traceback on KeyboardInterrupt use --full-trace)

Results (7.71s):
pytest.main() 返回 pytest.ExitCode.INTERRUPTED: True
```

- 例二：传递选项和参数：pytest.main(["-qq"], plugins=[MyPlugin()])指定一个插件。

代码如下：

```
# Author: lindafang
# Date: 2020-05-13 17:47
# File: test_main_plugin.py
```

```python
import pytest


class MyPlugin:
    def pytest_sessionfinish(self):
        import allure
        allure.title("笔者导入了allure")
        print("allure test run reporting finishing")

# 因为没有选择要执行的文件,执行的范围是配置的 path 所在的范围
# 静默执行,并且不显示测试用例所收集的结果消息
pytest.main(["-qq"], plugins=[MyPlugin()])
```

由于 pytest.main 不在 if __name__ == '__main__': 的下面,选择在终端执行,执行结果参考如下,因为没有选择要执行的文件,执行的范围是配置的 path 所在的范围或者所在的目录下的所有用例。有可能只想执行插件加载,结果执行了所有测试用例。

执行结果如下:

```
/usr/local/bin/python3.6
/Users/lindafang/PyCharmProjects/pytest_book/src/chapter-2/test_main_plugin.py
FFFFF.FF.FFFF...FFF.ssss.xFF....ssss...F.FsF.
[100%]allure test run reporting finishing

========================= FAILURES =========================
_____ test_f _____

    def test_f():
>       assert f() == 4
E       assert 3 == 4
E        +  where 3 = f()

test_assert_1.py:12: AssertionError
_____ test_long_str_comparison _____

    def test_long_str_comparison():
        str3 = 'abcdef'
        str4 = 'adcdef'
>       assert str3 == str4
E       AssertionError: assert 'abcdef' == 'adcdef'
E         - abcdef
E         ?  ^
E         + adcdef
E         ?  ^
```

```
test_assert_2.py:8: AssertionError
_____ test_eq_list _____

    def test_eq_list():
>       assert [0, 1, 2] == [0, 1, 3]
E       assert [0, 1, 2] == [0, 1, 3]
E         At index 2 diff: 2 != 3
E         Full diff:
E         - [0, 1, 2]
E         ?         ^
E         + [0, 1, 3]
E         ?         ^

test_assert_2.py:12: AssertionError

.......略过一些结果


Process finished with exit code 0
```

- 例三:传递选项和参数:pytest.main(["-x","mytestdir"]),执行到1个失败便停止。这是"-x"参数的作用,遇到失败就不再往下执行了。

在test_assert_2.py中加入后执行。

代码如下:

```python
# Author: lindafang
# Date: 2020-04-30 15:34
# File: test_assert_2.py
import pytest
def test_long_str_comparison():
    str3 = 'abcdef'
    str4 = 'adcdef'
    assert str3 == str4


def test_eq_list():
    assert [0, 1, 2] == [0, 1, 3]


def test_dict_comparison():
    dict1 = {
        'name': 'linda',
        'age': 18,
    }
```

```python
    dict2 = {
        'name': 'linda',
        'age': 88,
    }
    assert dict1 == dict2


def test_set_comparison():
    set1 = set("1308")
    set2 = set("8035")
    assert set1 == set2


if __name__ == '__main__':
    pytest.main(["-x", "test_assert_2.py"])
```

执行结果如下：4个测试用例都有错误，收集到4个，在第一个测试用例执行之后就停止。

```
/usr/local/bin/python3.6
/Users/lindafang/PyCharmProjects/pytest_book/src/chapter-2/test_assert_2.py
======================= test session starts =======================
platform darwin -- Python 3.6.8, pytest-5.2.1, py-1.8.0, pluggy-0.13.1
rootdir: /Users/lindafang/PyCharmProjects/pytest_book/src/chapter-2, inifile: pytest.ini
plugins: rerunfailures-5.0, forked-1.0.2, sugar-0.9.2, assume-1.2.2, xdist-1.28.0,
ordering-0.6, metadata-1.8.0, bdd-3.2.1
collected 4 items

test_assert_2.py F

============================ FAILURES =============================
_____ test_long_str_comparison _____

    def test_long_str_comparison():
        str3 = 'abcdef'
        str4 = 'adcdef'
>       assert str3 == str4
E       AssertionError: assert 'abcdef' == 'adcdef'
E         - abcdef
E         ?  ^
E         + adcdef
E         ?  ^

test_assert_2.py:8: AssertionError
======================= 1 failed in 0.03s =========================

Process finished with exit code 0
```

注意：调用 pytest.main() 会引入测试文件及其引用的所有模块。由于 Python 引入机制的缓存特性，当这些文件发生变化时，后续再调用 pytest.main()（在同一个程序执行过程中）时，并不会响应这些文件的变化。

基于这个原因，一般不推荐在同一个程序中多次调用 pytest.main()（例如：为了重新执行测试。如果确实有这个需求，则可以考虑 pytest-repeat 插件）。

2.4.5 pytest 执行结束时返回的状态码

pytest 命令执行结束，可能会返回以下 6 种状态码：
ExitCode 0：（OK）所有收集到的用例测试通过；
ExitCode 1：（TESTS_FAILED）用例测试失败；
ExitCode 2：（INTERRUPTED）用户打断测试执行；
ExitCode 3：（INTERNAL_ERROR）测试执行的过程中，发生内部错误；
ExitCode 4：（USAGE_ERROR）pytest 命令使用错误；
ExitCode 5：（NO_TESTS_COLLECTED）没有收集到测试用例。

它们在枚举类 _pytest.main.ExitCode 中声明，并且其作为公开 API 的一部分，能够直接引入和访问，代码如下：

```
from pytest import ExitCode
```

2.4.6 输出代码中的控制台信息

pytest -s 可以输出代码中的控制台信息。

2.4.7 显示详细信息

pytest -v 可以显示更详细的信息，可以多个参数连在一起使用。

显示详细信息包括控制台输出信息，代码如下：

```
pytest -s -v 文件名
```

执行结果如下：

```
lindafang@linda chapter-2 % pytest -s -v test_raise.py
Test session starts (platform: darwin, Python 3.6.8, pytest 5.2.1, pytest-sugar 0.9.2)
cachedir: .pytest_cache
metadata: {'Python': '3.6.8', 'Platform': 'Darwin-19.3.0-x86_64-i386-64bit', 'Packages':
{'pytest': '5.2.1', 'py': '1.8.0', 'pluggy': '0.13.1'}, 'Plugins': {'rerunfailures': '5.0', 'forked':
'1.0.2', 'sugar': '0.9.2', 'assume': '1.2.2', 'xdist': '1.28.0', 'ordering': '0.6', 'metadata':
'1.8.0', 'bdd': '3.2.1'}, 'JAVA_HOME': '/Library/Java/JavaVirtualMachines/JDK1.8.0_201.JDK/
Contents/Home'}
```

```
rootdir: /Users/lindafang/PyCharmProjects/pytest_book/src/chapter-2
plugins: rerunfailures-5.0, forked-1.0.2, sugar-0.9.2, assume-1.2.2, xdist-1.28.0,
ordering-0.6, metadata-1.8.0, bdd-3.2.1
collecting ... 如果确实是解释器请求退出,则这个原因所引起的异常会被测试认定为通过

 test_raise.py::test_mytest √
50% ■■■■■          如果不是解释器请求退出,则这个原因所引起的异常会被测试认定为不通过


————————————————————— test_mytest_error —————————————————————

    def test_mytest_error():
        with pytest.raises(ImportError):
            print("如果不是解释器请求退出,则这个原因所引起的异常会被测试认定为不通过")
>           f()

test_raise.py:22:
_____

    def f():
        # 解释器请求退出
>       raise SystemExit(1)
E       SystemExit: 1

test_raise.py:10: SystemExit

 test_raise.py::test_mytest_error ×
100% ■■■■■■■■■■

Results (0.04s):
       1 passed
       1 failed
         - test_raise.py:19 test_mytest_error
```

2.4.8 不显示详细信息

pytest -q 命令不显示太详细的信息。q 是 quiet 的缩写,表示静默。

2.4.9 显示简单总结结果

-r 选项可以在执行结束后打印一个简短的总结报告。当执行的测试用例很多时,可以对结果有个清晰的了解:在终端输入:pytest -q -rA。

执行结果如下:

```
=============== short test summary info =====================
PASSED test_assert_1.py::test_f
PASSED test_assert_1.py::test_f
PASSED test_assert_2.py::test_long_str_comparison
PASSED test_assert_2.py::test_long_str_comparison
PASSED test_assert_2.py::test_eq_list
......#略过部分
PASSED test_raise.py::test_mytest
PASSED test_raise.py::test_mytest
PASSED test_raise.py::test_mytest_error
PASSED test_raise.py::test_mytest_error
SKIPPED [1] test_assert_skip.py:10: unconditional skip
SKIPPED [1] test_assert_skip.py:19: android平台没有这个功能
xfail test_assert_skip.py::test_xfail
FAILED test_assert_1.py::test_f - assert 3 == 4
FAILED test_assert_2.py::test_long_str_comparison - AssertionError: assert 'abcdef' == 
'adcdef'
FAILED test_assert_2.py::test_eq_list - assert [0, 1, 2] == [0, 1, 3]
FAILED test_assert_2.py::test_dict_comparison - AssertionError: assert {'age': 18, 'name':
'linda'} == {'age':...
FAILED test_assert_2.py::test_set_comparison - AssertionError: assert {'0', '1', '3', '8'} ==
{'0', '3', '5', ...
FAILED test_assert_3.py::test_true - assert (5 > 3) is False
FAILED test_assert_3.py::test_in_dict - assert 'linda1' in "I'm linda"
FAILED test_assert_3.py::test_long_list - AssertionError: assert ['0', '1', '2...'4', '5', ...] ==
['0', '2', ...
FAILED test_assert_3.py::test_long_copy - AssertionError: assert [12, 12, 12, 12, 12, 12,
...] == [11, 11, 11,...
FAILED test_assert_except.py::test_mytest - SystemExit: 1
FAILED test_assert_except.py::test_match - AssertionError: Pattern '.*40011.*' not found
in '返回40013支付错误'
FAILED test_assert_sample.py::test_approx_simple_fail - assert (0.1 + 0.2) == 0.35 ± 3.5e-07
FAILED test_assert_sample.py::test_warrior_long_description - AssertionError: assert 'A
seasoned v...
FAILED test_assert_sample.py::test_get_starting_equiment - AssertionError: 装备不符
FAILED test_fail_cus.py::test_foo_compare - assert 比较两个Foo实例:
FAILED test_fail_cus.py::test_num - AssertionError: 这两个值不相等
FAILED test_raise.py::test_mytest_error - SystemExit: 1
```

-r选项后面要紧接这个参数,用于过滤显示测试用例的结果。

以下是所有有效的字符参数:

- f:失败的;
- E:出错的;
- s:跳过执行的;

- x：跳过执行，并标记为 xfailed 的；
- X：跳过执行，并标记为 xpassed 的；
- p：测试通过的；
- P：测试通过，并且有输出信息的，即用例中有 print 等；
- a：除了测试通过的，其他所有的，即除了 p 和 P 的；
- A：所有的。

上述字符参数可以叠加使用，例如：执行后期望过滤掉失败的和跳过执行的(-rfs)。
代码如下：

```
pytest -rfs

================ short test summary info ==============
FAILED test_assert_1.py::test_f - assert 3 == 4
FAILED test_assert_2.py::test_long_str_comparison - AssertionError: assert 'abcdef' == 'adcdef'
FAILED test_assert_2.py::test_eq_list - assert [0, 1, 2] == [0, 1, 3]
FAILED test_assert_2.py::test_dict_comparison - AssertionError: assert {'age': 18, 'name': 'linda'} == {'age':...
FAILED test_assert_2.py::test_set_comparison - AssertionError: assert {'0', '1', '3', '8'} == {'0', '3', '5', ...
FAILED test_assert_3.py::test_true - assert (5 > 3) is False
FAILED test_assert_3.py::test_in_dict - assert 'linda1' in "I'm linda"
FAILED test_assert_3.py::test_long_list - AssertionError: assert ['0', '1', '2...'4', '5', ...] == ['0', '2', ...
FAILED test_assert_3.py::test_long_copy - assert [12, 12, 12, 12, 12, 12, ...] == [11, 11, 11, 11, 11, 11, ...]
FAILED test_assert_except.py::test_mytest - SystemExit: 1
FAILED test_assert_except.py::test_match - AssertionError: Pattern '.*40011.*' not found in '返回40013支付错误'
FAILED test_assert_sample.py::test_approx_simple_fail - assert (0.1 + 0.2) == 0.35 ± 3.5e-07
FAILED test_assert_sample.py::test_warrior_long_description - AssertionError: assert 'A seasoned v...
FAILED test_assert_sample.py::test_get_starting_equiment - AssertionError: 装备不符
FAILED test_fail_cus.py::test_foo_compare - assert 比较两个 Foo 实例：
FAILED test_fail_cus.py::test_num - AssertionError: 这两个值不相等
FAILED test_raise.py::test_mytest_error - SystemExit: 1
SKIPPED [1] test_assert_skip.py:10: unconditional skip
SKIPPED [1] test_assert_skip.py:19: Android平台没有这个功能
```

2.4.10 执行指定的测试用例

pytest 支持多种方式执行特定的测试用例，这些丰富的形式让测试用例的运行及管理变得更灵活实用。

执行指定模块（文件）中的测试用例，代码如下：

```
pytest 文件名
```

2.4.11　执行指定目录下所有的测试用例

如果在某个文件夹下执行该文件夹下所有测试用例，则可以直接输入 pytest 或者输入 pytest ../文件夹名。

如果在某个文件夹上一级执行该文件夹下所有测试用例，则可以直接输入 pytest 文件夹名/，代码如下：

```
pytest chapter-2/
```

2.4.12　-k 参数执行包含特定关键字的测试用例

-k 参数主要搜索测试方法函数类名中包含与不包含的名字，并可进行与或非的组合逻辑搜索。

执行当前目录下名字包含_class 但不包含 two 的测试用例：

```
pytest -k "Case and not three"
```

执行结果如下：

```
lindafang@linda chapter-1 % pytest -vk "Case and not three"
.....
collecting ...
 test_frame.py::TestCase.test_four √
50% ■■■■■■
 test_frame_2.py::TestCase.test_four √
100% ■■■■■■■■■■

Results (0.05s):
       2 passed
       8 deselected
```

执行指定文件中名字包含 TestCase(类名) 和 three(方法名) 的测试用例：

```
pytest -v 文件名 -k "TestCase and three"
```

执行结果如下：

```
lindafang@linda chapter-2 % pytest ../chapter-1/test_frame.py -k "TestCase and three"
Results (0.06s):
       1 passed
       3 deselected
```

执行当前路径(chapter-1)下名字包含 TestCase 或者 test_one 的函数名、文件名、方法名、类名的用例：

```
pytest -v -k "TestCase or test_one"
```

执行结果如下：

```
lindafang@linda src % pytest -v -k "TestCase or test_one"
........................
collecting ...
 chapter-1/test_frame.py::test_one √
17%  ■■
 chapter-1/test_frame.py::TestCase.test_three √
33%  ■■■■
 chapter-1/test_frame.py::TestCase.test_four √
50%  ■■■■■■
 chapter-1/test_frame_1.py::test_one √
67%  ■■■■■■■■
 chapter-1/test_frame_2.py::TestCase.test_three √
83%  ■■■■■■■■■■
 chapter-1/test_frame_2.py::TestCase.test_four √
100% ■■■■■■■■■■■■

Results (0.38s):
       6 passed
      75 deselected
```

执行指定文件中名字包含 comparison 的用例：

```
pytest -v 文件名 -k "comparison"
```

执行结果如下：

```
Results (0.06s):
       3 failed
         - test_assert_2.py:5 test_long_str_comparison
         - test_assert_2.py:15 test_dict_comparison
         - test_assert_2.py:27 test_set_comparison
       1 deselected
lindafang@linda chapter-2 % pytest test_assert_2.py -k "comparison"
```

注意：Python 的关键字不可以应用在 -k 选项中，例如，class、def 等。

2.4.13 执行指定 nodeid 的测试用例

pytest 为每个收集到的测试用例指定一个唯一的 nodeid。其由模块名加说明符构成，中间以::间隔。其中，说明符可以是类名、函数名。可以通过这种用例唯一的 nodeid 执行指定的测试用例。

示例代码如下，下面创建了 3 个测试用例，分别对应不同的说明符。

```python
# Author: lindafang
# File: test_nodeid.py

import pytest


def test_one():
    print('test_one')
    assert 1


class TestNodeId:
    def test_one(self):
        print('TestNodeId::test_one')
        assert 1

    # 这里是参数化,将在第 4 章中详细讲解
    # 循环执行两遍,第一遍取 x = 1, y = 1,第二遍取 x = 3, y = 4
    @pytest.mark.parametrize('x,y', [(1, 1), (3, 4)])
    def test_two(self, x, y):
        print(f'TestNodeId::test_two::{x} == {y}')
        assert x == y
```

test_one 测试方法是函数，是 TestNodeId 类中的 test_one 类中的方法，此外 test_two 是带有参数化的可执行多次的测试方法。通常使用类名和方法名进行标识，参数化的测试方法使用参数数据进行标识。

执行结果如图 2-8 所示。

指定某个函数名执行，命令如下：

```
pytest -q -s test_nodeid.py::test_one
```

执行结果如下：

```
lindafang@linda chapter-2 % pytest -q -s test_nodeid.py::test_one
test_one
```

第2章　pytest的测试用例管理及运行管理　59

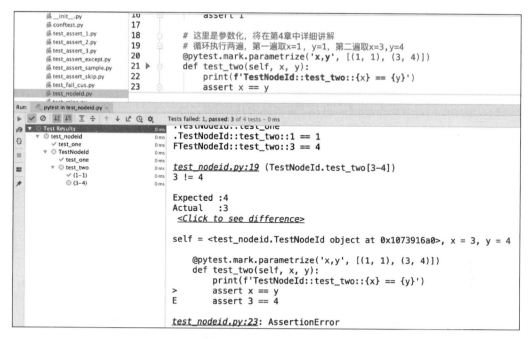

图 2-8　nodeid 执行结果

```
test_nodeid.py √
100% ▇▇▇▇▇▇▇▇▇▇

Results (0.06s):
       1 passed
```

指定某个文件中的某个类名＋函数名执行，命令如下：

pytest －q －s test_nodeid.py::TestNodeId::test_one

执行结果如下：

```
chapter-2 % pytest -q -s test_nodeid.py::TestNodeId::test_one
TestNodeId::test_one

Results (0.04s):
       1 passed
```

注意：这里也可以使用-k 选项以便达到同样的效果。
首先，可以使用--collect-only 选项查看用例名。
执行及结果如下：

```
pytest -q -s --collect-only test_nodeid.py
test_nodeid.py::test_one
test_nodeid.py::TestNodeId::test_one
test_nodeid.py::TestNodeId::test_two[1-1]
test_nodeid.py::TestNodeId::test_two[3-4]
```

然后，使用-k 执行符合规则的用例，例如：执行 test_nodeid.py::test_one：

```
 pytest -q -s -k 'test_one and not TestNodeId' test_nodeid.py
test_one
.
1 passed, 3 deselected in 0.02s
```

结果和执行 pytest -q -s test_nodeid.py::test_one 相同。

2.4.14 -m 参数执行指定标记的用例

-m 参数全称为 marker，可以通过方法名称的表达式灵活选择测试某种方法和不测试一些方法，举例说明如下。

创建一个新的 Python 文件 test_markers.py，代码如下：

```
# Author: lindafang
# File: test_markers.py
"""
Markers 可以实现选择某种方法测试或不选择某些方法测试

命令行:
pytest -v test_markers.py
pytest -v -m 'p0' test_markers.py
pytest -v -m 'not p0' test_markers.py
pytest -v -m 'not p0 and not func' test_markers.py
pytest -v -m 'func or postgres' test_markers.py

"""
import pytest
  @pytest 是一组装饰器的方法,用于在测试方法中添加一些功能,这些细节将在第3章中详细讲解
# 通过 pytest 中的 mark 标记标注每个测试方法的含义
@pytest.mark.p0
def test_something_p0():
    pass


@pytest.mark.func
def test_something_func():
```

```
    pass


@pytest.mark.postgres
def test_something_postgres():
    pass
```

运行脚本中所有方法,输入下面的命令并执行。

```
py.test -v test_markers.py
```

运行脚本中标记为 p0 的测试方法,输入下面的命令并执行。

```
pytest -v -m 'p0' test_markers.py
```

执行结果中会有 pytestUnknownMarkWarning 的警告信息,可以忽略,此外可以在配置文件中配置自定义的标记,配置后警告信息会消失。

配置自定义标记的步骤如下:
(1) 新建 pytest.ini 文件。
(2) 在文件中输入以下内容。

```
[pytest]
markers =
    p0: 冒烟测试用例
    func: 功能测试用例
    postgres: 数据库测试用例
```

(3) p0、func、postgres 应与 test_markers.py 文件测试方法中所标记的意义一致,并写清自定义的标记所表示的意思。
(4) 再次执行 pytest -v -m 'p0' test_markers.py。
(5) 执行,正常情况下结果中不再显示警告信息。

```
lindafang@linda chapter-2 % pytest -v -m 'p0' test_markers.py
collecting ...
 test_markers.py::test_something_p0 √
100% ■■■■■■■■■■

Results (0.04s):
       1 passed
       2 deselected
```

运行脚本中不是 p0 的所有测试方法,输入 pytest -v -m 'not p0' test_markers.py 命令

并执行，其结果如下：

```
lindafang@linda chapter-2 % pytest -v -m 'not p0' test_markers.py
……
collecting ...
 test_markers.py::test_something_func √
50%  ■■■■■
 test_markers.py::test_something_postgres √
100% ■■■■■■■■■■

Results (0.04s):
       2 passed
       1 deselected
```

运行脚本中不是 p0 也不是 func 的所有测试方法，输入 pytest -v -m 'not p0 and not func' test_markers.py 命令并执行，其结果如下：

```
lindafang@linda chapter-2 % pytest -v -m 'not p0 and not func' test_markers.py

collecting ...
 test_markers.py::test_something_postgres √
100% ■■■■■■■■■■

Results (0.04s):
       1 passed
       2 deselected
```

运行脚本中 func 或是 postgres 的测试方法，输入 pytest -v -m 'func or postgres' test_markers.py 命令并执行，其结果如下：

```
lindafang@linda chapter-2 % pytest -v -m 'func or postgres' test_markers.py
collecting ...
 test_markers.py::test_something_func √
50% ■■■■■
 test_markers.py::test_something_postgres √
100% ■■■■■■■■■■

Results (0.04s):
       2 passed
       1 deselected
```

2.4.15 执行指定包中的测试用例

有时要测试的用例不在常用的文件夹中,此时可以通过指定包来执行包所能收集到的测试用例。执行命令 pytest --pyargs pkg.testing；此命令中的 pytest 会引入 pkg.testing 包,并在它的系统目录下搜寻测试用例并执行。

Python 包的布局及建立步骤如下。

先在 pytest-book 下创建文件 setup.py,代码如下：

```python
# Author: lindafang
# Date: 2020 - 05 - 13 21:02
# File: setup.py

from setuptools import setup, find_packages

setup(name = "utils", packages = find_packages())
```

(1) 在 src 同级目录下建立 utils 的包。

在下面创建 max.py 文件,代码如下：

```python
# Author: lindafang
# Date: 2020 - 05 - 13 21:03
# File: max.py

def max(values):

    _max = values[0]

    for val in values:
        if val > _max:
            _max = val

    return _max
```

(2) 在 utils 下创建 tests 的包。

在 tests 文件夹创建测试 max.py 的测试文件 test_max.py,代码如下：

```python
# Author: lindafang
# Date: 2020 - 05 - 13 21:05
# File: test_max.py


def test_max():
    values = (2, 3, 1, 4, 6, 7)
```

```
    val = max(values)
    assert val == 6
```

(3) 整体的布局如下:

```
setup.py
src
└── chapter-1
    │   test_frame.py
    │
    utils
    │   max.py
    │   __init__.py
    │
    └──tests
           test_max.py
            __init__.py
```

(4) 最后通过执行包 pytest --pyargs utils 执行测试用例:

```
lindafang@linda src % pytest -- pyargs utils
Test session starts (platform: darwin, Python 3.6.8, pytest 5.2.1, pytest-sugar 0.9.2)
rootdir: /Users/lindafang/PyCharmProjects/pytest_book
plugins: rerunfailures-5.0, forked-1.0.2, sugar-0.9.2, assume-1.2.2, xdist-1.28.0, ordering-0.6, metadata-1.8.0, bdd-3.2.1
collecting ...

―――――――――――――――――― test_max ――――――――――――――――――

    def test_max():
        values = (2, 3, 1, 4, 6, 7)

        val = max(values)
>       assert val == 6
E       assert 7 == 6

utils/tests/test_max.py:11: AssertionError

 src/utils/tests/test_max.py ×
100% ██████████████

Results (0.07s):
       1 failed
         - src/utils/tests/test_max.py:7 test_max
```

2.4.16 修改回溯信息的输出模式

pytest 回溯信息的输出一共有 6 种模式：auto、long、short、line、native、no，用--tb 选项指定：

```
pytest - l, -- showlocals          # 打印本地变量
pytest -- tb = auto                # 默认模式
pytest -- tb = long                # 尽可能详细的输出
pytest -- tb = short               # 更简短的输出
pytest -- tb = line                # 每个失败信息总结在一行中
pytest -- tb = native              # Python 的标准输出
pytest -- tb = no                  # 不打印失败信息
```

--full-trace 是一种比--tb＝long 更详细的输出模式。它甚至能观察到用户打断执行（Ctrl＋C）时的回溯信息，而上述 6 种模式默认是不输出此类信息的。

输入命令 pytest --tb＝line test_assert_2.py 并执行，每个失败信息总结在一行中。

执行结果如下：

```
lindafang@linda chapter - 2 % pytest -- tb = line test_assert_2.py
Test session starts (platform: darwin, Python 3.6.8, pytest 5.2.1, pytest - sugar 0.9.2)
rootdir: /Users/lindafang/PyCharmProjects/pytest_book/src/chapter - 2, inifile: pytest.ini
plugins: rerunfailures - 5.0, forked - 1.0.2, sugar - 0.9.2, assume - 1.2.2, xdist - 1.28.0,
ordering - 0.6, metadata - 1.8.0, bdd - 3.2.1
collecting ...
/Users/lindafang/PyCharmProjects/pytest_ book/src/chapter - 2/test _ assert _ 2. py: 8:
AssertionError: assert 'abcdef' == 'adcdef'

 test_assert_2.py ×
25 % ■■■
/Users/lindafang/PyCharmProjects/pytest_book/src/chapter - 2/test_assert_2.py:12: assert
[0, 1, 2] == [0, 1, 3]

 test_assert_2.py ×
50 % ■■■■■■
/Users/lindafang/PyCharmProjects/pytest_ book/src/chapter - 2/test _ assert _ 2. py: 24:
AssertionError: assert {'age': 18, 'name': 'linda'} == {'age': 88, 'name': 'linda'}

 test_assert_2.py ×
75 % ■■■■■■■■■
/Users/lindafang/PyCharmProjects/pytest_ book/src/chapter - 2/test _ assert _ 2. py: 31:
AssertionError: assert {'0', '1', '3', '8'} == {'0', '3', '5', '8'}
```

```
test_assert_2.py ×
100% ■■■■■■■■■

Results (0.04s):
     4 failed
         - test_assert_2.py:5 test_long_str_comparison
         - test_assert_2.py:11 test_eq_list
         - test_assert_2.py:15 test_dict_comparison
         - test_assert_2.py:27 test_set_comparison
```

选项--tb=no 不打印具体失败信息,输入下面命令执行:

```
pytest --tb=no test_assert_2.py
```

执行结果如下:

```
lindafang@linda chapter-2 % pytest --tb=no test_assert_2.py
Test session starts (platform: darwin, Python 3.6.8, pytest 5.2.1, pytest-sugar 0.9.2)
rootdir: /Users/lindafang/PyCharmProjects/pytest_book/src/chapter-2, inifile: pytest.ini
plugins: rerunfailures-5.0, forked-1.0.2, sugar-0.9.2, assume-1.2.2, xdist-1.28.0, ordering-0.6, metadata-1.8.0, bdd-3.2.1
collecting ...

 test_assert_2.py ×
25% ■■

 test_assert_2.py ×
50% ■■■■

 test_assert_2.py ×
75% ■■■■■■

 test_assert_2.py ×
100% ■■■■■■■■■

Results (0.04s):
     4 failed
         - test_assert_2.py:5 test_long_str_comparison
         - test_assert_2.py:11 test_eq_list
         - test_assert_2.py:15 test_dict_comparison
         - test_assert_2.py:27 test_set_comparison
```

2.5 运行的失败管理

2.5.1 最多允许失败的测试用例数

当达到最大上限时,退出执行。如未配置,则没有上限。

命令 pytest -x 遇到第一个失败时,退出执行。

```
Results (0.07s):
       1 failed
         - test_assert_1.py:11 test_f
lindafang@linda chapter-2 % pytest -x
```

命令 pytest --maxfail=2 遇到第 2 个失败时,退出执行。同理,命令 pytest --maxfail=3 遇到第 3 个失败时退出执行。

执行结果如下:

```
Results (0.09s):
       3 failed
         - test_assert_1.py:11 test_f
         - test_assert_2.py:5 test_long_str_comparison
         - test_assert_2.py:11 test_eq_list
lindafang@linda chapter-2 % pytest --maxfail=3
```

2.5.2 失败运行管理的原理

用例运行失败后,通常希望再次执行失败的用例,看是环境问题还是脚本问题,因此测试方法运行失败需被记录。

这部分内容与第 12 章缓存目录设置配置内容相一致。cacheprovider 插件将执行状态写入缓存文件夹,pytest 会将本轮测试的执行状态写入 .pytest_cache 文件夹,这个行为是由自带的 cacheprovider 插件实现的。

pytest 默认将测试执行的状态写入根目录中的 .pytest_cache 文件夹,也可以通过在 pytest.ini 中配置 cache_dir 选项来自定义缓存的目录,它可以是相对路径,也可以是绝对路径,相对路径指的是相对于 pytest.ini 文件所在的目录。

例如,我们想把第 12 章执行的状态的缓存和源码放在一起,该如何实现呢?实现步骤如下。

在 src/chapter12/pytest.ini 中添加如下配置:

```
[pytest]
cache_dir = .pytest-cache
```

这样，即使在项目的根目录下执行 src/chapter12/中的用例，也只会在 pytest_book/src/chapter12/.pytest_cache 中生成缓存，而不是 pytest_book/.pytest_cache 中。

（1）在 chapter12 中复制 chapter-2 中的 test_assert_2.py 并改名为 test_assert_12.py，需要将内容进行适当修改。

（2）在 src 路径下执行 pytest src/chapter12。

（3）执行状态未写在根目录下，而是写在测试文件所在的目录中（.pytest_cache）。

执行结果如图 2-9 所示。

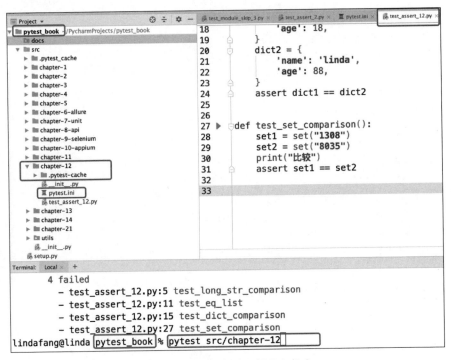

图 2-9　执行状态写入缓存文件夹

cacheprovider 插件实现失败管理的准备：

（1）在 chapter12 下创建 pytest.ini 文件，代码如下：

```
[pytest]
cache_dir = .pytest-cache
```

（2）在 chapter12 下创建 test_failed.py 文件，代码如下：

```
src/chapter12/test_failed.py

import pytest
```

```python
@pytest.mark.parametrize('num', [1, 2])
def test_failed(num):
    assert num == 1
```

(3)在 chapter12 下创建 test_pass.py 文件,代码如下:

```
src\chapter12\test_pass.py
```

```python
def test_pass():
    assert 1
```

(4)如果大家是从前到后进行实践,则需要删除 test_assert_12.py 和 .pytest_cache。
(5)再在根目录下执行 pytest src/chapter12。
(6)chapter12 下会自动创建缓存文件夹及文件 .pytest_cache。

执行结果如下,可以看到一共收集到 3 个测试用例,其中有一个失败,另外两个成功,并且两个执行成功的用例分属不同的测试模块。

执行结果如下:

```
lindafang@linda pytest_book % pytest --tb=line src/chapter12
collecting ...
 test_failed.py √
33% ■■■■
/Users/lindafang/PyCharmProjects/pytest_book/src/chapter12/test_failed.py:10: assert 2 == 1

 test_failed.py ×
67% ■■■■■■■
 test_pass.py √
100% ■■■■■■■■■■

Results (0.04s):
       2 passed
       1 failed
         - test_failed.py:8 test_failed[2]
```

同时,pytest 也在 src/chapter12/目录下生成缓存文件夹(.pytest_cache),具体的目录结构如图 2-10 所示。

现在结合上面的组织结构,具体介绍一下 cacheprovider 插件的功能。
--lf,--last-failed:只执行上一轮失败的用例。

缓存中的 lastfailed 文件记录了上次失败的用例 ID,可以通过--cache-show 命令查看它的内容。

--cache-show 命令是 cacheprovider 提供的新功能,它不会导致任何用例的执行。
执行结果如下:

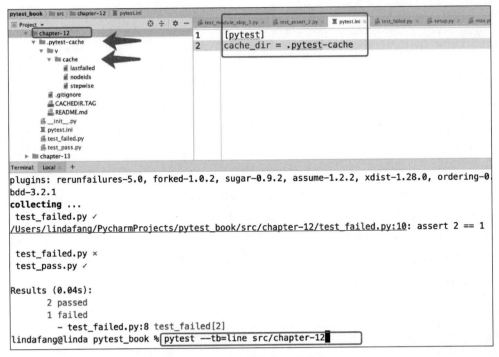

图 2-10 生成缓存的目录结构

```
lindafang@linda pytest_book % pytest src/chapter12/ -q --cache-show 'lastfailed'
Test session starts (platform: darwin, Python 3.6.8, pytest 5.2.1, pytest-sugar 0.9.2)
cachedir: .pytest-cache
rootdir: /Users/lindafang/PyCharmProjects/pytest_book/src/chapter12, inifile: pytest.ini
plugins: rerunfailures-5.0, forked-1.0.2, sugar-0.9.2, assume-1.2.2, xdist-1.28.0,
ordering-0.6, metadata-1.8.0, bdd-3.2.1
cachedir: /Users/lindafang/PyCharmProjects/pytest_book/src/chapter12/.pytest-cache
------------------- cache values for 'lastfailed' ------------------------
cache/lastfailed contains:
  {'test_failed.py::test_failed[2]': True}


Results (0.02s):
```

可以看到，它记录了一个用例，此用例为上次失败的测试用例的 ID：test_failed.py::test_failed[2]。

下次执行，当使用--lf 选项时，pytest 在收集阶段只会选择这个失败的用例，而忽略其他的用例。命令 pytest --lf --collect-only src/chapter12/中的 collect-only 选项只收集用例而不执行。

执行结果如下：

```
lindafang@linda pytest_book % pytest --lf --collect-only src/chapter12/
Test session starts (platform: darwin, Python 3.6.8, pytest 5.2.1, pytest-sugar 0.9.2)
cachedir: .pytest-cache
rootdir: /Users/lindafang/PyCharmProjects/pytest_book/src/chapter12, inifile: pytest.ini
plugins: rerunfailures-5.0, forked-1.0.2, sugar-0.9.2, assume-1.2.2, xdist-1.28.0,
ordering-0.6, metadata-1.8.0, bdd-3.2.1
collecting ... <Module test_failed.py>
  <Function test_failed[2]>
run-last-failure: rerun previous 1 failure (skipped 5 files)


Results (0.03s):
       1 deselected
```

仔细观察一下上面的回显，collecting ... <Module test_failed.py> 只收集到 test_fail.py 文件中 <Function test_failed[2]>，表示第二个执行用例失败了，需要重新执行（跳过 5 个文件）。1 deselected（没有选择），此处的"没有选择"是 <Module test_failed.py> 中执行通过的第一个用例。跳过 5 个文件包括 pytest.ini 和 test_pass.py 在内的 5 个文件。

实际上，--lf 复写了用例收集阶段的两个钩子方法：pytest_ignore_collect(path, config) 和 pytest_collection_modifyitems(session, config, items)。

先来看一看 pytest_ignore_collect(path, config)，如果它的结果返回值为 true，就忽略 path 路径中的用例。

cacheprovider.py 中 pytest_ignore_collect 方法的源码如下：

```python
_pytest/cacheprovider.py

    def last_failed_paths(self):
        """Returns a set with all Paths()s of the previously failed nodeids (cached).
        """
        try:
            return self._last_failed_paths
        except AttributeError:
            rootpath = Path(self.config.rootdir)
            result = {rootpath / nodeid.split("::")[0] for nodeid in self.lastfailed}
            result = {x for x in result if x.exists()}
            self._last_failed_paths = result
            return result

    def pytest_ignore_collect(self, path):
        """
        Ignore this file path if we are in --lf mode and it is not in the list of
        previously failed files.
        """
```

```python
            if self.active and self.config.getoption("lf") and path.isfile():
                last_failed_paths = self.last_failed_paths()
                if last_failed_paths:
                    skip_it = Path(path) not in self.last_failed_paths()
                    if skip_it:
                        self._skipped_files += 1
                    return skip_it
```

可以看到,如果当前收集的文件不在上一次失败的路径集合内,就会忽略这个文件,所以这次执行就不会到 test_pass.py 中收集用例了,故而只收集到两个用例,并且 pytest.ini 也在忽略的名单上,所以实际上是跳过了。

至于 pytest_collection_modifyitems(session,config,items)钩子方法,接下来和--ff 命令一起讲解。

--ff,--failed-first:先执行上一轮失败的用例,再执行其他的用例。

先通过实践看一看这个命令的效果,再去分析它的实现,命令如下:

```
pytest --collect-only -s --ff src/chapter12/
```

执行结果如下:

```
lindafang@linda pytest_book % pytest --collect-only -s --ff src/chapter12/
Test session starts (platform: darwin, Python 3.6.8, pytest 5.2.1, pytest-sugar 0.9.2)
cachedir: .pytest-cache
rootdir: /Users/lindafang/PyCharmProjects/pytest_book/src/chapter12, inifile: pytest.ini
plugins: rerunfailures-5.0, forked-1.0.2, sugar-0.9.2, assume-1.2.2, xdist-1.28.0, ordering-0.6, metadata-1.8.0, bdd-3.2.1
collecting ... <Package /Users/lindafang/PyCharmProjects/pytest_book/src/chapter12>
  <Module test_failed.py>
    <Function test_failed[2]>
    <Function test_failed[1]>
  <Module test_pass.py>
    <Function test_pass>
run-last-failure: rerun previous 1 failure first



Results (0.03s):
```

可以看到一共收集到 3 个测试用例,和正常所收集到的测试用例的顺序相比,上一轮失败的 test_failed.py::test_failed[2]用例在最前面,将优先执行。

实际上,-ff 只复写了钩子方法:pytest_collection_modifyitems(session,config,items),它可以过滤或者重新排序所收集到的用例。

cacheprovider.py 中 pytest_collection_modifyitems 方法的源码如下:

```
_pytest/cacheprovider.py
    def pytest_collection_modifyitems(self, session, config, items):
        ...
            if self.config.getoption("lf"):
                items[:] = previously_failed
                config.hook.pytest_deselected(items = previously_passed)
            else:    -- failedfirst
                items[:] = previously_failed + previously_passed
……
```

可以看到，如果使用的是 lf，就把之前成功的用例状态置为 deselected，这轮执行就会忽略它们。如果使用的是-ff，只是将之前失败的用例按顺序调到前面。

另外，也可以看到 lf 的优先级要高于 ff，所以它们同时使用，ff 是不起作用的。

--nf，--new-first：先执行新加的或修改的用例，再执行其他的用例。

缓存中的 nodeids 文件记录了上一轮执行的所有用例，命令如下：

```
pytest src/chapter12 -- cache-show 'nodeids'
```

执行结果如下：

```
lindafang@linda pytest_book % pytest src/chapter12 -- cache-show 'nodeids'
Test session starts (platform: darwin, Python 3.6.8, pytest 5.2.1, pytest-sugar 0.9.2)
cachedir: .pytest-cache
rootdir: /Users/lindafang/PyCharmProjects/pytest_book/src/chapter12, inifile: pytest.ini
plugins: rerunfailures-5.0, forked-1.0.2, sugar-0.9.2, assume-1.2.2, xdist-1.28.0, ordering-0.6, metadata-1.8.0, bdd-3.2.1
cachedir: /Users/lindafang/PyCharmProjects/pytest_book/src/chapter12/.pytest-cache
------------------ cache values for 'nodeids' ------------------
cache/nodeids contains:
  ['test_failed.py::test_failed[1]',
   'test_failed.py::test_failed[2]',
   'test_pass.py::test_pass']


Results (0.02s):
```

从执行结果可以看到上一轮共执行了 3 个测试用例。

现在可以在 test_pass.py 中新加一个用例，并修改一下 test_failed.py 文件中的用例（但是不添加新用例）。

代码如下：

```
    src\chapter12\test_pass.py

def test_pass():
    assert 1


def test_new_pass():
    assert 1
```

再来执行一下收集命令,命令如下:

```
pytest --collect-only -s --nf src/chapter12/
```

执行结果如下:

```
lindafang@linda pytest_book % pytest --collect-only -s --nf src/chapter12/
Test session starts (platform: darwin, Python 3.6.8, pytest 5.2.1, pytest-sugar 0.9.2)
cachedir: .pytest-cache
rootdir: /Users/lindafang/PyCharmProjects/pytest_book/src/chapter12, inifile: pytest.ini
plugins: rerunfailures-5.0, forked-1.0.2, sugar-0.9.2, assume-1.2.2, xdist-1.28.0,
ordering-0.6, metadata-1.8.0, bdd-3.2.1
collecting ... <Package /Users/lindafang/PyCharmProjects/pytest_book/src/chapter12>
  <Module test_pass.py>
    <Function test_new_pass>
    <Function test_pass>
  <Module test_failed.py>
    <Function test_failed[1]>
    <Function test_failed[2]>



Results (0.06s):
```

可以看到,新加的用例排在最前面,其次修改过的测试用例紧接其后,最后才是旧的用例,这个行为在源码中有所体现。

cacheprovider.py 中 pytest_collection_modifyitems 方法的源码如下:

```
    _pytest/cacheprovider.py

    def pytest_collection_modifyitems(self, session, config, items):
        if self.active:
            new_items = OrderedDict()
            other_items = OrderedDict()
            for item in items:
                if item.nodeid not in self.cached_nodeids:
```

```
                    new_items[item.nodeid] = item
                else:
                    other_items[item.nodeid] = item

        items[:] = self._get_increasing_order(
            new_items.values()
        ) + self._get_increasing_order(other_items.values())
        self.cached_nodeids = [x.nodeid for x in items if isinstance(x, pytest.Item)]

    def _get_increasing_order(self, items):
        return sorted(items, key=lambda item: item.fspath.mtime(), reverse=True)
```

item.fspath.mtime()代表用例所在文件的最后修改时间,reverse＝True表明按倒序进行排列。

items[:]＝self._get_increasing_order(new_items.values())＋self._get_increasing_order(other_items.values())保证新加的用例永远在最前面。

--cache-clear:先清除所有缓存,再执行用例。

源码如下:

```
_pytest/cacheprovider.py

class Cache:

    ...

    @classmethod
    def for_config(cls, config):
        cachedir = cls.cache_dir_from_config(config)
        if config.getoption("cacheclear") and cachedir.exists():
            rm_rf(cachedir)
            cachedir.mkdir()
        return cls(cachedir, config)
```

可以看到,它会先把已有的缓存文件夹删除(rm_rf(cachedir)),再创建一个空的同名文件夹(cachedir.mkdir()),这样会导致上述的功能失效,所以一般不使用这个命令。

如果上一轮没有失败的用例,则需要先清除缓存,再执行 test_pass.py 模块(它的用例都是能测试成功的),如图 2-11 所示。

执行及结果如下:

```
pytest -- cache-clear -q -s src/chapter12/test_pass.py
.
1 passed in 0.01s
```

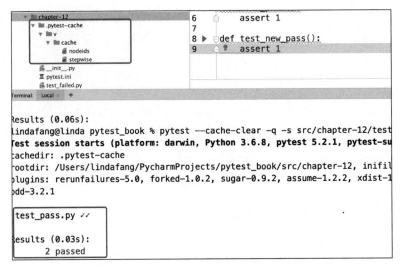

图 2-11　执行 test_pass.py 模块结果

其结果是不是少了点什么呢？对！因为没有失败的用例，所以不会生成 lastfailed 文件，那么这个时候再使用 --lf 和 --ff 会发生什么呢？我们来试试。

注意：如果我们观察得足够仔细，就会发现现在的缓存目录和之前相比不仅少了 lastfailed 文件，还少了 CACHEDIR.TAG、.gitignore 和 README.md 这 3 个文件。这是一个 Bug，现在 pytest 的版本是 5.2.1，预计会在之后的版本得到修复。

2.6　跳过 skip 测试用例的执行

实际工作中，测试用例的执行可能会依赖于一些外部条件，例如：只能运行在某个特定的操作系统（Windows），或者我们本身期望它们测试失败，例如：被某个已知的 Bug 所阻塞。如果我们能为这些用例提前做上标记，那么 pytest 就可以相应地预处理它们，并提供一个更加准确的测试报告。

在这种场景下，常用的标记有以下几种。

skip：只有当某些条件得到满足时，才执行测试用例，否则跳过整个测试用例的执行。例如，在非 Windows 平台上跳过只支持 Windows 系统的用例。

xfail：因为一个确切的原因，我们知道这个用例会失败。例如，对某个未实现的功能的测试，或者阻塞于某个已知 Bug 的测试。

pytest 默认不显示 skip 和 xfail 用例的详细信息，我们可以通过 -r 选项来自定义这种行为。

通常，我们使用一个字母作为一种类型的代表，具体的规则如下：

(f)ailed,(E)rror,(s)kipped,(x)failed,(X)passed,(p)assed,(P)assed with output,(a)ll except passed(p/P),or(A)ll。

下面通过几种不同的实现方式对不同层次的用例进行跳过操作。

2.6.1 @pytest.mark.skip 装饰器

跳过执行某个用例最简单的方式就是使用@pytest.mark.skip装饰器，并且可以设置一个可选参数 reason，表明跳过的原因，使用装饰器的原因是在不修改测试方法函数代码的前提下就可达到目的。

例如：如果要直接跳过 test_cakan() 方法，则可以在方法的上面直接加上@pytest.mark.skip，这样这种方法就不会被执行了。

```
def f():
    return 3

@pytest.mark.skip
def test_cakan():
    pass
    assert f() == 4
```

加上跳过的原因，其代码可以这样写。

```
@pytest.mark.skip(reason = "当前没有这个办法")
def test_the_unknown():
    ...
```

执行的结果如下，如果显示 s，则表示跳过，第一种方法没写原因，则显示 unconditional skip，第二种方法写了原因，则直接显示具体原因。

```
test_assert_skip_1.py s
Skipped: unconditional skip
s
Skipped: 当前没有这个办法
```

2.6.2 pytest.skip 方法

如果我们想在测试执行期间（也可以在 SetUp/TearDown 期间）强制跳过后续的步骤，则可以考虑 pytest.skip() 方法，它同样可以设置一个参数 msg，表明跳过的原因。

例如：从配置文件中读取的数值与系统返回的值比对。如果读取的是 32 位系统的数值，但配置所返回的数值是 64 位的系统的数值，比较后不符合，所以返回 0。返回 1 表示有效配置，返回 0 表示无效配置。这个可以测试不同配置在不同环境下是否配置成功。

代码如下，其他功能代码需要自己补全：

```
def valid_config():
    #返回 1 表示有效配置,返回 0 表示无效配置
    #这里可以从配置文件中读取的数值,与系统返回的值比对
    #例如,读取的数值是 32 位系统的数值,但配置所返回的数值是 64 位系统的数值,
    #比较后所以返回 0
    #这个可以测试不同配置在不同环境下是否配置成功
    return 0


def test_function():
    if not valid_config():
        pytest.skip("不支持此项配置")
```

执行的结果如下:

```
test_assert_skip_1.py s
Skipped: 不支持此项配置
```

另外,还可以为其设置一个布尔型的参数 allow_module_level(默认为 False),表明是否允许在模块中调用这种方法,如果置为 True,则跳过模块中剩余的部分,也就是说其值为 True 时这个模块中所有测试方法都被跳过。

例如,在 Mac 平台下,sys.platform 返回 darwin,测试这个模块。

代码如下:

```
#Author: lindafang
#File: test_Windows_skip.py

'''
sys.platform 返回值
系统            返回值
Windows     :   'win32'
Linux       :   'Linux'
Windows/Cygwin  :   'cygwin'
Mac OS X    :   'darwin'    本机是这个

'''
import sys
import pytest


    #如果参数换成 win32,则表示 Mac 系统不执行,而 Windows 系统执行
    #大家可以更换一下试试
if not sys.platform.startswith("darwin"):
```

```
        print(sys.platform)
        pytest.skip("如果不是 Mac 系统,就跳过而不执行,"
                    "如果是 Mac 系统,则执行,因为后面参数是模块级参数,所以是整个文件的用例",
allow_module_level = True)


def f():
    print("\n" + sys.platform)
    return 3


def test_cakan():
    assert f() == 4


def test_2():
    print("查看这个用例是不是也被执行,当参数是 win32 时不执行")
    pass
```

将参数换成 win32 后,运行结果如下:因为笔者的计算机是 Mac 系统的,所以就跳过不执行。

```
==================== test session starts =============================
 Skipped: 如果不是 Windows 系统,就跳过不执行,如果是 Windows 系统,则执行,因为后面参数是模块级参数所以是整个文件的用例

collected 0 items / 1 skipped

====================== 1 skipped in 0.03s =============================
Process finished with exit code 0
```

注意:当在用例中设置 allow_module_level 参数时,并不会生效,因为这个参数是模块级专用参数。

在代码中增加语句 allow_module_level=True,由于这个是模块级专用的参数,所以放在某个测试方法中起不到把整个模块也就是文件都跳过的作用。

代码如下:

```
# Author: lindafang
# Date: 2020 - 05 - 09 14:20
# File: test_Windows_skip_1.py

import sys
import pytest
```

```python
def test_login():
    pytest.skip("跳出", allow_module_level = True)
    assert 'linda' == 'linda'

def f():
    print("\n" + sys.platform)
    return 3


def test_cakan():
    assert f() == 4

def test_2():
    print("查看这个用例是不是也被执行,当参数是win32时不执行")
    pass
```

执行的结果如下,只跳过这个用例,而不跳过这个用例下面所有的用例。也就是说, allow_module_level=True 这个参数虽然影响范围是文件模块级但放在某种方法用例中,它就不起作用了,如图 2-12 所示。第 1 种方法跳过,第 2 种方法失败,第 3 种方法成功了。

后两种方法并未跳过,而是都执行了,所以用例中的 allow_module_level=True 这个参数未影响到整个模块。

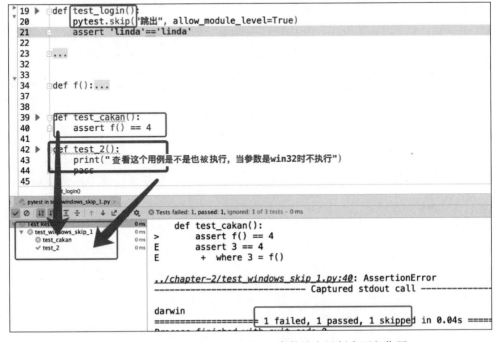

图 2-12　allow_module_level=True 参数放在用例中不起作用

2.6.3 @pytest.mark.skipif 装饰器

带条件的跳过执行：满足条件就跳过，不满足条件就不跳过。在单个用例中使用有条件的跳过，以及在单个用例中使用两种有条件的跳过，体会这两个条件的逻辑组合所产生的不同的结果。使用装饰器可以将条件跳过在模块级共享。

1. 在单个用例中使用 pytest.mark.skipif 标记

如果想有条件地跳过某些测试用例的执行，可以使用@pytest.mark.skipif 装饰器。

例如：不同的环境执行不同的用例。

代码如下：

```
environment = 'android'


@pytest.mark.skipif('environment == "android"', reason = 'android平台没有这个功能')
def test_cart_3():
    pass
```

2. 一个用例上用两个不同的 pytest.mark.skipif 标记

例如，当 Python 的版本低于 3.6 时，跳过用例，并且一个用例上面可以叠加不同的跳过条件。

代码如下：

```
# Author: lindafang
# File: test_assert_skip_2.py

import pytest
import sys

environment = 'android'

@pytest.mark.skipif('environment == "android"', reason = 'android平台没有这个功能')
def test_cart_3():
    pass


@pytest.mark.skipif(sys.platform == 'win32', reason = '不在 Windows 下运行')
@pytest.mark.skipif(sys.version_info < (3,6), reason = '3.6版本以下不执行,你需要更高版本')
def test_cart():
    print("当 Python 版本大于或等于 3.6 时执行,Windows 下不执行")
    pass
```

执行结果如下，第一种方法跳过，第二种方法执行了。因为 Python 的版本为 3.6.8，并且不是 Windows 平台,所以执行了。

```
============== test session starts ==============
test_assert_skip_2.py s
Skipped: Android 平台没有这个功能
.当 Python 版本大于或等于 3.6 时执行,Windows 平台下不执行
                                                    [100%]

========== 1 passed, 1 skipped in 0.02s ==========
Process finished with exit code 0
```

你也可以把版本中子版本 6 改成 9,这样就应该跳过了。提示信息未改,执行结果如下:

```
s
Skipped: 3.6 版本以下不执行,你需要更高版本
```

注意:sys.version_info < (3,6),这个 3 后面是逗号",",而不是点。

当一个用例指定了多个 skipif 条件时,只需满足其中一个,就可以跳过这个用例的执行。

3. 两个模块之间共享 pytest.mark.skipif 标记

例如:我们新建 test_module 文件夹,创建两个文件:test_module_skip_1 和 test_module_skip_2,在文件 test_module_skip_1 中定义 minversion,表明支持的 Python 最低版本。

代码如下:

```python
# Author: lindafang
# Date: 2020-05-09 16:22
# File: test_module_skip_2.py
import sys

import pytest

minversion = pytest.mark.skipif(sys.version_info < (3, 8),
                                reason = '请使用 Python 3.8 或者更高的版本.')

@minversion
def test_one():
    assert True
```

此外,在 test_module_skip_2.py 中引入 minversion,代码如下:

```
# Author: lindafang
# Date: 2020-05-09 16:22
# File: test_module_skip_2.py

from .test_module_skip_1 import minversion


@minversion
def test_two():
    assert True
```

现在,我们来执行这两个用例(当前的 Python 版本为 3.6.8),执行结果如下:

```
lindafang@linda test_module % pytest -rs -k "module" .
...
collecting ...
 test_module/test_module_skip_1.py s
 50%  ██████
 test_module/test_module_skip_2.py s
100%  ████████████
================= short test summary info =============================
SKIPPED [1] test_module/test_module_skip_1.py:13: 请使用 Python 3.8 或者更高的版本.
SKIPPED [1] test_module/test_module_skip_2.py:8: 请使用 Python 3.8 或者更高的版本.

Results (0.03s):
        2 skipped

====================== 2 skipped in 0.03s =============================
```

可以看到,minversion 在两个测试模块中都生效了。

因此,在大型的测试项目中,可以在一个文件中定义所有的执行条件,需要时再引入模块中。

注意:不存在 pytest.skipif() 的方法。

2.6.4　pytest.importorskip 方法

当引入某个模块失败或者引入的版本不符合时,我们同样可以跳过后续部分的执行。

1. 引入失败时跳过

代码如下:

```
# Author: lindafang
# Date: 2020-05-09 16:48
# File: test_importorskip.py
```

```python
import pytest

docutils = pytest.importorskip("docutils")

@docutils
def test_importorskip():
    pass


def test_severn():
    print("7")
```

执行结果如下:

```
Skipped: could not import 'docutils': No module named 'docutils'
collected 0 items / 1 skipped

================ 1 skipped in 0.03s ==============================
Process finished with exit code 0
```

2. 引入模块版本不符合时跳过

我们也可以为其指定一个最低满足要求的版本,判断的依据是检查引入模块的 __version__ 属性,代码如下:

```python
# Author: lindafang
# Date: 2020 - 05 - 09 16:48
# File: test_import_minversion.py
import pytest

csv1 = pytest.importorskip("csv", minversion = "2.0")
import csv

@csv1
def test_import_minversion():
    pass
```

执行结果如下,csv 的版本是 1.0,要获得的最小版本是 2.0,所以跳过而未导入。
执行结果如下:

```
Skipped: module 'csv' has __version__ '1.0', required is: '2.0'
collected 0 items / 1 skipped

============================== 1 skipped in 0.03s =========
```

我们还可以再为其指定一个 reason 参数，表明跳过的原因。

我们注意到，pytest.importorskip 和 pytest.skip(allow_module_level=True) 都可以在模块的引入阶段跳过剩余部分。实际上，在源码中它们抛出的都是同样的异常，代码如下：

```
pytest.skip(allow_module_level = True)

raise Skipped(msg = msg, allow_module_level = allow_module_level)
pytest.importorskip()

raise Skipped(reason, allow_module_level = True) from None
```

只是 importorskip 额外增加了 minversion 参数，代码如下：

```
_pytest/outcomes.py
if minversion is None:
        return mod
    verattr = getattr(mod, "__version__", None)
    if minversion is not None:
        if verattr is None or Version(verattr) < Version(minversion):
            raise Skipped(
                "module %r has __version__ %r, required is: %r"
                % (modname, verattr, minversion),
                allow_module_level = True,
            )
```

从代码中也证实了，它实际检查的是模块的 __version__ 属性。

3. 使用 pytest.skip() 实现导入不成功时跳过

对于一般场景下，使用下面的方法可以实现同样的效果：

```
try:
    import docutils
except ImportError:
    pytest.skip("could not import 'docutils': No module named 'docutils'",
                allow_module_level = True)
```

2.6.5 跳过测试类

跳过的层次也是对应的：跳过函数、跳过方法、跳过模块、跳过类。那么在类上如何应用 @pytest.mark.skip 或 @pytest.mark.skipif 呢？

代码如下：

```
# Author: lindafang
# Date: 2020-05-09 20:42
```

```
#File: test_skip_class.py

import pytest

@pytest.mark.skip("作用于类中的每个用例,所以 pytest 共收集到两个 SKIPPED 的用例。")
class TestMyClass():
    def test_one(self):
        assert True

    def test_two(self):
        assert True
```

2.6.6 跳过测试模块

在模块中定义 pytestmark 变量的方法以便跳过测试模块。

代码如下:

```
# Author: lindafang
# Date: 2020-05-09 20:45
# File: test_module_skip_3.py

import pytest

pytestmark = pytest.mark.skip('作用于模块中的每个用例,所以 pytest 共收集到两个 SKIPPED 的用例。')


def test_one():
    assert True


def test_two():
    assert True
```

或者,在模块中调用 pytest.skip 方法,并设置 allow_module_level=True。

代码如下:

```
# Author: lindafang
# Date: 2020-05-09 20:45
# File: test_module_skip_3.py

import pytest
```

```
pytest.skip('在用例收集阶段就已经跳出了,所以不会收集到任何用例。', allow_module_level = 
True)


def test_one():
    assert True


def test_two():
    assert True
```

2.6.7 跳过指定文件或目录

通过在 conftest.py 中配置 collect_ignore_glob 项,可以在用例的收集阶段跳过指定的文件和目录。

例如:在 chapter-2 文件夹下执行,跳过 test_module 测试目录中文件名匹配 test_*.py 规则的文件和以 test_assert 开头的测试用例的收集,也就是不执行,代码如下:

```
# 配置 test_module 下所有以 test_开关的 Python 文件,以及 test_assert 开关的 Python
# 文件不参与收集,也就是不执行
collect_ignore_glob = ['test_module/test_*.py','test_assert*.py']
```

执行结果如下,未执行 test_module 文件下的文件,也未执行以 test_assert 开头的文件。

```
lindafang@linda chapter-2 % pytest
...
collecting ...

------------------------- test_foo_compare -------------------------

    def test_foo_compare():
        f1 = Foo(1)
        f2 = Foo(2)
>       assert f1 == f2 , '第一个实例中的内容与第二实例中的内容不相等'
E       assert 1 == 2

test_fail_cus.py:19: AssertionError

 test_fail_cus.py ×
7%

----------------- test_num -------------------------
```

```
        def test_num():
>           assert 1 == 2, '这两个值不相等'
E           AssertionError: 这两个值不相等
E           assert 1 == 2

test_fail_cus.py:23: AssertionError

 test_fail_cus.py ×
14 % ■■
 test_markers.py √√√
36 % ■■■■
 test_nodeid.py √√√
57 % ■■■■■■

_____ TestNodeId.test_two[3-4] _____

self = <test_nodeid.TestNodeId object at 0x10656c0f0>, x = 3, y = 4

    @pytest.mark.parametrize('x,y', [(1, 1), (3, 4)])
    def test_two(self, x, y):
        print(f'TestNodeId::test_two::{x} == {y}')
>       assert x == y
E       assert 3 == 4

test_nodeid.py:23: AssertionError
_____ Captured stdout call _____
TestNodeId::test_two::3 == 4

 test_nodeid.py ×
64 % ■■■■■■■
 test_raise.py √
71 % ■■■■■■■

_____ test_mytest_error _____

    def test_mytest_error():
        with pytest.raises(ImportError):
            print("如果不是解释器请求退出,则这个原因引起的异常被认定为测试不通过")
>           f()

test_raise.py:22:
_____

    def f():
        #解释器请求退出
>       raise SystemExit(1)
```

```
E       SystemExit: 1

test_raise.py:10: SystemExit
-------------------- Captured stdout call --------------------
如果不是解释器请求退出,则这个原因引起的异常被认定为测试不通过

 test_raise.py ×
79 % ■■■■■■■■
 test_Windows_skip_1.py s
86 % ■■■■■■■■■


------------------ test_cakan ------------------

    def test_cakan():
>       assert f() == 4
E       assert 3 == 4
E        +  where 3 = f()

test_Windows_skip_1.py:40: AssertionError
---------------------------- Captured stdout call ---------------------
darwin

 test_Windows_skip_1.py ×√
100 % ■■■■■■■■■■

Results (0.11s):
       8 passed
       5 failed
         - test_fail_cus.py:16 test_foo_compare
         - test_fail_cus.py:22 test_num
         - test_nodeid.py:20 TestNodeId.test_two[3-4]
         - test_raise.py:19 test_mytest_error
         - test_Windows_skip_1.py:39 test_cakan
       1 skipped
```

如果注释本设置,则执行 chapter-2 下的全部用例,执行统计结果如下:

```
Results (0.18s):
      15 passed
      19 failed
         - test_assert_1.py:11 test_f
         - test_assert_2.py:5 test_long_str_comparison
         - test_assert_2.py:11 test_eq_list
         - test_assert_2.py:15 test_dict_comparison
```

```
        - test_assert_2.py:27 test_set_comparison
        - test_assert_3.py:15 test_true
        - test_assert_3.py:21 test_in_dict
        - test_assert_3.py:33 test_long_list
        - test_assert_3.py:39 test_long_copy
        - test_assert_except.py:13 test_mytest
        - test_assert_except.py:26 test_match
        - test_assert_sample.py:29 test_approx_simple_fail
        - test_assert_sample.py:45 test_warrior_long_description
        - test_assert_sample.py:65 test_get_starting_equiment
        - test_fail_cus.py:16 test_foo_compare
        - test_fail_cus.py:22 test_num
        - test_nodeid.py:20 TestNodeId.test_two[3-4]
        - test_raise.py:19 test_mytest_error
        - test_Windows_skip_1.py:39 test_cakan
1 xfailed
9 skipped
```

2.6.8 各种跳过小结

通过小结看到不同跳过的使用层次,如表2-1所示。

表2-1 各种跳过小结

名称	pytest.mark.skip	pytest.mark.skipif	pytest.skip	pytest.importorskip	conftest.py
用例	@pytest.mark.skip()	@pytest.mark.skipif()	pytest.skip(msg='')	/	/
类	@pytest.mark.skip()	@pytest.mark.skipif()	/	/	/
模块	pytestmark=pytest.mark.skip()	pytestmark=pytest.mark.skipif()	pytest.skip(allow_module_level=True)	pytestmark=pytest.importorskip()	/
文件或目录	/	/	/	/	collect_ignore_glob

2.7 标记用例为预期失败

在测试过程中有时会遇到这种情况,在执行某个用例后找出了用例中的Bug,开发人员会修改,但最近提交的版本中这个Bug并未被修改,开发人员可能在下个版本才会修改,这时我们知道即使执行这个用例也会失败。

这时可以使用@pytest.mark.xfail标记用例,表示期望这个用例执行失败。用例会正常执行,只是失败时不再显示堆栈信息,最终的结果有两个:用例执行失败(XFAIL:符合预期的失败)、用例执行成功(XPASS:不符合预期的成功)。也就是说执行成功反倒不正确了。

在用例执行中可以使用 pytest.xfail 实现这种效果。

如何取消设置的标识呢?通过下面 3 节内容进行说明。

2.7.1 @pytest.mark.xfail 标记用例

首先看一下@pytest.mark.xfail()的用法,xfail 有几个不同的参数,参数的不同组合会有不同的结果。condition 是条件,reason 是原因,raises 是引起异常,run 参数表明是否执行,strict 是失败的用例是否显示为 FAILED 的开关,然后看一下源码,再通过具体实例体会不同参数的组合及执行的结果标记。

源码如下:

```
def xfail(self,condition = None, reason = None, raises = None, run = True, strict = False):
    """mark the the test function as an expected failure if eval(self,condition)
    has a True value.

    Optionally specify a reason for better reporting and run = False if
    you don't even want to execute the test function.

    See http://doc.pytest.org/en/latest/skipping.html
    """
```

(1) 预期失败的用例在执行失败时显示 XFAIL 标记,包含 reason 参数。

condition 为位置参数,默认值为 None。reason 为关键字参数,默认值为 None,可以加字符串写清原因。

代码如下:

```
# Author: lindafang
# Date: 2020 - 05 - 14 14:27
# File: test_xfail.py
import pytest

class Test_pytest():

    @pytest.mark.xfail(reason = '该功能有 Bug')
    def test_one(self, ):
        print(" ---- start ------ ")
        print("test_one 方法执行")
        assert 2 == 1

    def test_two(self):
```

```
            print("test_two 方法执行")
            assert "o" in "love"

        def test_three(self):
            print("test_three 方法执行")
            assert 3 - 2 == 1

#r 参数是总结报告,s 参数是输出控制台打印信息
if __name__ == "__main__":
    pytest.main(['-sr', 'test_xfail.py'])
```

执行的结果如下,2 个 passed 通过,1 个 xfailed 是预期失败,原因是断言失败。

```
Testing started at 15:07 ...
============ test session starts ==============================
...
collected 3 items

test_xfail.py x ---- start ------
test_one 方法执行

self = <test_xfail.Test_pytest object at 0x109ba0048>

    @pytest.mark.xfail(reason = '该功能有 Bug')
    def test_one(self, ):
        print(" ---- start ------ ")
        print("test_one 方法执行")
>       assert 2 == 1
E       assert 2 == 1

test_xfail.py:19: AssertionError
.test_two 方法执行
.test_three 方法执行                                              [100%]

==================== 2 passed, 1 xfailed in 0.04s =========================
Process finished with exit code 0
```

(2) 预期失败的用例在执行通过时显示 XPASS 标记。

如果将 assert 2 == 1 改成 assert 1==1,则该功能是成功的,也就是说本来功能是有 Bug 的,但断言成功了,说明断言错误或功能有问题,这时显示的标记为 XPASS。

执行的结果如下:2 passed,1 xpassed。

```
test_xfail.py X ---- start ------
test_one 方法执行
```

```
.test_two 方法执行
.test_three 方法执行
                                                        [100%]

===================== 2 passed, 1 xpassed in 0.03s =========================
Process finished with exit code 0
```

(3) 带条件的标记用例 XFAIL,包含 condition 参数。

和@pytest.mark.skipif 一样,它也可以接收一个 Python 表达式,表明只有满足条件时才标记用例。例如,只在 pytest 3.6 及以上版本标记用例,如果用例返回失败,则显示 xfail。如果用例返回正确,则显示 xpass。

代码如下:

```python
# Python 3.6 及以上版本标记该测试用例,如果用例返回失败,则显示 xfail.如果用例返回正确,
# 则显示 xpass
@pytest.mark.xfail(sys.version_info >= (3, 6), reason = "Python 3.6 API 变更了")
def test_function():
    print("不同版本下执行不同测试用例")
    assert 2 == 1
```

执行结果一,断言失败显示 XFAIL:

```
 test_xfail_raise.py::test_xfail x
50% ■■■■■        # 不同版本下执行不同测试用例

 test_xfail_raise.py::test_function x
100% ■■■■■■■■■■

Results (0.07s):
       2 xfailed
```

执行结果二,断言成功显示 XPASS:

```
 test_xfail_raise.py::test_xfail x
50% ■■■■■        # 不同版本下执行不同测试用例

 test_xfail_raise.py::test_function X
100% ■■■■■■■■■■

Results (0.06s):
       1 xpassed
       1 xfailed
```

(4) 将不符合预期的成功 XPASS 标记为失败,使用 strict 参数。

strict 为关键字参数,默认值为 False。

当 strict=False 时,如果用例执行失败,则结果标记为 XFAIL,表示符合预期的失败。如果用例执行成功,则结果标记为 XPASS,表示不符合预期的成功。

当 strict=True 时,如果用例执行成功,则结果将标记为 FAILED,而不再标记为 XPASS。

我们也可以在 pytest.ini 文件中配置,代码如下:

```
[pytest]
xfail_strict = true
```

如果用例的失败不是因为所期望的异常导致的,pytest 将会把测试结果标记为 FAILED。上面的代码将让 assert 1==1 断言成功,执行结果如下:

```
collecting ...
 test_xfail_raise.py::test_xfail x
50% ■■■■■         # 不同版本下执行不同测试用例
--------------- test_function ------------
[XPASS(strict)] Python 3.6 API 变更了

 test_xfail_raise.py::test_function ×
100% ■■■■■■■■■■

Results (0.08s):
       1 failed
         - test_xfail_raise.py:13 test_function
       1 xfailed
```

(5) 标记为 XFAIL 的用例不再执行,包含 run 参数。

run 为关键字参数,默认值为 True。

当 run=False 时,pytest 不会再执行测试用例,而直接将结果标记为 XFAIL。

执行结果如下,有个 xx,Results 为 1 xfailed。也就是一个被执行,而另一个未被执行。

```
collecting ...
 test_xfail_raise.py xx
100% ■■■■■■■■■■

Results (0.06s):
       1 xfailed
```

使用 PyCharm 中的 pytest 运行工具执行,其结果有两个 xfailed,并且显示其中一个是跳过执行 NOTRUN。

```
Testing started at 15:46 ...
/usr/local/bin/python3.6 "/Applications/PyCharm
CE.app/Contents/helpers/PyCharm/_jb_pytest_runner.py" -- path
/Users/lindafang/PyCharmProjects/pytest_book/src/chapter-2/test_xfail_raise.py
Launching pytest with arguments /Users/lindafang/PyCharmProjects/pytest_book/src/chapter-2/test_
xfail_raise.py in /Users/lindafang/PyCharmProjects/pytest_book/src/chapter-2

==================== test session starts ============================
platform darwin -- Python 3.6.8, pytest-5.2.1, py-1.8.0, pluggy-0.13.1
rootdir: /Users/lindafang/PyCharmProjects/pytest_book/src/chapter-2, inifile: pytest.ini
plugins: rerunfailures-5.0, forked-1.0.2, sugar-0.9.2, assume-1.2.2,
xdist-1.28.0, ordering-0.6, metadata-1.8.0, bdd-3.2.1collected 2 items

test_xfail_raise.py x
    @pytest.mark.xfail()
    def test_xfail():
>       print(broken_fixture())

test_xfail_raise.py:6:
_____

    def broken_fixture():
>       raise Exception("Sorry, it's 中断异常")
E       Exception: Sorry, it's 中断异常

test_xfail_raise.py:10: Exception
x
cls = <class '_pytest.runner.CallInfo'>
func = <function call_runtest_hook.<locals>.<lambda> at 0x10700d950>
when = 'setup'
reraise = (<class '_pytest.outcomes.Exit'>, <class 'KeyboardInterrupt'>)

    @classmethod
    def from_call(cls, func, when, reraise=None):
        #: context of invocation: one of "setup", "call",
        #: "teardown", "memocollect"
        start = time()
        excinfo = None
        try:
>           result = func()

/Library/Frameworks/Python.framework/Versions/3.6/lib/python3.6/site-packages/_pytest/
runner.py:229:
_____
/Library/Frameworks/Python.framework/Versions/3.6/lib/python3.6/site-packages/_pytest/
runner.py:201: in <lambda>
    lambda: ihook(item=item, **kwds), when=when, reraise=reraise
```

```
/Library/Frameworks/Python.framework/Versions/3.6/lib/python3.6/site-packages/pluggy/
hooks.py:286: in __call__
    return self._hookexec(self, self.get_hookimpls(), kwargs)
/Library/Frameworks/Python.framework/Versions/3.6/lib/python3.6/site-packages/pluggy/
manager.py:93: in _hookexec
    return self._inner_hookexec(hook, methods, kwargs)
/Library/Frameworks/Python.framework/Versions/3.6/lib/python3.6/site-packages/pluggy/
manager.py:87: in <lambda>
    firstresult=hook.spec.opts.get("firstresult") if hook.spec else False,
/Library/Frameworks/Python.framework/Versions/3.6/lib/python3.6/site-packages/_pytest/
skipping.py:87: in pytest_runtest_setup
    check_xfail_no_run(item)
_ _ _ _ _ _ _ _ _ _ _ _ _ _ _ _ _ _ _ _ _ _ _ _ _ _ _ _ _ _ _ _ _ _ _ _ _ _ _ _ _ _ _ _

item = <Function test_function>

    def check_xfail_no_run(item):
        """check xfail(run=False)"""
        if not item.config.option.runxfail:
            evalxfail = item._evalxfail
            if evalxfail.istrue():
                if not evalxfail.get("run", True):
>                   xfail("[NOTRUN] " + evalxfail.getexplanation())
E                   _pytest.outcomes.XFailed: [NOTRUN] Python3.6 api 变更了

/Library/Frameworks/Python.framework/Versions/3.6/lib/python3.6/site-packages/_pytest/
skipping.py:105: XFailed
                                                                            [100%]

================ 2 xfailed in 0.08s =============================
Process finished with exit code 0
```

（6）引起正确异常的失败，包含 raises 参数。

raises 为关键字参数，默认值为 None。可以指定为一个异常类或者多个异常类的元组，表明我们期望用例上报指定的异常。pytest.mark.xfail() 也可以接收一个 raises 参数，用来判断用例是否因为一个具体的异常而导致失败，代码如下：

```
@pytest.mark.xfail(raises=Exception)
def test_xfail():
    print(broken_fixture())



def broken_fixture():
    raise Exception("Sorry, it's 中断异常")
```

执行结果如下,如果 test_xfail() 触发一个 Exception 异常,则用例标记为 xfailed。如果没有,则正常执行 test_xfail()。

执行结果一,引起的异常相同。

```
test_xfail_raise.py x
100%  ■■■■■■■■■■

Results (0.08s):
       1 xfailed
```

修改 raise=IndexError 后再执行,正常执行,显示 XPASS,由于设置了 strict 为 true,这种直接结果为失败,即 1 failed,与预期引起的异常不同。

```
collecting ...  #如果没有异常或异常引起的不同才执行这个
-------------------- test_xfail ----------------
[XPASS(strict)]

test_xfail_raise.py x
100%  ■■■■■■■■■■

Results (0.07s):
       1 failed
         - test_xfail_raise.py:5 test_xfail
```

注意:如果 test_xfail() 测试成功,则用例的结果是 xpass,而不是 pass。pytest.raises 适用于检查由代码故意引发的异常,而 @pytest.mark.xfail() 更适合于记录一些未修复的 Bug。

(7) 小结。

我们以表 2-2 来总结不同参数组合对测试结果的影响(其中 xfail=pytest.mark.xfail)。

表 2-2 参数对结果的影响

测试结果	@xfail()	@xfail(strict=True)	@xfail(raises=Exception)	@xfail(strict=True, raises=Exception)	@xfail(..., run=False)
用例测试成功	XPASS	FAILED	XPASS	FAILED	XFAIL
用例测试失败	上报 AssertionError	XFAIL	XFAIL	FAILED	FAILED
用例上报 Exception	XFAIL	XFAIL	XFAIL	XFAIL	XFAIL

2.7.2 使用 pytest.xfail 标记用例

可以通过 pytest.xfail 方法在用例执行过程中直接将用例结果标记为 XFAIL,并跳过剩余的部分,也就是该用例中的后续代码不会被执行。

应用的场景是功能未完成、已知有问题。除此之外,用例的执行需要前置条件或操作,如果前置条件或操作失败,那么我们就可以直接将该用例设为失败,也就是 xfail。

同样可以为 pytest.xfail 指定一个 reason 参数,表明原因。

(1) 功能未完成,已知有 Bug 未修改,执行预期失败。

代码如下:

```python
# Author: lindafang
# Date: 2020-05-14 14:27
# File: test_xfail.py
import pytest


class Test_pytest():

    def test_one(self, ):
        print("----start------")
        pytest.xfail(reason='该功能尚未完成')
        print("test_one 方法执行")
        assert 2 == 1

    def test_two(self):
        print("test_two 方法执行")
        assert "o" in "love"

    def test_three(self):
        print("test_three 方法执行")
        assert 3 - 2 == 1


if __name__ == "__main__":
    # 如果加上 r,则有些总结信息看得更清楚
    pytest.main(['-sr', 'test_xfail.py'])
```

执行结果如图 2-13 所示。

(2) 无效配置时直接显示预期失败,不进行后续测试。

在代码中增加下面测试方法,为大家展示无效配置时显示失败的效果。

```python
def valid_config():
    # 如果返回值为 true,则是有效配置.如果返回值为 false,则是无效配置
```

第2章　pytest的测试用例管理及运行管理

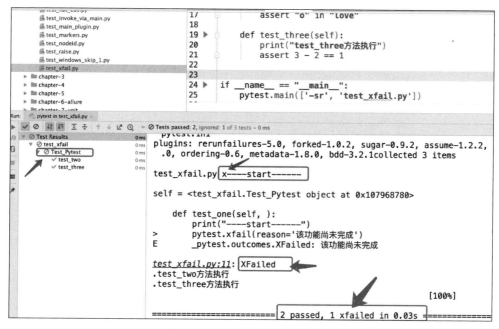

图 2-13　pytest.xfail 的执行结果

```
        return False

#在类中增加
    def test_function(self):
        if not valid_config():
            pytest.xfail("这是个无效的配置,无法进行后续测试")
```

执行结果如下：

```
Testing started at 14:39 ...
...
============== test session starts ==============================
...
collected 4 items

test_xfail.py x ---- start ------

self = <test_xfail.Test_pytest object at 0x108ba4a58>

    def test_one(self, ):
        print(" ---- start ------ ")
>       pytest.xfail(reason = '该功能尚未完成')
E       _pytest.outcomes.XFailed: 该功能尚未完成
```

```
test_xfail.py:16: XFailed
.test_two 方法执行
.test_three 方法执行
x
self = <test_xfail.Test_pytest object at 0x108b9b940>

    def test_function(self):
        if not valid_config():
>           pytest.xfail("这是个无效的配置,无法进行后续测试")
E           _pytest.outcomes.XFailed: 这是个无效的配置,无法进行后续测试

test_xfail.py:30: XFailed
                                                                    [100%]

================ 2 passed, 2 xfailed in 0.05s =========================
Process finished with exit code 0
```

2.7.3　xfail 标记如何失效

(1) 通过 --runxfail 参数让 xfail 标记失效。

通过命令行选项 pytest --runxfail 让 xfail 标记失效,使这些用例变成正常执行的用例,仿佛没有被标记过一样。同样,pytest.xfail()方法也将失效。

执行命令 pytest --runxfail test_xfail.py 会有以下结果,显示 2 个通过,1 个失败。

执行结果如下:

```
lindafang@linda chapter-2 % pytest --runxfail test_xfail.py
Test session starts (platform: darwin, Python 3.6.8, pytest 5.2.1, pytest-sugar 0.9.2)
rootdir: /Users/lindafang/PyCharmProjects/pytest_book/src/chapter-2, inifile: pytest.ini
plugins: rerunfailures-5.0, forked-1.0.2, sugar-0.9.2, assume-1.2.2, xdist-1.28.0,
ordering-0.6, metadata-1.8.0, bdd-3.2.1
collecting ...

---------------- Test_pytest.test_one --------------------
[XPASS(strict)] 该功能有 Bug
--------------- Captured stdout call ----------------
----start------
test_one 方法执行

 test_xfail.py ×√√
 100% ██████████████

 Results (0.04s):
       2 passed
       1 failed
         - test_xfail.py:14 Test_pytest.test_one
```

(2) pytest.param 用于将参数化标记为失败。

这部分知识与第 4 章的知识有交叉，建议学习完第 4 章内容之后回头学习这部分知识会更好理解。

@pytest.mark.parametrize 功能将在第 4 章详细讲解。

pytest.param 方法可用于为 @pytest.mark.parametrize 或者参数化的 fixture 指定一个具体的实参，它有一个关键字参数 marks，可以接收一个或一组标记，用于标记这轮测试的用例。

代码如下：

```python
# Author: lindafang
# Date: 2020-05-14 16:57
# File: test_xfail_param.py

import pytest
import sys


@pytest.mark.parametrize(
    ('n', 'expected'),
    [(2, 1),
     pytest.param(2, 1, marks=pytest.mark.xfail(), id='XPASS'),
     pytest.param(0, 1, marks=pytest.mark.xfail(raises=ZeroDivisionError), id='XFAIL'),
     pytest.param(1, 2, marks=pytest.mark.skip(reason='无效的参数,跳过执行')),
     pytest.param(1, 2, marks=pytest.mark.skipif(sys.version_info <= (3, 8), reason='请使用3.8及以上版本的Python.'))])
def test_params(n, expected):
    assert 2 / n == expected
```

执行 pytest -rA test_xfail_param.py，结果如下：

```
lindafang@linda chapter-2 % pytest -rA test_xfail_param.py
Test session starts (platform: darwin, Python 3.6.8, pytest 5.2.1, pytest-sugar 0.9.2)
rootdir: /Users/lindafang/PyCharmProjects/pytest_book/src/chapter-2, inifile: pytest.ini
plugins: rerunfailures-5.0, forked-1.0.2, sugar-0.9.2, assume-1.2.2, xdist-1.28.0,
ordering-0.6, metadata-1.8.0, bdd-3.2.1
collecting ...
 test_xfail_param.py √Xxss
100% ■■■■■■■■■■■
============================== PASSES ==============================
================ short test summary info ===============
PASSED test_xfail_param.py::test_params[2-1]
PASSED test_xfail_param.py::test_params[XPASS]
```

```
PASSED test_xfail_param.py::test_params[XFAIL]
PASSED test_xfail_param.py::test_params[1-20]
PASSED test_xfail_param.py::test_params[1-21]
SKIPPED [1] test_xfail_param.py:9: 无效的参数,跳过执行
SKIPPED [1] test_xfail_param.py:9: 请使用3.8及以上版本的Python.
xfail test_xfail_param.py::test_params[XFAIL]
XPASS test_xfail_param.py::test_params[XPASS]

Results (0.07s):
       1 passed
       1 xpassed
       1 xfailed
       2 skipped
```

2.8 中断调试及错误处理

2.8.1 失败时加载 PDB 环境

在工具栏中会有一个小虫子的图标 ,一般笔者在调试时使用这个查看变量等的情况。PDB(Python Debugger)是 Python 内建的调试器,如果习惯使用它也是一个不错的选择。pytest 允许通过以下命令在执行失败时进入这个调试器模式,代码如下:

```
pytest -- pdb                      # 将在每次失败时调用 Python 调试器
pytest - x -- pdb                  # 只在第一次失败的测试中执行此操作
pytest -- pdb -- maxfail = 3       # 在前三次失败时调用
```

举例说明:

pytest 会在测试用例失败(或者 Ctrl+C)时,调用这个调试器,可以访问测试用例的本地变量 x。失败的信息存储在 sys.last_value、sys.last_type、sys.last_traceback 变量中,可以在交互环境中访问它们。使用 exit 命令,即可退出 PDB 环境。

代码如下:

```
# Author: lindafang
# Date: 2020 - 05 - 15 10:40
# File: test_pdb.py

def test_fail():
    x = 1
    assert x == 0
```

执行 pytest --pdb test_pdb.py,结果如下:

```
lindafang@linda chapter-2 % pytest --pdb test_pdb.py
Test session starts (platform: darwin, Python 3.6.8, pytest 5.2.1, pytest-sugar 0.9.2)
rootdir: /Users/lindafang/PyCharmProjects/pytest_book/src/chapter-2, inifile: pytest.ini
plugins: rerunfailures-5.0, forked-1.0.2, sugar-0.9.2, assume-1.2.2, xdist-1.28.0,
ordering-0.6, metadata-1.8.0, bdd-3.2.1
collecting ...

————————— test_fail —————————————————————

    def test_fail():
        x = 1
>       assert x == 0
E       assert 1 == 0

test_pdb.py:7: AssertionError


>>>>>>>>>>>>>>>> traceback >>>>>>>>>>>>>>>>>>>>>>>>>>>>>>>>>>>>

    def test_fail():
        x = 1
>       assert x == 0
E       assert 1 == 0

test_pdb.py:7: AssertionError
>>>>>>>>>>>>>>>> entering PDB >>>>>>>>>>>>>>>>>>>

>>>>>>>>>> PDB post_mortem (IO-capturing turned off) >>>>>>>>>>>>>>>>>>>
>
/Users/lindafang/PyCharmProjects/pytest_book/src/chapter-2/test_pdb.py(7)test_fail()
-> assert x == 0
#以下(Pdb)是提示,后面的内容是输入进去的,也就是笔者输入 x 这个变量,系统返回的值为1
(Pdb) x
1
(Pdb) import sys
(Pdb) sys.last_value
AssertionError('assert 1 == 0',)
(Pdb) sys.last_type
<class 'AssertionError'>
(Pdb) sys.last_traceback
<traceback object at 0x106e50748>
(Pdb) exit

!!!!!!!!!!!!!!!!! _pytest.outcomes.Exit: Quitting debugger !!!!!!!!!!!!!!!

Results (109.33s):
       1 failed
         - test_pdb.py:5 test_fail
```

2.8.2　开始执行时就加载 PDB 环境

通过以下命令，pytest 允许在每个测试用例开始执行时就加载 PDB 环境：

```
pytest -- trace
```

2.8.3　设置断点

通常可以通过在代码行号附近单击具体行设置一个断点，在单击 ![] 后运行会停下来。如图 2-14 所示。

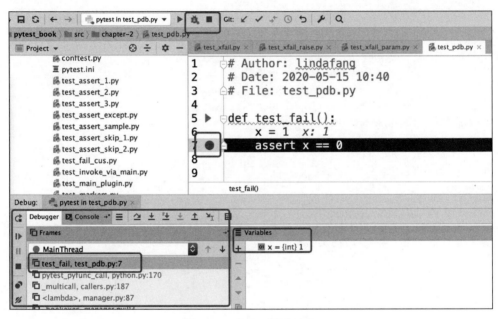

图 2-14　Debug 调试工具

也可以在测试用例代码中添加 import pdb;pdb.set_trace()，当其被调用时，pytest 会停止这条用例的输出，其他用例不受影响。通过 continue 命令，退出 PDB 环境，并继续执行用例。

代码如下：

```
# Author: lindafang
# Date: 2020-05-15 10:40
# File: test_pdb.py

def test_fail():
    x = 1
```

```
    import pdb;
    pdb.set_trace()
    assert x == 0
```

执行结果如下,在进行(Pdb)后,输入 x,返回 1。输入 import sys,无返回。输入 sys.path,返回当前系统路径。输入 continue,用例接着运行,输出结果如下。

```
Testing started at 12:06 ...
/usr/local/bin/python3.6 "/Applications/PyCharm
CE.app/Contents/helpers/PyCharm/_jb_pytest_runner.py" -- target test_pdb.py::test_fail
Launching pytest with arguments test_pdb.py::test_fail in
/Users/lindafang/PyCharmProjects/pytest_book/src/chapter-2

============ test session starts =================
platform darwin -- Python 3.6.8, pytest-5.2.1, py-1.8.0, pluggy-0.13.1
rootdir: /Users/lindafang/PyCharmProjects/pytest_book/src/chapter-2, inifile: pytest.ini
plugins: rerunfailures-5.0, forked-1.0.2, sugar-0.9.2, assume-1.2.2, xdist-1.28.0,
ordering-0.6, metadata-1.8.0, bdd-3.2.1collected 1 item

test_pdb.py
>>>>>>>>>>>>>> PDB set_trace (IO-capturing turned off) >>>>>>>>>>>>>>>>>>>
> /Users/lindafang/PyCharmProjects/pytest_book/src/chapter-2/test_pdb.py(9)test_fail() ->
assert x == 0
(Pdb) 1
(Pdb) (Pdb) ['/Users/lindafang/PyCharmProjects/pytest_book/src/chapter-2',
'/Users/lindafang/PyCharmProjects/pytest_book',
'/Users/lindafang/PyCharmProjects/pytest_book/src',
'/Applications/PyCharm CE.app/Contents/helpers/PyCharm',
'/Library/Frameworks/Python.framework/Versions/3.6/lib/python36.zip',
'/Library/Frameworks/Python.framework/Versions/3.6/lib/python3.6',
'/Library/Frameworks/Python.framework/Versions/3.6/lib/python3.6/lib-dynload',
'/Users/lindafang/Library/Python/3.6/lib/python/site-packages',
'/Library/Frameworks/Python.framework/Versions/3.6/lib/python3.6/site-packages',
'/Library/Frameworks/Python.framework/Versions/3.6/lib/python3.6/site-packages/selenium_
page_objects-0.2.1-py3.6.egg',
'/Users/lindafang/Downloads/code/pytest-book-code/tasks_proj/src',
'/Library/Frameworks/Python.framework/Versions/3.6/lib/python3.6/site-packages/pywinauto-0.6.
8-py3.6.egg']
(Pdb) (Pdb)
>>>>>>>>>>>>>>>>>>>> PDB continue (IO-capturing resumed) >>>>>>>>>>>>>
F
test_pdb.py:4 (test_fail)
1 != 0

Expected :0
```

```
Actual   :1
<Click to see difference>

def test_fail():
        x = 1
        import pdb;
        pdb.set_trace()
>       assert x == 0
E       assert 1 == 0

test_pdb.py:9: AssertionError
                                                                    [100%]

=========== FAILURES ====================================
------------ test_fail ----------------------------------

    def test_fail():
        x = 1
        import pdb;
        pdb.set_trace()
>       assert x == 0
E       assert 1 == 0

test_pdb.py:9: AssertionError
============= 1 failed in 41.13s ============================
Process finished with exit code 0
```

2.8.4 使用内置的中断函数

Python 3.7 中有一个内置 breakpoint() 函数。pytest 可以在以下场景中使用：

当 breakpoint() 被调用，并且 PYTHONBREAKPOINT 为 None 时，pytest 会使用内部自定义的 PDB 代替系统的 PDB，测试执行结束时，自动切换回系统自带的 PDB。

当加上 --pdb 选项时，breakpoint() 和测试发生错误时，都会调用内部自定义的 PDB，--pdbcls 选项允许指定一个用户自定义的 PDB 类。

2.8.5 错误句柄

这是 pytest 5.0 版本新增特性，在测试中发生段错误或者超时的情况下，faulthandler 标准模块可以转存 Python 的回溯信息，它在 pytest 的执行中默认为已加载，使用-p no:faulthandler 选项可以关闭它。同样，faulthandler_timeout=X 配置项可用于当测试用例的完成时间超过 X 秒时，转存所有线程的 Python 回溯信息。

举例说明，在配置文件中设置测试执行的超时时间为 5s，代码如下：

```
src/chapter-2/pytest.ini

[pytest]
faulthandler_timeout = 5
```

在测试用例中添加等待 7s 的操作，代码如下：

```
test_fault_handler.py

import time


def test_faulthandler():
    time.sleep(7)
    assert 1
```

当默认已加载 faulthandler 时，输入 pytest -q test_fault_handler.py，执行结果如下，显示 Timeout (0:00:05)。在执行刚超过 5s 的时候会打印出回溯信息，但不会中断测试的执行。

执行结果如下：

```
lindafang@linda chapter-2 % pytest -q test_fault_handler.py
Test session starts (platform: darwin, Python 3.6.8, pytest 5.2.1, pytest-sugar 0.9.2)
rootdir: /Users/lindafang/PyCharmProjects/pytest_book/src/chapter-2, inifile: pytest.ini
plugins: rerunfailures-5.0, forked-1.0.2, sugar-0.9.2, assume-1.2.2, xdist-1.28.0,
ordering-0.6, metadata-1.8.0, bdd-3.2.1
Timeout (0:00:05)!
Thread 0x00007fff5e7aadc0 (most recent call first):
  File "/Users/lindafang/PyCharmProjects/pytest_book/src/chapter-2/test_fault_handler.
py", line 9 in test_faulthandler
  File
……#此处略过部分执行结果
"/Library/Frameworks/Python.framework/Versions/3.6/lib/python3.6/site-packages/_pytest/
config/__init__.py", line 90 in main
  File "/Library/Frameworks/Python.framework/Versions/3.6/bin/pytest", line 8 in <module>

 test_fault_handler.py √
100% ▇▇▇▇▇▇▇▇▇▇

Results (7.06s):
       1 passed
```

接下来看一下去掉 faulthandler 的情况，超时并不会触发回溯信息的打印，输入 pytest -q -p no:faulthandler test_fault_handler.py 之后的结果如下：

```
lindafang@linda chapter-2 % pytest -q -p no:faulthandler test_fault_handler.py
...
 test_fault_handler.py √
100%■■■■■■■■■■

Results (7.03s):
       1 passed
```

注意：这个功能是从 pytest-faulthandler 插件合并而来的，但是有两点不同：去掉此功能时，使用-p no:faulthandler 代替原来的--no-faulthandler；使用 faulthandler_timeout 配置项代替--faulthandler-timeout 命令行选项来配置超时时间。当然，也可以使用-o faulthandler_timeout=X 在命令行进行配置。

2.9 结果分析及报告

2.9.1 分析测试执行时间

获取执行最慢的 10 个测试用例：

```
pytest --durations=10
```

默认情况下，可以使用-vv 选项查看它们。

```
pytest -vv --durations=10

========= slowest 10 test durations =========================================
10.00s call     test_invoke_via_main.py::test_one
7.00s call      test_fault_handler.py::test_faulthandler
0.12s call      test_assert_3.py::test_long_copy
0.00s call      test_assert_3.py::test_long_list
0.00s call      test_assert_sample.py::test_warrior_long_description
0.00s call      test_assert_2.py::test_set_comparison
0.00s call      test_assert_2.py::test_dict_comparison
0.00s call      test_xfail_param.py::test_params[XFAIL]
0.00s setup     test_assert_1.py::test_f
0.00s call      test_assert_1.py::test_f
```

2.9.2 创建及定制 JUnitXML 格式的测试报告

测试报告是我们在测试过程中必不可少的一个环节，pytest 通过这种形式提供报告的输出和定制。同样，因为重要，pytest 的第三方插件 allure 已经将报告做到极致，并且与 Jenkins 有很好的集成。因此以下内容在实际工作中用得不多。如果真要用到的这里没有说明的方法请直接到pytest官网上阅读。

使用pytest --junitxml=path命令,可以在指定的path中创建一个能被Jenkins或者其他CI工具读取的XML格式的测试报告。

具体步骤如下:

(1)可以在项目的pytest.ini文件输入下面的信息:

```
src/chapter-2/pytest.ini

[pytest]
junit_suite_name = pytest_chinese_doc
```

通过设置junit_suite_name的值,可以自定义xml文件中testsuite根节点的name信息,junit_suite_name是4.0版本新增的配置项。

(2)输入pytest -q --junitxml=report/test_one.xml test_nodeid.py::test_one并执行。

(3)在当前路径下的report下面可找到test_one.xml文件,代码如下:

```
<?xml version = "1.0" encoding = "UTF-8"?>
<testsuites>
    <testsuite errors = "0" failures = "0" hostname = "linda.local" name = "pytest_chinese_doc" skipped = "0" tests = "1"
              time = "0.015" timestamp = "2020-05-15T12:34:16.368873">
        <testcase classname = "test_nodeid" file = "test_nodeid.py" line = "7" name = "test_one" time = "0.001">
            <system-out>test_one
            </system-out>
        </testcase>
    </testsuite>
</testsuites>
```

(4)<testsuite>节点的name属性的值,变为我们所期望的pytest_chinese_doc。

(5)time属性表明测试用例执行的全部耗时,包含setup和teardown中的操作。如果只想记录测试用例执行的时间,只需进行如下配置,大家可以自己实验:

```
src/chapter-2/pytest.ini

junit_duration_report = call
```

1. 在报告中为测试用例附加额外的子节点信息

使用record_property fixture为test_record_property用例添加一个额外的test_id,代码如下:

```
src/chapter-2/test_xml_report.py
```
```python
def test_record_property(record_property):
    record_property("test_id", 10010)
    assert 1
```

在报告中的表现为<property name="test_id" value="10010" />：

```xml
<?xml version="1.0" encoding="UTF-8"?>
<testsuites>
    <testsuite errors="0" failures="0" hostname="linda.local" name="pytest_chinese_doc" skipped="0" tests="1"
               time="0.068" timestamp="2020-05-15T13:00:02.436844">
        <testcase classname="test_xml_report" file="test_xml_report.py" line="4" name="test_record_property"
                  time="0.000">
            <properties>
                <property name="test_id" value="10010"/>
            </properties>
        </testcase>
    </testsuite>
</testsuites>
```

注意：变动后的报告可能不符合最新的JUnitXML模式检查规则，从而可能导致在某些CI工具上发生未知的错误。

2. 在报告中为测试用例附加额外的属性信息

可以通过record_xml_attribute fixture为测试用例附加额外的属性，而不是通过record_property为其添加子节点。为测试用例添加一个test_id属性，并修改原先的classname属性，代码如下：

```
src/chapter-2/test_xml_report.py
```
```python
def test_record_property2(record_xml_attribute):
    record_xml_attribute('test_id', 10010)
    record_xml_attribute('classname', 'custom_classname')
    assert 1
```

在报告中的表现为<testcase classname="custom_classname" test_id="10010" …：

```xml
<?xml version="1.0" encoding="UTF-8"?>
<testsuites>
    <testsuite errors="0" failures="0" hostname="linda.local" name="pytest_chinese_doc" skipped="0" tests="1"
```

```xml
            time = "0.062" timestamp = "2020 - 05 - 15T13:03:29.847495">
      < testcase classname = "custom_classname" file = "test_xml_report.py" line = "8" name =
"test_record_property2"
                 test_id = "10010" time = "0.000">
      </testcase>
    </testsuite>
</testsuites> testsuites >
  < testsuite errors = "0" failures = "0" hostname = "NJ - LUYAO - T460" name = "pytest_chinese_
doc" skipped = "0" tests = "1"
    time = "0.028" timestamp = "2019 - 09 - 27T15:35:47.093494">
    < testcase classname = "custom_classname" file = "test_xml_report.py" line = "34" name =
"test_record_property2"
        test_id = "10010" time = "0.001"></testcase>
  </testsuite>
</testsuites>
```

注意：record_xml_attribute 目前是一个实验性的功能，未来可能被更强大的 API 所替代，但功能本身会被保留。变动后的报告可能不符合最新的 JUnitXML 模式检查规则，从而可能导致在某些 CI 工具上发生未知的错误。

3. 在报告中为测试集附加额外的子节点信息

这部分内容涉及第 3 章内容，建议学完第 3 章后再参考查看。

这是 pytest 4.5 版本新增功能，可以通过自定义一个 session 作用域级别的 fixture 为测试集添加子节点信息，并且会作用于所有的测试用例，这个自定义的 fixture 需要调用另外一个 record_testsuite_property fixture。

record_testsuite_property 接收两个参数 name 和 value 以构成< property >标签，其中 name 必须为字符串，value 会转换为字符串并进行 XML 转义。

代码如下：

```python
src/chapter - 2/test_xml_report.py

@pytest.fixture(scope = "session")
def log_global_env_facts(record_testsuite_property):
    record_testsuite_property("EXECUTOR", "lindafang")
    record_testsuite_property("LOCATION", "HRB")

def test_record_property3(log_global_env_facts):
    assert 1
```

生成的测试报告表现在 testsuite 节点中，多了一个 properties 子节点，包含所有新增的属性节点，而且，它和所有的 testcase 节点是平级的：

```xml
<?xml version = "1.0" encoding = "UTF-8"?>
<testsuites>
    <testsuite errors = "0" failures = "0" hostname = "linda.local" name = "pytest_chinese_doc" skipped = "0" tests = "1"
               time = "0.047" timestamp = "2020-05-15T13:07:54.708452">
        <properties>
            <property name = "EXECUTOR" value = "lindafang"/>
            <property name = "LOCATION" value = "HRB"/>
        </properties>
        <testcase classname = "test_xml_report" file = "test_xml_report.py" line = "19" name = "test_record_property3"
                  time = "0.000">

        </testcase>
    </testsuite>
</testsuites>
```

注意：这样生成的xml文件符合最新的xUnit标准，这点和record_property、record_xml_attribute正好相反。

2.10 不稳定测试用例处理

"不稳定"测试用例是指，有时候用例通过，有时候用例失败。用例间歇或偶发性出现故障，并且不清楚原因。pytest提供了处理这些不确定测试用例的功能。这些功能可以帮助快速确定、修复或缓解这些问题，或提供一些其他的常规策略。

2.10.1 为什么不稳定测试是个问题

当使用连续集成（CI）服务器时，不稳定测试尤其麻烦，因此在合并新代码之前必须通过所有测试。如果测试结果不稳定，则意味着代码更改破坏了测试，开发人员可能会对测试结果产生不信任，这可能导致忽略真正的失败。它也是浪费时间的一个问题，因为开发人员必须重新运行测试套件并调查虚假故障。

2.10.2 潜在的根本原因是什么

1. 系统状态

从广义上讲，不稳定测试用例表明测试依赖于一些未被适当控制的系统状态，即测试环境没有充分被隔离。更高级别的测试更有可能依赖更多的状态。

当测试套件并行运行时（例如使用pytest-xdist），有时会出现不稳定测试用例。这可以表明测试依赖于测试顺序。

也许不同的测试在完成自身之后无法清理并留下导致后面测试失败的数据。

不稳定测试依赖于先前测试的数据，该测试不会自行清理，并且会并行运行以前的测试数据，但以前的测试数据并不总是存在的。

修改全局状态的测试通常不能并行运行。

2．过于严格的断言

过于严格的断言可能会导致浮点比较及时序问题。pytest.approx 在这里很有用。

2.10.3　pytest 为我们提供的解决策略

Xfail 严格模式

pytest.mark.xfail 和 strict=False 可用于标记测试，以便当其失败时不会导致整个构建中断。这一般用于临时手动隔离一些问题，但不推荐永久使用。

2.10.4　pytest_CURRENT_TEST

pytest_CURRENT_TEST 环境变量可用于确定"哪个测试卡住了"。

2.10.5　可以重新运行的插件

重新运行任何失败的用例，可以通过给予它们额外的机会来减轻不稳定测试的负面影响，这样整体构建就不会失败。以下几个 pytest 插件可以支持用例重新运行：

- flasky；
- pytest-flakefinder-blog post；
- pytest-rerunfailures；
- pytest-replay：这个插件有助于重现 CI 运行期间观察到的局部崩溃或片状测试；
- 使用随机化测试的插件，打乱运行顺序也可以帮助暴露测试状态依赖问题；
- pytest-random-order；
- pytest-randomly。

2.10.6　测试人员采用的解决策略

为解决不稳定测试用例的执行处理问题，可以采用将测试套件拆分的方法，对于了解测试失败的原因可通过保存失败视频和截图方法进行分析，如果多个测试都包括某个容易失败的函数，则可以先把它从执行中删除，或重新单独编写。

1．拆分测试套件

将单个测试套件拆分为两个套件是常见的方法，例如单元与集成，并且仅将单元测试套件用作 CI。这也有助于保持构建时间的可管理性，因为高级别测试往往更慢，但是，这意味着打破构建的代码可能会合并，因此需要额外警惕并监视集成测试结果。

2．失败的视频/截图

对于 UI 测试，这些对于了解测试失败时 UI 的状态非常重要。pytest-splinter 可以与 pytest-bdd 插件一起使用，并且可以在测试失败时保存屏幕截图，这有助于查看原因。

3. 删除或重写测试

如果其他测试涵盖了该函数,则可以先从套件中删除该测试。或在本地文件中,使其来源更明显。

2.11 本章小结

本章是 pytest 中最重要的一章,重点内容是关于断言和运行管理的。以下是本章的要点:

(1) pytest 的运行顺序。

(2) pytest 中的断言管理。

(3) pytest 各种执行方式与结果。

(4) pytest 执行中的跳过运行。

(5) pytest 执行中的预期失败的结果。

(6) pytest 失败结果。

(7) pytest 中断调试。

(8) pytest 的定制 xml 报告。

(9) pytest 运行时出错解决策略。

第 3 章 pytest 中最闪亮的 fixture 功能

第 2 章介绍了使用 pytest 运行各种范围的测试用例、跳过和标记失败用例、运行中断错误管理及运行结果和报告设置。本章主要针对 pytest 中的 fixture 功能进行讲解。基本原理是通过装饰器的方式把所有测试方法进行有效组织的过程。

3.1 fixture 介绍

fixture 是 pytest 特有的功能，它用 @pytest.fixture 标识，定义在函数前面。在编写测试函数的时候，可以将此函数的名称作为传入参数，pytest 会以依赖注入方式将该函数的返回值作为测试函数的传入参数。

fixture 的管理从简单的单元测试扩展到复杂的功能测试，允许通过配置和组件选项参数化 fixture 和测试用例，或者跨功能、类、模块，甚至整个测试会话复用 fixture。

此外，pytest 继续支持经典的 xUnit 风格的测试。可以根据自己的喜好，混合使用两种风格，或者逐渐过渡到新的风格，也可以从已有的 unittest.TestCase 项目中执行测试。

3.2 fixture 目标

fixture 主要的目的是为了提供一种可靠和可重复性的手段去运行那些最基本的测试内容。例如在测试网站的功能时，每个测试用例都要登录和退出，利用 fixture 就可以只执行一次，否则每个测试用例都要执行这两步也是冗余。

对比 xUnit 经典的 setup/teardown 形式，fixture 拥有一个明确的名称，可以不必是 setup 这些固定的方法名，通过声明使其能够在函数、类、模块，甚至整个测试会话中被激活使用。

假如把 fixture 看作资源，在测试用例执行之前需要去配置这些资源，执行完后需要去释放资源。例如 module 级别的 fixture，适合于那些测试用例都只需执行一次的操作。

fixture 以一种模块化的方式实现，因为每个 fixture 的名字都能触发一个 fixture 函数，而这个函数本身又能调用其他的 fixture，这就是 fixture 的嵌套功能。

fixture 还提供了参数化功能，根据配置和不同组件来选择不同的参数。

3.3 fixture 基本的依赖注入功能

fixture 允许测试用例可以轻松地接收和处理特定的需要预初始化操作的应用对象,而不用过分关心导入/设置/清理的细节,这是一个典型的依赖注入的实践,其中,fixture 扮演着注入者(injector)的角色,而测试用例扮演着消费者(client)的角色。

测试用例执行时,有些模块必须先执行,这样才能进行后续的测试,但并不是所有的模块都是如此。以前使用 setup 的方式把先执行的用例放在 setup 中,这样所有测试方法都先执行 setup 方法,但有些测试方法并不需要。fixture 的依赖注入功能可以帮你灵活解决此类问题。

最典型的使用场景:

登录功能,有些功能需要登录才能使用,例如支付功能和查看购物车,而浏览商品功能不需要登录。fixture 实现步骤如下:

(1) 导入 pytest。
(2) 创建 login() 函数。
(3) 在 login() 函数上加@pytest.fixture()。
(4) 在要使用的测试方法中传入(登录函数名称),也就是先执行 login() 函数再执行本测试方法。
(5) 不传入参数表明不需要登录,此时可以直接执行测试方法。

代码如下:

```python
# Author: lindafang
# Date: 2020 - 08 - 16 14:25
# File: test_fixture.py

import pytest

# 不带参数时 scope = "function"
# 有依赖关系,先登录,与其他功能有依赖


@pytest.fixture()
def login():
    print("\n用户名 linda 密码登录!")


def test_cart(login):
    print('\n用例 1,登录后执行查看购物车其他功能 1')

def test_find_goods():
    print('\n用例 2,不登录,执行浏览商品功能 2')

def test_pay(login):
    print('\n用例 3,登录后执行支付功能 3')
```

执行结果如下：

首先 login() 函数没有以 test_ 开头，所以不执行。

执行用例 test_cart 时，发现参数是 login，就开始从本用例中查找是否有这个参数或函数。找到 login() 的函数后，执行它，返回值传入 test_cart，之后再执行 test_cart 测试函数。

执行用例 test_find_goods 时，未发现参数，直接执行测试用例 2。

执行用例 test_pay 时，发现参数是 login，就开始从本用例中查找是否有这个参数或函数。找到 login() 的函数后，执行它，之后再执行 test_pay 测试函数。

```
==================== test session starts ====================
...
test_fixture.py::test_cart
用户名 linda 密码登录！
PASSED                                              [ 33%]
用例 1,登录后执行查看购物车其他功能 1

test_fixture.py::test_find_goods PASSED             [ 66%]
用例 2,不登录后执行浏览商品功能 2

test_fixture.py::test_pay
用户名 linda 密码登录！
PASSED                                              [100%]
用例 3,登录后执行支付功能 3


================== 3 passed in 0.03s ========================
Process finished with exit code 0
```

注意：@pytest.fixture() 中不带参数时范围默认 scope = "function"，也就是共享数据为函数级，就是在本用例函数中搜索是否有需要的共享数据。"\n"是换行符。

3.4 fixture 应用在初始化设置

初始化过程一般进行数据初始化、连接初始化等。

常用场景：测试用例执行时，有的用例的数据是可读取的，需要把数据读进来再执行测试用例。setup 和 teardown 可以实现。fixture 可以灵活命名实现。

具体实现步骤：

(1) 导入 pytest。

(2) 创建 data() 函数。

(3) 在 data() 函数上加 @pytest.fixture()。

(4) 在要使用的测试方法 test_login 中传入(data 函数名称)，也就是先执行 data() 函数再执行测试方法。

(5) 不传入参数表明可以直接执行测试方法。

代码如下：

```python
# Author: lindafang
# File: test_fixture_data.py
import pytest
import csv


@pytest.fixture()
def data():
    test_data = {'name': 'linda', 'age': 18}
    return test_data


def test_login(data):
    name = data['name']
    age = data['age']
    print("笔者的名字叫：{}，今年{}。".format(name, age))
```

如果测试数据是从 csv 文件中读取的，执行操作步骤如下：

(1) 新建 userinfo.csv 文件，代码如下：

```
username,age
linda,18
tom,8
steven,28
```

(2) 在 test_fixture_data.py 中增加代码如下：

```python
@pytest.fixture()
def read_data():
    with open('userinfo.csv') as f:
        row = csv.reader(f, delimiter=',')
        next(row)                    # 不读取首行，跳到下一行
        users = []
        for r in row:
            users.append(r)          # 读取的字段均为 str 类型
    return users


def test_logins(read_data):
    name = read_data[0][0]           # 只取第 1 组数据中的第 1 个
```

```
        age = read_data[0][1]
        print("s笔者的名字叫:{},今年{}岁.".format(name, age))
```

(3) 鼠标右击选择 pytest 执行。

```
test_fixture_data.py .笔者的名字叫:linda,今年18岁。
.s笔者的名字叫:linda,今年18。
                                                        [100%]

===================== 2 passed in 0.02s =================================
```

3.5 fixture 应用在配置销毁

3.5.1 使用 yield 代替 return

已经可以将测试方法前要执行的或依赖的问题解决了,但测试方法后需要销毁并清除的数据该如何处理呢？范围是模块级别的,类似 setupClass。解决方法是通过在同一模块中加入 yield 关键字,yield 调用第一次返回结果,第二次执行它下面的语句并返回结果。

步骤如下：
(1) 添加@pytest.fixture(scope=module)语句。
(2) 在登录的方法中添加 yield,之后添加销毁清除的步骤。
代码如下：

```
# Author: lindafang
# Date: 2020-08-16 14:25
# File: test_fixture_module_yield.py
import pytest


# 在整个模块只做一次,范围是模块级别的


@pytest.fixture(scope="module")
def open():
    print("打开浏览器,打开百度首页")

    yield

    print('执行teardown')
    print('最后关闭浏览器')


# open在第一种方法中使用,其他方法可以不添加
```

```
def test_s7():
    print('用例 7,')

#如果第一种方法中没有 open,则第一种方法单独执行,再执行下面的第二种方法
def test_s8(open):
    print('用例 8,')

#也就是说,open 添加在哪个测试方法中,就从该方法以后实现 open 依赖注入效果
def test_s9(open):
    print('用例 9,')
```

3.5.2 使用 with 写法

对于支持 with 写法的对象,程序也可以隐式地执行它的清理销毁操作。当使用 with 出现问题时,由 Python 进行处理以便销毁。

代码如下:

```
# Author: lindafang
# Date: 2020 - 12 - 01 10:56
# File: test_fixture_with.py
import smtplib

import pytest

@pytest.fixture()
def smtp_connection_yield():
    with smtplib.SMTP("smtp.163.com", 25, timeout = 5) as smtp_connection:
        print(" --- start connection")
        yield smtp_connection
        print(" --- end connection")

def test_send_mail(smtp_connection_yield):
    print("发邮件")
```

3.5.3 使用 addfinalizer 方法

fixture 函数能够接收一个 request 参数,表示测试请求的上下文。本书使用 request.addfinalizer 方法为 fixture 添加清理销毁函数。

代码如下:

```
# Author: lindafang
# Date: 2020 - 08 - 11 13:54
# File: test_addfinalizer.py
```

```python
import pytest

@pytest.fixture()
def demo_fixture(request):
    print("\n这个 fixture 在每个用例前执行一次")

    def demo_finalizer():
        print("\n在每个用例完成后执行的 teardown")

    # 注册 demo_finalizer 为终结函数
    request.addfinalizer(demo_finalizer)


def test_01(demo_fixture):
    print("\n=== 执行了用例：test_01 === ")

def test_02(demo_fixture):
    print("\n=== 执行了用例：test_02 === ")

def test_03(demo_fixture):
    print("\n=== 执行了用例：test_03 === ")
```

执行结果如图 3-1 所示。

图 3-1　addfinalizer 的函数使用

注意：如果在 yield 之前或者 addfinalizer 注册之前代码发生错误并退出，则不会再执行后续的清理操作。

3.5.4　yield 与 addfinalizer 的区别

那么，除了在使用上的区别之外，yield 与 addfinalizer 还有什么不同呢？

addfinalizer 可以注册多个终结销毁函数，而 yield 无法实现多个。在执行 yield 的过程中出现错误无法再执行 yield 后面的函数，而 addfinalizer 函数在执行的过程中即使出错也会执行后面的函数。

代码如下:

```python
# Author: lindafang
# Date: 2020-08-11 14:10
# File: test_addfinalizer02.py
import pytest


@pytest.fixture()
def demo_fixture(request):
    print("\n 这个 fixture 在每个用例前执行一次")

    def demo_finalizer():
        print("\n 在每个用例完成后执行的 teardown")

    def demo_finalizer2():
        print("\n 在每个用例完成后执行的 teardown2")

    def demo_finalizer3():
        print("\n 在每个用例完成后执行的 teardown3")

    # 注册 demo_finalizer 为终结函数
    request.addfinalizer(demo_finalizer)
    request.addfinalizer(demo_finalizer2)
    request.addfinalizer(demo_finalizer3)


def test_01(demo_fixture):
    print("\n=== 执行了用例: test_01 ===")


def test_02(demo_fixture):
    print("\n=== 执行了用例: test_02 ===")


def test_03(demo_fixture):
    print("\n=== 执行了用例: test_03 ===")
```

运行结果如图 3-2 所示,可以看到,注册的 3 个函数都被执行了,但是要注意执行顺序,可以看出与注册的顺序恰好相反。

当 setUp 的代码执行出错时,addfinalizer 依旧会被执行。

代码如下:

```python
@pytest.fixture
def equipments(request):
```

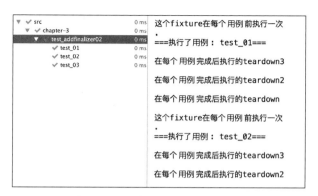

图 3-2 yield 与 addfinalizer 的区别

```
r = []
for port in ('C1', 'C3', 'C28'):
    equip = connect(port)
    request.addfinalizer(equip.disconnect)
    r.append(equip)
return r
```

例如,C1、C3、C28 这 3 个端口连接,如果 C28 这个端口失败了,这时会抛出一个连接异常,但是在执行 teardown 关闭连接的时候,C1 和 C3 依然可以正常关闭。

3.6　fixture 方法源码详细讲解

通过上面两个例子大家对 fixture 的功能有了些了解,下面是@pytest.fixture()方法的源码,笔者只将注释部分按句翻译,并会在后面不同的例子中将其功能讲解清楚。

源码及部分翻译如下:

```
def fixture(
    callable_or_scope = None,
    * args,
    scope = "function",
    params = None,
    autouse = False,
    ids = None,
    name = None
):
    """Decorator to mark a fixture factory function.

    使用装饰器标记 fixture 的功能

    This decorator can be used, with or without parameters, to define a
```

fixture function.
可以使用此装饰器(带或不带参数)来定义fixture功能

　　The name of the fixture function can later be referenced to cause its
　　invocation ahead of running tests: test
　　modules or classes can use the "pytest.mark.usefixtures(fixturename)"
marker.

　　fixture功能的名称可以在以后使用,引用它会在运行测试之前调用它: test模块或类可以使用pytest.mark.usefixtures(fixturename)标记

　　Test functions can directly use fixture names as input
　　arguments in which case the fixture instance returned from the fixture
　　function will be injected.

　　测试功能可以直接使用fixture名称作为输入参数,在这种情况下,fixture实例的返回值将通过参数形式被注入

Fixtures can provide their values to test functions using "return" or "yield" statements. When using "yield" the code block after the "yield" statement is executed as teardown code regardless of the test outcome, and must yield exactly once.

　　多个fixture能给测试方法提供数据,通过return和yield的方式。当我们应用yield时,我们执行代码初始化在遇到yield语句之后我们开始执行测试方法,执行测试方法结束后必须执行yield后面的代码一次。

　　:arg scope: the scope for which this fixture is shared, one of
　　　　　　　　"function" (default), "class", "module",
　　　　　　　　"package" or "session".

　　　　　　　　"package" is considered **experimental** at this time.

:参数作用范围: fixture被分享的范围 , "function" (default), "class", "module", "package" or "session".
"package"被认为是实验性的。

　　:arg params: an optional list of parameters which will cause multiple
　　　　　　　　invocations of the fixture function and all of the tests
　　　　　　　　using it.
　　　　　　　　The current parameter is available in "request.param".

　　一个可选的参数列表,它将导致多个参数调用fixture功能和所有测试使用它。当前的参数可能在request.param中调用

　　:arg autouse: if True, the fixture func is activated for all tests that
　　　　　　　　can see it. If False (the default) then an explicit

```
                        reference is needed to activate the fixture.
        如果值为 true,则为所有测试激活 fixture func 以便可以看到它。如果值为 false(默认值)则
    显式定义来激活 fixture

        :arg ids: list of string ids each corresponding to the params
                    so that they are part of the test id. If no ids are provided
                    they will be generated automatically from the params.
        每个字符串 id 的列表,每个字符串对应于 params,这样它们就是测试 id 的一部分。如果没有提
    供 id 它们将从 params 自动生成
        :arg name: the name of the fixture. This defaults to the name of the
                    decorated function. If a fixture is used in the same module in
                    which it is defined, the function name of the fixture will be
                    shadowed by the function arg that requests the fixture; one way
                    to resolve this is to name the decorated function
                    ``fixture_<fixturename>`` and then use
                    ``@pytest.fixture(name='<fixturename>')``.
        fixture 的名称,默认为装饰函数的名称。如果 fixture 在定义它的同一模块中使用,fixture 的
    功能名称将被请求 fixture 的功能 arg 遮蔽,解决这个问题的一种方法是将装饰函数命名"fixture_
    <fixturename>"然后使用"@ pytest.fixture(name = '<fixturename>')"。
        """
```

3.7 不同层级 scope 使用 fixture 实例

上述 3.6 节源码中可以看到:fixture(scope="function",params=None,autouse=False,ids=None,name=None):scope 有 5 个级别参数 function(默认)、class、module、package 和 session。package 被认为是实验性的。
- function:每个函数或方法都会调用;
- class:每个类调用一次,一个类可以有多种方法;
- module:每个.py 文件调用一次,该文件内又有多个 function 和 class;
- Session:多个文件调用一次,可以跨.py 文件调用,每个.py 文件就是 module。

如何使用 fixture 功能实现不同层级的数据共享呢?默认的 function 级别已经举例说明了,也就是每种方法都可以调用 fixture 标记的方法,通过传参的方式。

现在来举个 module 层级的例子,module 是模块级,也就是每个文件只调用一次。

3.7.1 模块(module)级别使用 fixture 实例

常用使用场景:当进行测试执行时,有些动作只在这个文件中的开始或结束执行一次。这就需要运用层级 scope 的设置了,当把 scope 参数设置为 module 时,只在文件开始执行一次。例如,在进行 WebUI 自动化测试时,需要在测试用例前打开浏览器这个动作。

实现步骤：

（1）导入 pytest。

（2）创建 open()函数。

（3）在 open()函数上添加@pytest.fixture(scope="module")。

（4）在测试方法中传入参数(open 函数名称)。

具体代码如下：

```python
# Author: lindafang
# Date: 2020-08-16 14:25
# File: test_fixture_module.py
import pytest


# 在整个模块只做一次事,范围是模块级别的


@pytest.fixture(scope="module")
def open():
    print("打开浏览器,打开百度首页")


# open 在第一种方法中使用,其他方法可以不添加
def test_s7():
    print('用例 7,')


# 如果第一种方法中没有 open,则第一种方法单独执行,再执行下面的第二种方法
def test_s8(open):
    print('用例 8,')


# 也就是说,open 添加在哪个测试方法中,就从该方法以后实现 open 依赖注入效果
def test_s9(open):
    print('用例 9,')
```

全部添加 open 参数时执行结果如下：

```
test_fixture_module.py 打开浏览器,打开百度首页
.用例 7,
.用例 8,
.用例 9,
```

s7()函数不添加 open 参数时执行结果如下：

```
test_fixture_module.py .用例 7,
打开浏览器,打开百度首页
.用例 8,
.用例 9,
```

其他执行结果,读者可以自行体会。因为 module 作用于全文件,可以分几种情况: open 在第一种方法中使用,其他方法可以不添加。open 方法在全文件开始前只执行一次。如果第一种方法中没有 open,则第一种方法单独执行,再执行下面的第二种方法,也就是说,open 添加在哪个测试方法中,就从该方法以后实现 open 依赖注入效果。

3.7.2 类(class)级别使用 fixture 实例

当 fixture 为 class 级别的时候,如果一个 class 里面有多个用例,则都调用了此 fixture,那么当此 fixture 只在该 class 中时所有用例开始前执行一次。

代码如下:

```python
# Author: lindafang
# Date: 2020-12-01 11:10
# File: test_fixture_class.py

import pytest

@pytest.fixture(scope="class")
def first():
    print("\n 获取用户名,scope 为 class 级别只运行一次")
    a = "linda"
    return a

def test_3():
    print("不在类中的测试方法")

class TestCase():
    def test_1(self, first):
        '''用例传 fixture'''
        print("测试账号: %s" % first)
        assert first == "linda"

    def test_2(self, first):
        '''用例传 fixture'''
        print("测试账号: %s" % first)
        assert first == "linda"
```

执行的结果如图 3-3 所示。

3.7.3 会话(session)级别使用 fixture 与 conftest.py 配合

fixture 为 session 级别是可以跨.py 模块调用的,也就是当有多个.py 文件用例的时候,如果多个用例只需调用一次 fixture,那就可以设置为 scope="session"。

既然已经是跨模块,需要在.py 模块之上。因此采用一个单独的文件 conftest.py,文件名称是固定的，pytest 会自动识别该文件。放到工程的根目录下就可以全局调用了，如果放到某个 package 包下，那就只在该 package 内有效。

图 3-3　class 级别 fixture 的使用

常用使用场景：当用户与其他测试工程师合作开发时，公共的模块要存放在不同文件中，要存放在大家都能访问的地方。通常是一些公用的配置。公共模块也可以，例如登录模块是大家公用的，因此应放在项目或包的路径下。

具体执行步骤如下：

（1）在本目录下创建 conftest.py 文件（文件名必须是这个）。

（2）将登录模块带@pytest.fixture 写在 conftest.py 文件中。

```
# Author: lindafang
# File: conftest.py
import pytest

@pytest.fixture()
def login():
    print("\n用户名 linda 密码登录！")
```

（3）在原来的 test_fixture.py 文件中删除部分代码（login 方法），代码如下：

```
import pytest

def test_cart(login):
    print('\n用例 1,登录后执行查看购物车其他功能 1')

def test_find_goods():
    print('\n用例 2,不登录后执行浏览商品功能 2')

def test_pay(login):
    print('\n用例 3,登录后执行支付功能 3')
```

（4）右击鼠标，选择 pytest 执行 test_fixture.py。

在执行过程中当读到 login 时，如果在本用例中没找到，则去本目录下 conftest.py 中查找。如果找到就执行，如果找不到就报错。同时其他工程师也可以在本目录中新建文件，并使用 login 函数，可以跟上述代码类似。

执行结果如图 3-4 所示。

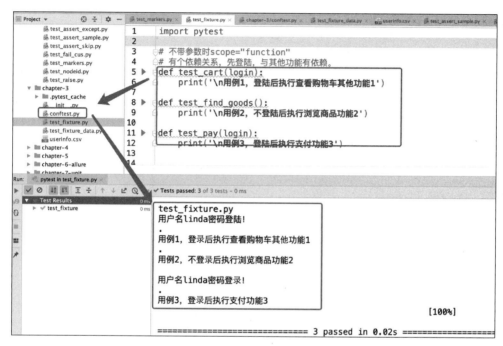

图 3-4　conftest.py 文件共享数据执行结果

3.7.4　session 级别实例

需要使用网络接入的 fixture 往往依赖于网络的连通性，并且创建过程一般非常耗时。放在全局 conftest.py 文件中，类似一个单例模式的方法，提前建立连接，所有人可以使用这个连接。

（1）创建 conftest.py 文件，把网络连接的函数写在这个文件中，代码如下：

```
# Author: lindafang
# Date: 2020-05-07 15:05
# File: conftest.py
import pytest
import smtplib


@pytest.fixture()
def login():
```

```
        print("\n用户名 linda 密码登录!")


@pytest.fixture(scope = 'module')
def smtp_connection():
    return smtplib.SMTP("smtp.163.com", 25, timeout = 5)
```

(2) 在相同的目录下,新建一个测试模块 test_module.py,将 smtp_connection 作为形参传入每个测试用例,它们共享同一个 smtp_connection()的返回值。

```
# Author: lindafang
# Date: 2020 - 12 - 01 11:22
# File: test_module_1.py


def test_ehlo(smtp_connection):
    response, _ = smtp_connection.ehlo()
    assert response == 250
    smtp_connection.extra_attr = 'test'
    assert 0     # 为了展示,强制置为失败


def test_noop(smtp_connection):
    response, _ = smtp_connection.noop()
    assert response == 250
    assert smtp_connection.extra_attr == 0    # 为了展示,强制置为失败
```

(3) 执行这个测试模块,结果如下:

```
F
src/chapter3/test_module_1.py:5 (test_ehlo)
smtp_connection = < smtplib.SMTP object at 0x1104bbe48 >

    def test_ehlo(smtp_connection):
        response, _ = smtp_connection.ehlo()
        assert response == 250
        smtp_connection.extra_attr = 'test'
>       assert 0    # 为了展示,强制置为失败
E       assert 0

test_module_1.py:10: AssertionError
F
src/chapter3/test_module_1.py:12 (test_noop)
test != 0
```

```
Expected : 0
Actual   : test
< Click to see difference >

smtp_connection = < smtplib.SMTP object at 0x1104bbe48 >

    def test_noop(smtp_connection):
        response, _ = smtp_connection.noop()
        assert response == 250
>       assert smtp_connection.extra_attr == 0   # 为了展示,强制置为失败
E       AssertionError: assert 'test' == 0
E        + where 'test' = < smtplib.SMTP object at 0x1104bbe48 >.extra_attr

test_module_1.py:16: AssertionError
```

可以看到：两个测试用例使用的 smtp_connection 实例都是< smtplib.SMTP object at 0x1104bbe48 >，说明 smtp_connection 只被调用了一次。在前一个用例 test_ehlo 中修改 smtp_connection 实例（上述例子中，为 smtp_connection 添加 extra_attr 属性），也会反映到 test_noop 用例中，如图 3-5 所示。

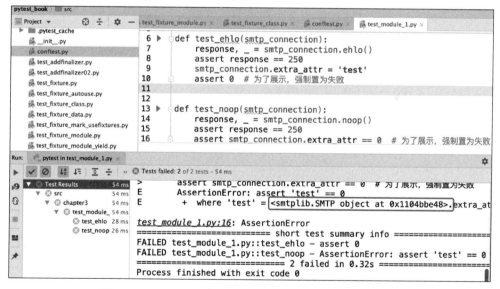

图 3-5　conftest.py 中 session 级别文件共享数据执行结果

如果期望拥有一个会话级别作用域的 fixture，则可以简单地将其声明为 session。

```
@pytest.fixture(scope = 'session')
def smtp_connection():
    return smtplib.SMTP("smtp.163.com", 25, timeout = 5)
```

注意：pytest每次只缓存一个fixture实例,当使用参数化的fixture时,pytest可能会在声明的作用域内多次调用这个fixture。

3.8 使用params传递不同数据

从上述3.6节源码中可以看到：fixture(scope="function", params=None, autouse=False, ids=None, name=None),params是参数,默认可以不选。

如果需要在一系列测试用例的执行中,每轮执行都使用同一个fixture,但是有不同的依赖场景,则可以考虑对fixture进行参数化,这种方式适用于对多场景的功能模块进行详尽测试。

可以通过指定params关键字参数创建两个fixture实例,每个实例供一轮测试使用,所有的测试用例执行两遍,在fixture的声明函数中,可以使用request.param获取当前使用的入参。

3.8.1 测试方法使用两个简单测试数据

每次执行应使用不同数据,一次应使用一个。可以通过fixture自带的params准备不同的数据。下面是一个测试方法使用2个简单测试数据。

代码如下：

```python
# Author: lindafang
# Date: 2020-05-16 14:25
# File: test_fixture_params.py
import pytest

# 每次执行应使用不同数据,一次应使用一个。这种方法应准备不同的数据。
@pytest.fixture(params=['apple', 'banana'])
def fruit(request):
    # 将列表中的数据依次返回
    return request.param

def test_fruit(fruit):
    print("\n笔者今天吃{}".format(fruit))
    assert True


def test_cook_fruit(fruit):
    print("笔者今天做{}派。".format(fruit))
    assert True
```

执行结果如图3-6所示。

第3章　pytest中最闪亮的fixture功能

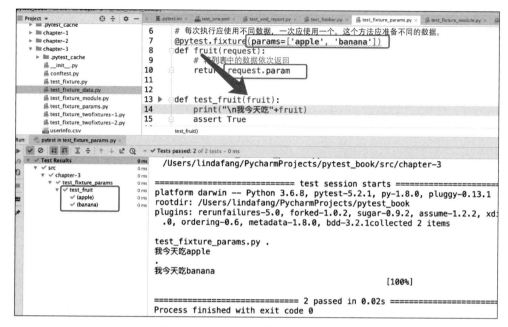

图 3-6　fixture 中参数 params 的使用

3.8.2　二(多)个测试方法共用两个简单测试数据

在代码中增加下述代码,实现多个测试使用一组不同数据:

```
def test_cook_fruit(fruit):
    print("笔者今天做{}派。".format(fruit))
    assert True
```

执行结果如图 3-7 所示。

3.8.3　有效测试数据与预期失败 xfail 的测试数据

在测试某些功能时,如果步骤一致,则可以使用数据驱动方式执行测试,也就是说多组数据使用一个测试方法。通常正确有效的数据分得一个测试方法,因为错误的步骤通常简单,而正确的步骤会多些,必须分开验证,但当步骤相同而数据不同时,预期失败和预期正确的数据可以在一个测试方法中实现。这样就只有一组或几组数据会引起预期的失败,但不能在测试方法中标记预期失败,而应该在通过参数传递的测试数据中使用 pytest.param 标记预期失败。

代码如下:

```
# Author: lindafang
# Date: 2020-05-16 14:25
```

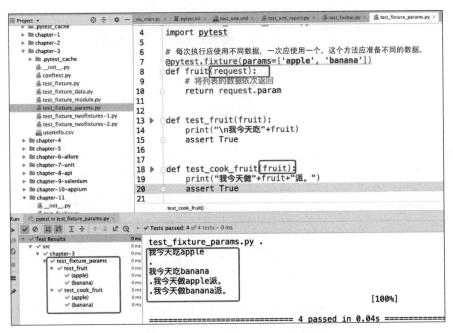

图 3-7　fixture 中参数 params 的使用

```
#File: test_fixture_param_marks.py

import pytest

#第二个数据做个预期失败标记
@pytest.fixture(params = [('3 + 5', 8),
                          pytest.param(('6 * 9', 42),
                                       marks = pytest.mark.xfail,
                                       ids = 'failed')])
def data_set(request):
    return request.param


def test_data(data_set):
    #eval 可以将字符串中的表达式进行计算,也就是 3 + 5 计算后与 8 比对,6 * 9
    #计算后与 42 比对
    assert eval(data_set[0]) == data_set[1]
```

在 data_set 方法中使用 request 传递 fixture 中的参数,使用 request.param 解析每个参数。在方法上面加上 pytest.fixture 装饰器,通过参数传递不同的测试数据。一组是正确的,3+5 和 8 比对,另一组将 6*9 和 42 比对,并且在 test_data 中通过 data_set 的依赖注入的方式传入数据,通过 eval 方法将表达式,例如 3+5 从字符串转换成可执行的算式,计算的结果与传递参数的第二列 data_set[1] 数据比对是否相等。

最重要的是第二组数据预期结果不正确，通过 marks 标记成失败 pytest.mark.xfail，并用 ids 显示出 failed，执行的结果如图 3-8 所示。

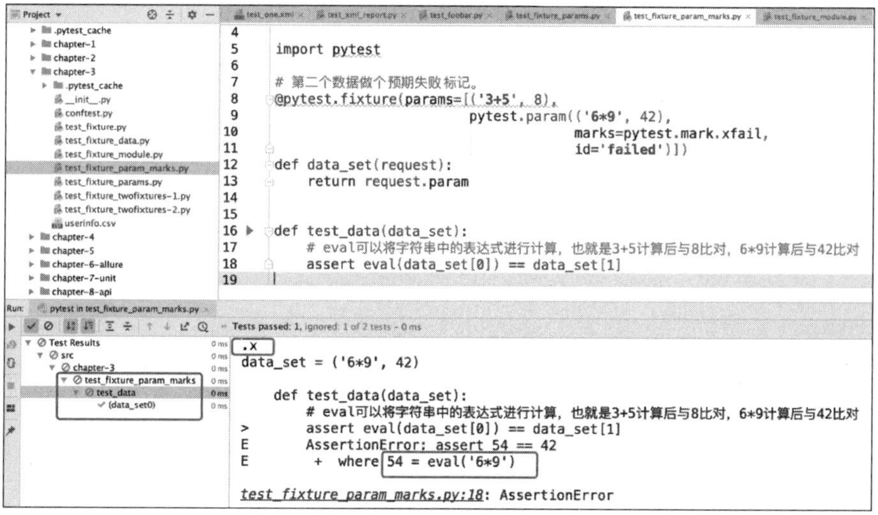

图 3-8　预期失败的执行结果

3.8.4　params 与 ids 的应用

对于复杂类型的测试数据通常加上 id 或 name 来表明数据的含义，并标记测试要点。测试数据除了字符串以外，还可以是表达式，以及元组、字典、类等类型。使用 ids 关键字参数，自定义测试 ID。

代码如下：

```python
# Date: 2020-05-16 14:25
# File: test_fixture_params_ids.py
import pytest

@pytest.fixture(params=[0, 'a'], ids=['number', 'charactor'])
def a(request):
    return request.param

def test_a(a):
    print(a)
    pass

def idfn(fixture_value):
    if fixture_value == 0:
        return "eggs"
    elif fixture_value == 1:
        return False
    elif fixture_value == 2:
```

```
            return None
        else:
            return fixture_value


@pytest.fixture(params = [0, 1, 2, 3], ids = idfn)
def b(request):
    return request.param

def test_b(b):
    print(b)
    pass

class C:
    pass

@pytest.fixture(params = [(1, 2), {'d': 1}, C()])
def c(request):
    return request.param

def test_c(c):
    print(c)
    pass
```

ids=idfn,idfn 是前面使用的方法,在 fixture 中将 params 的每个数据传入 idfn,ids 的值就是其返回值。当 params=0 时,返回值 eggs;当 params=1 时,返回值 False;当 params=2 时,返回值 None,ids 显示 params 的值;当 params=3 时,返回值直接显示为 3。

执行结果如图 3-9 所示。

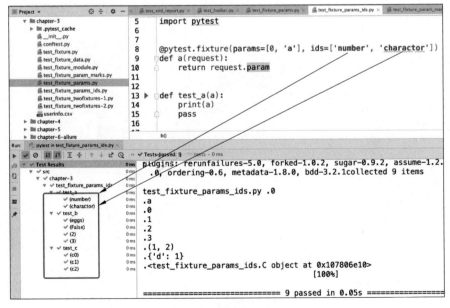

图 3-9　参数化与 ids

从执行结果可以看出：ids 可以接收一个函数，用于生成测试 ID。当测试 ID 指定为 None 时，使用的是 params 原先对应的值。

注意：当测试 params 中包含元组、字典或者对象时，测试 ID 使用的是 fixture 函数名＋param 的下标。

如果使用 CIL 终端执行，则结果如下：

```
lindafang@linda chapter-3 % pytest -s -v test_fixture_params_ids.py
...
  collected 9 items

test_fixture_params_ids.py::test_a[number] 0
PASSED
test_fixture_params_ids.py::test_a[charactor] a
PASSED
test_fixture_params_ids.py::test_b[eggs] 0
PASSED
test_fixture_params_ids.py::test_b[False] 1
PASSED
test_fixture_params_ids.py::test_b[2] 2
PASSED
test_fixture_params_ids.py::test_b[3] 3
PASSED
test_fixture_params_ids.py::test_c[c0] (1, 2)
PASSED
test_fixture_params_ids.py::test_c[c1] {'d': 1}
PASSED
test_fixture_params_ids.py::test_c[c2] < test_fixture_params_ids.C object at 0x107951ef0 >
PASSED
```

3.8.5　params 综合实例

这个测试用例使用不同的 SMTP 服务器，共执行了 3 次。在参数化的 fixture 中，pytest 为每个 fixture 实例自动指定一个测试 ID。

```
# Author: lindafang
# Date: 2020-12-01 12:51
# File: test_request.py

import pytest
import smtplib

@pytest.fixture(scope='module', params=['smtp.163.com','smtp.126.com',"mail.python.org"])
```

```python
def smtp_connection_params(request):
    server = request.param
    with smtplib.SMTP(server, 587, timeout=5) as smtp_connection:
        yield smtp_connection

def test_parames(smtp_connection_params):
    response, _ = smtp_connection_params.ehlo()
    assert response == 250
```

可以看到执行显示的 ID 是 test_parames[smtp.163.com]、test_parames[smtp.126.com]、test_parames[mail.python.org]，如图 3-10 所示。

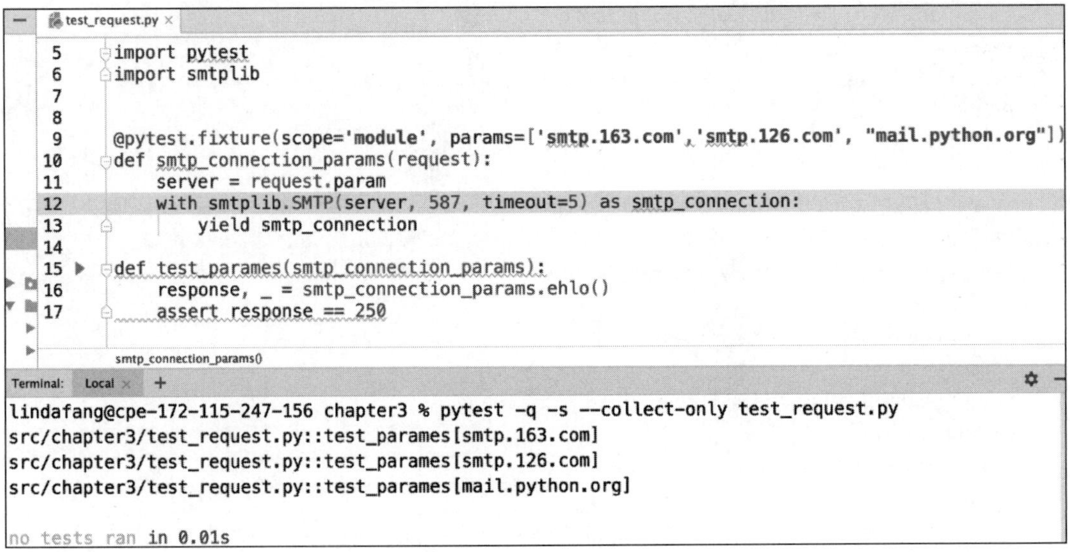

图 3-10　param 的综合实例

3.9　自动调用 fixture

如果每次使用 fixture 都要通过传参的方式，则应改变原来测试方法的结构。如何不通过注入的方式让测试方法执行呢？有 2 种方式可选，第一种在 fixture 的参数中将 autouse 参数设置为 True，这样便会自动应用所作用的范围。第二种使用 @pytest.mark.usefixtures，在需要的测试方法上添加。

3.9.1　使用 fixture 中参数 autouse＝True 实现

在方法上面加 @pytest.fixture(autouse＝True)，因为 scope 参数未设置，所以使用默认，其作用范围是 function，也就是说每个测试方法都默认应用了。

代码如下:

```python
#File: test_fixture_autouse.py
import pytest


# autouse = True 不需要在测试方法中传参调用,默认全使用
@pytest.fixture(autouse = True)
def open_browser():
    print('打开首页!')


def test_login():
    print('case1: 登录')


def test_search():
    print('case2:搜索')


def test_exit():
    print('case3,退出')
```

执行结果如下,可以看到 3 个测试方法前都自动先执行了打开首页的方法。

```
test_fixture_autouse.py 打开首页!
.case1: 登录
打开首页!
.case2:搜索
打开首页!
.case3,退出
                                    [100%]

==================== 3 passed in 0.03s ====================
```

autouse=True 的 fixture 在其他级别作用域中的工作流程:autouse fixture 遵循 scope 关键字的定义,如果其含有 scope='session',则不管它在哪里定义的,都将只执行一次。scope='class' 表示每个测试类执行一次。

如果在测试模块中定义 autouse fixture,那么这个测试模块所有的用例自动使用它。

如果在 conftest.py 中定义 autouse fixture,那么同文件夹和子文件夹中的所有测试模块中的用例都将自动使用它。

如果在插件中定义 autouse fixture,那么安装这个插件的项目中的所有用例都将自动使用它。

3.9.2 使用@pytest.mark.usefixtures

上面全部自动使用的方式类似单元测试框架中 setup/teardown 方法，就是在每个测试方法前后执行，而有时只有部分测试方法需要这个，也就是更灵活的方式。那么我们可以在测试方法上添加@pytest.mark.usefixtures("start")实现。

代码如下：

```python
#File: test_fixture_mark_usefixtures.py
import pytest


@pytest.fixture(scope="function")
def start():
    print("\n----begin-登录--")


@pytest.mark.usefixtures("start")
def test_soso():
    print('\ncase1:登录后执行搜索')


def test_cakan():
    print('\ncase2:不登录就看')


@pytest.mark.usefixtures("start")
def test_cart():
    print('\ncase3,登录,加购物车')


@pytest.mark.skipif(reason="不想退出")
def test_quit():
    print('case3,登录,退出')
```

执行结果如下，只有测试方法上面带有标记 mark.usefixtures("start")的测试方法才执行 start 方法中的内容。

```
test_fixture_mark_usefixtures.py
----begin-登录--
.
case1:登录后执行搜索
.
```

```
case2:不登录就看

----begin-登录--
.
case3,登录,加购物车
s
Skipped: 不想退出
                                    [100%]

==================== 3 passed, 1 skipped in 0.03s ====================
```

3.9.3 数据库自动应用的实例

有时候,想在测试用例中自动使用 fixture,而不是作为参数使用或者 usefixtures 标记。设想,现在有一个与数据库相关的 fixture,包含 begin/rollback/commit 的体系结构,希望通过 begin/rollback 包裹每个测试用例。

不想对原测试方法进行任何改动,或全部实现自动应用,当没有特例,也都不需要返回值时可以选择自动应用。此种情况该如何解决呢?

代码如下:

```python
#File: test_db_transact.py

import pytest


class DB:
    def __init__(self):
        self.intransaction = []

    def begin(self, name):
        self.intransaction.append(name)

    def rollback(self):
        self.intransaction.pop()


@pytest.fixture(scope="module")
def db():
    return DB()


class TestClass:
    @pytest.fixture(autouse=True)
```

```
    def transact(self, request, db):
        db.begin(request.function.name)
        yield
        db.rollback()

    def test_method1(self, db):
        assert db.intransaction == ["test_method1"]

    def test_method2(self, db):
        assert db.intransaction == ["test_method2"]
```

3.10 第三方插件通过文件夹共享测试数据

如果想让多个测试共享同样的测试数据文件,则下面的两个好方法可以实现。

(1)把这些数据加载到 fixture 中,在测试中再使用这些 fixture。

(2)把这些数据文件放到 tests 文件夹中,一些第三方插件能帮助你管理这方面的测试,例如:pytest-datadir 和 pytest-datafiles。

3.11 fixture 的并列与嵌套调用

常用使用场景:相互依赖的 fixture 可能是多个,或提前准备的数据也可能是多个,或数据准备也可能有依赖和先后。

3.11.1 并列使用 fixture

常用使用场景:前提条件有多个,可以并列使用多个 fixture,多个@注解,在一个测试方法中使用时将传入的参数写多个就能解决此问题。

具体执行步骤:

(1)新建文件 test_fixture_twofixtures-1.py。

(2)创建 open_browser()函数。在函数上添加@pytest.fixture(scope="module")。

(3)创建 login()函数。在函数上添加@pytest.fixture()。

(4)创建 3 个测试函数。在第一个测试函数 test_soso 的参数中传入 open_browser、login 的函数名,表示调用这两个函数。

代码如下:

```
#File: test_fixture_twofixtures-1.py
import pytest

#fixture 可以并列使用
```

```python
@pytest.fixture()
def login():
    print('这是个登录模块!')


@pytest.fixture(scope = "module")
def open_browser():
    print('打开首页!')


def test_soso(open_browser,login):
    print('case1:登录后执行搜索')


def test_cakan():
    print('case2:不登录就看')


def test_cart(login):
    print('case3,登录,加购物车')
```

执行结果如下,由于 open_browser 函数上的 fixture 的层级范围是 module,所以先执行 open_browser,也就是最先执行的是测试方法 test_soso,读到输入参数时先执行 open_browser,然后执行 login 方法,再执行 test_soso 中的代码。之后正常执行 test_cakan 和 test_cart 方法,执行 test_cart 方法时先执行 login 方法。

```
test_fixture_twofixtures-1.py 打开首页!
这是个登录模块!
.case1:登录后执行搜索
.case2:不登录就看
这是个登录模块!
.case3,登录,加购物车
```

3.11.2 嵌套调用 fixture

常用使用场景:有先后关系的嵌套调用,多个@注解依赖可以通过传参方式,测试方法调用时只需调用最后的那种方法 fixture,然后调用其他的 fixture。

在上面代码的基础上复制并修改,将 login 函数的参数中加入 open_browser,所以下面的测试方法就不用添加 open_browser 了。

代码如下:

```
#File: test_fixture_twofixtures-2.py

import pytest
```

```python
# fixture可以嵌套使用
@pytest.fixture()
def login(open_browser):
    print('这是个登录模块!')


@pytest.fixture(scope = "module")
def open_browser():
    print('打开首页!')


def test_soso(login):
    print('case1：登录后执行搜索')


def test_cakan():
    print('case2:不登录就看')


def test_cart(login):
    print('case3,登录,加购物车')
```

执行结果如下：

```
test_fixture_twofixtures-2.py 打开首页!
这是个登录模块!
.case1：登录后执行搜索
.case2:不登录就看
这是个登录模块!
.case3,登录,加购物车
```

3.11.3　多个 fixture 的实例化顺序

由于存在 fixture 的并列关系、嵌套关系及不同作用域，所以需要仔细体会这些复杂关系及执行顺序，因此多个 fixture 的实例化顺序遵循以下原则：

- 高级别作用域的实例化（例如：session）先于低级别作用域的（例如：class 或者 function）实例化；
- 相同级别作用域的实例化，其实例化顺序遵循它们在测试用例中被声明的顺序，也就是形参的顺序，或者 fixture 之间的相互调用关系；
- 自动应用 autouse 的 fixture，先于其同级别的其他 fixture 实例化。

代码如下：

```python
# Author: lindafang
# Date: 2020 - 12 - 01 13:32
# File: test_fixture_order.py

import pytest

order = []

@pytest.fixture(scope = "session", autouse = True)
def session1():
    order.append("session1")


@pytest.fixture(scope = "module", autouse = True)
def module1():
    order.append("module1")


# function1 嵌套调用 function3,所以先执行 function3,再执行 function1
@pytest.fixture(autouse = True)
def function1(function3):
    order.append("function1")


@pytest.fixture
def function3():
    order.append("function3")


# autouse 在所在 function 之前使用
@pytest.fixture(autouse = True)
def autouse1():
    order.append("autouse1")


@pytest.fixture(autouse = True)
def function2():
    order.append("function2")


def test_order(module1):
    assert order == ["session1", "module1", "autouse1", "function3", "function1", "function2"]
    # 特意断言错误以便查看执行顺序
    assert order == ["session1"]
```

执行结果如下:

```
F
src/chapter3/test_fixtures_order.py:39 (test_order)
['session1', ..., 'function2'] != ['session1']
```

```
Expected :['session1']
Actual   :['session1', ..., 'function2']
<Click to see difference>

module1 = None

    def test_order(module1):
        assert order == ["session1", "module1", "autouse1", "function3", "function1", "function2"]
                #特意断言错误以便查看执行顺序
>       assert order == ["session1"]
E       AssertionError: assert ['session1', ..., 'function2'] == ['session1']
E         Left contains 5 more items, first extra item: 'module1'
E         Use -v to get the full diff

test_fixtures_order.py:43: AssertionError
```

session1 拥有最高级的作用域（session），即使在测试用例 test_order 中最后被声明，它也是第 1 个被实例化的参数（参照第 1 条原则）。

module1 拥有仅次于 session 级别的作用域（module），所以它是第 2 个被实例化的参数（参照第 1 条原则）。

function1、function2、function3、autouse1 同属于 function 级别的作用域。

从 test_order（function1，module1，function2，session1）形参的声明顺序中可以看出 function1 比 function2 先被实例化（参照第 2 条原则）。

function1 的定义中又显式地调用了 function3，所以 function3 比 function1 先被实例化（参照第 2 条原则）。

autouse1 的定义中使能了 autouse 标记，所以它会在同级别的 fixture 之前被实例化，也就是在 function3、function1、function2 之前被实例化（参照第 3 条原则）。

所以在这个例子中 fixture 被实例化的顺序为 session1、module1、autouse1、function3、function1、function2。

注意：除了 autouse 的 fixture，需要测试用例显示声明（形参），不声明的参数不会被实例化。多个相同作用域的 autouse fixture，其实例化顺序遵循 fixture 函数名的排序。

3.11.4 fixture 返回工厂函数

常用使用场景：如果需要在一个测试用例中，多次使用同一个 fixture 实例，相对于直接返回数据，更好的方法是返回一个产生数据的工厂函数，并且，对于工厂函数产生的数据，也可以在 fixture 中对其管理。这里介绍的是一种思路，代码如下：

```python
@pytest.fixture
def make_customer_record():

    # 记录产生的数据
    created_records = []

    # 工厂
    def _make_customer_record(name):
        record = models.Customer(name=name, orders=[])
        created_records.append(record)
        return record

    yield _make_customer_record

    # 销毁数据
    for record in created_records:
        record.destroy()

def test_customer_records(make_customer_record):
    customer_1 = make_customer_record("Lisa")
    customer_2 = make_customer_record("Mike")
    customer_3 = make_customer_record("Meredith")
```

3.11.5 高效地利用 fixture 实例

在测试期间,pytest 只激活最少个数的 fixture 实例。如果拥有一个参数化的 fixture,所有使用它的用例会在所创建的第一个 fixture 实例被销毁后,才会使用第二个实例。

下面这个例子使用了两个参数化的 fixture,其中一个是模块级别的作用域,另一个是用例级别的作用域,并且使用 print 方法打印出它们的 setup/teardown 流程。

代码如下:

```python
# File: test_minfixture.py

import pytest


@pytest.fixture(scope="module", params=["mod1", "mod2"])
def modarg(request):
    param = request.param
    print("  SETUP modarg", param)
    yield param
    print("  TEARDOWN modarg", param)
```

```python
@pytest.fixture(scope = "function", params = [1, 2])
def otherarg(request):
    param = request.param
    print("  SETUP otherarg", param)
    yield param
    print("  TEARDOWN otherarg", param)


def test_0(otherarg):
    print("  RUN test0 with otherarg", otherarg)


def test_1(modarg):
    print("  RUN test1 with modarg", modarg)


def test_2(otherarg, modarg):
    print("  RUN test2 with otherarg {} and modarg {}".format(otherarg, modarg))
```

执行结果如下：

```
pytest -q -s test_minfixture.py
  SETUP otherarg 1
  RUN test0 with otherarg 1
.  TEARDOWN otherarg 1
  SETUP otherarg 2
  RUN test0 with otherarg 2
.  TEARDOWN otherarg 2
  SETUP modarg mod1
  RUN test1 with modarg mod1
.  SETUP otherarg 1
  RUN test2 with otherarg 1 and modarg mod1
.  TEARDOWN otherarg 1
  SETUP otherarg 2
  RUN test2 with otherarg 2 and modarg mod1
.  TEARDOWN otherarg 2
  TEARDOWN modarg mod1
  SETUP modarg mod2
  RUN test1 with modarg mod2
.  SETUP otherarg 1
  RUN test2 with otherarg 1 and modarg mod2
.  TEARDOWN otherarg 1
  SETUP otherarg 2
  RUN test2 with otherarg 2 and modarg mod2
.  TEARDOWN otherarg 2
  TEARDOWN modarg mod2

8 passed in 0.02s
```

mod1 的 TEARDOWN 操作完成后,才开始 mod2 的 SETUP 操作。用例 test_0 独立完成测试,用例 test_1 和 test_2 都使用了模块级别的 modarg,同时 test_2 也使用了用例级别的 otherarg。它们执行的顺序是,test_1 先使用 mod1,接着 test_2 使用 mod1 和 otherarg 1/otherarg 2,然后 test_1 使用 mod2,最后 test_2 使用 mod2 和 otherarg 1/otherarg 2,也就是说 test_1 和 test_2 共用相同的 modarg 实例,最少化地保留 fixture 的实例个数。

3.12 在不同的层级上重写 fixture

在大型测试中,可能需要在本地覆盖项目级别的 fixture,以增加可读性和可维护性。也就是说可以通过在不同层级中重写 fixture 改变 fixture 中原始的内容。思路是重写同名(文件或函数)的方法。使用时优先调用重写后的 fixture。

3.12.1 在文件夹(conftest.py)层级重写 fixture

下一层的 conftest.py 中的 fixtures 可以覆盖和访问上一级的 fixture,实现过程:

1. 在不同层次建立文件夹

创建 tests 文件夹,以及子文件夹 subfolder。在 tests 文件夹下面创建 conftest.py 文件和 test_something.py 两个文件,在 subfolder 下同样创建这两个文件,文件结构如图 3-11 所示。

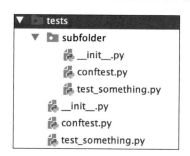

图 3-11 subfolder 文件夹下的文件重写上一级的 fixture 文件所建立的结构

2. 各个文件的代码

在 tests 文件夹下的 conftest.py 文件的 fixture 中创建 username 方法。

```
tests/
    __init__.py

    conftest.py
        # content of tests/conftest.py
        import pytest

        @pytest.fixture
        def username():
            return 'username'
```

在 tests 文件夹下的 test_something.py 文件中创建 test_username 方法,通过断言判断传入的是否为 username 方法所返回的 username 字符串。

```
test_something.py
    # content of tests/test_something.py
    def test_username(username):
        assert username == 'username'
```

在 subfolder 文件夹下的 conftest.py 文件的 fixture 中创建的 username 方法中调用 username,也就是通过这个参数调用上一级 conftest.py 中 username 的 fixture。通过同名的方法重写上一级的同名 fixture 方法。

```
subfolder/
__init__.py

conftest.py
    # content of tests/subfolder/conftest.py
    import pytest

    @pytest.fixture
    def username(username):
        return 'overridden-' + username
```

在 subfolder 文件夹下 test_something.py 文件的 test_username 测试方法中调用 username,这时调用本地的方法,也就是 subfolder 文件夹下的 conftest.py 文件的 username 方法中重写上一级的 usename 方法,并通过断言测试是否是本地重写的结果。

```
test_something.py
    # content of tests/subfolder/test_something.py
    def test_username(username):
        assert username == 'overridden-username'
```

执行结果如下:

```
collected 2 items

subfolder/test_something.py .
[ 50%]
test_something.py .
[100%]

===================== 2 passed in 0.08s =====================
```

子文件夹(subfolder)中 conftest.py 文件的 fixture 覆盖了上层文件夹中同名的 fixture。

子文件夹(subfolder)中 conftest.py 文件的 fixture 可以轻松地访问上层文件夹中同名的 fixture。

3.12.2 在模块层级重写 fixture

在 conftest.py 及自己同一文件下的 fixture 具有相同名字，由在同一文件中的 fixture 函数重写了 conftest.py 中的 fixture 函数。也就是说模块(文件中)中的 fixture 覆盖了 conftest.py 中同名的 fixture。模块(文件)中的 fixture 可以轻松地访问 conftest.py 中同名的 fixture。

代码如下：

```
tests/
    __init__.py

    conftest.py
        # content of tests/conftest.py
        import pytest

        @pytest.fixture
        def username():
            return 'username'

    test_something.py
        # content of tests/test_something.py
        import pytest

        @pytest.fixture
        def username(username):
            return 'overridden-' + username

        def test_username(username):
            assert username == 'overridden-username'

  # content of tests/test_something_else.py
  # Author: lindafang

import pytest

@pytest.fixture
def username(username):
    return 'overridden-else-' + username

def test_username(username):
    assert username == 'overridden-else-username'
```

3.12.3 在用例参数中重写 fixture

常用的使用场景：fixture 通常作为测试数据的初始化，而参数化是调用这些数据实现一个测试方法使用不同数据执行多次的效果。在 fixture 与参数 parametrize 组合在一起时，在 fixture 中读到的原始值可以被用例的参数 parametrize 所覆盖，这可体现数据层级的灵活性。

代码如下：

```
tests/
    __init__.py

    conftest.py
        # content of tests/conftest.py
        import pytest

        @pytest.fixture
        def username():
            return 'username'

        @pytest.fixture
        def other_username(username):
            return 'other-' + username

    test_something.py
        # content of tests/test_something.py
        import pytest

@pytest.mark.parametrize('username', ['directly-overridden-username'])
def test_username(username):
    assert username == 'directly-overridden-username'


@pytest.mark.parametrize('username', ['directly-overridden-username-other'])
def test_username_other(username, other_username):
    assert other_username == 'other-directly-overridden-username-other'
```

fixture 的值被用例的参数所覆盖，尽管用例 test_username_other 没有使用 username，但是 other_username 使用了 username，所以也同样受到了影响。

3.12.4 参数化的 fixture 可重写非参数化的 fixture，反之亦然

参数化的 fixture 和非参数化的 fixture 同样可以相互覆盖。其实就是在使用 fixture 的文件中建立同名的 fixture 方法重写上一级 conftest.py 文件中的同名方法。无论重写前的方法带参数，还是重写后的方法带参数都是可以的。也就是可以实现原来执行多次的方法

通过重写为非参数化的 fixture 而变成只执行一次。同样可以实现原来只执行一次的方法通过重写为参数化的 fixture 而变成执行多次的参数化 fixture。

代码如下：

```
tests/
    __init__.py

    conftest.py
        # content of tests/conftest.py
        import pytest

        @pytest.fixture(params = ['one', 'two', 'three'])
        def parametrized_username(request):
            return request.param

        @pytest.fixture
        def non_parametrized_username(request):
            return 'username'
```

在 test_something1.py 文件中，创建与 conftest.py 文件中同名的 parametrized_username 方法，重写为只返回一个值的非参数化方法。创建与 conftest.py 文件中同名的 non_parametrized_username 方法，添加参数及返回值，重写为参数化的方法。

```
test_something1.py
    # content of tests/test_something1.py
    import pytest

    @pytest.fixture
    def parametrized_username():
        return 'overridden-username'

    @pytest.fixture(params = ['one', 'two', 'three'])
    def non_parametrized_username(request):
        return request.param

    def test_username(parametrized_username):
        assert parametrized_username == 'overridden-username'

    def test_parametrized_username(non_parametrized_username):
        assert non_parametrized_username in ['one', 'two', 'three']
```

在 test_something_else1.py 方法中不需要重写，正常调用 conftest.py 文件中的 fixture。

```
test_something_else1.py
    # content of tests/test_something_else.py
```

```
def test_username(parametrized_username):
    assert parametrized_username in ['one', 'two', 'three']

def test_username1(non_parametrized_username):
    assert non_parametrized_username == 'username'
```

3.13 本章小结

本章是 pytest 框架最重要的章节,主要讲述了 fixture 的使用。
(1) fixture 基本的依赖注入功能,可以将多个功能关联起来。
(2) 可以实现初始化设置和配置销毁功能。
(3) fixture 还能在不同层面上使用,达到不同层面运行的依赖。
(4) 利用 params 传递不同数据。
(5) 自动调用 fixture。
(6) fixture 并列与嵌套调用。
(7) 在不同层级上重写 fixture。

第 4 章 pytest 的数据驱动和参数传递

4.1 参数化介绍

常见使用场景：简单注册功能，也就是输入用户名、输入密码、单击注册，而测试数据会有很多个，可以通过测试用例设计技术组织出很多测试数据，例如用户名都是字母，密码也都是字母，或者都是数字，也可是它们的组合，或是边界值长度的测试数据等。这时可以通过参数化技术实现测试数据驱动执行每组测试用例。

测试数据与测试用例是多对一的关系，所以完全可以把它们分开来看，把数据部分抽象成参数，通过对参数的赋值来驱动用例的执行。

参数化传递是实现数据驱动的一种技术，可以实现测试数据与测试用例分离。各个方面的参数化如下：

- 测试用例的参数化：使用@pytest.mark.parametrize 可以在测试用例、测试类甚至测试模块中标记多个参数或 fixture 的组合；
- 参数化的行为可以表现在不同的层级上；
- 多参数的参数化：一个以上参数与数据驱动结果；
- fixture 的参数化：相关知识可参考第 3.8 节；
- 自定义参数化：可以通过 pytest_generate_tests 这个钩子方法自定义参数化的方案；
- 使用第三方插件实现数据驱动 DDT。

4.2 参数化的应用

通过@pytest.mark.parametrize 可以实现数据驱动。@pytest.mark.parametrize 的根本作用是在收集测试用例的过程中，通过对指定参数的赋值来新增被标记对象的调用（执行）。下面以例说明具体参数化时如何使用不同数据。

4.2.1 单一参数化应用

通常使用场景：测试方法中只有一个数据是变化的，也就是通过一个参数把多组测试数据传递进去。执行时,每组数据都执行一遍。

实现的具体步骤如下:

(1) 在测试方法中输入@pytest.mark.parametrize。

(2) 其中有两个参数,一个是参数名,另一个是参数值,这个值可以是多个,可以是数字或字符等。

(3) 测试方法中的参数与 parametrize 中的参数名相同。

(4) 通过参数名在测试方法中调用这些数据。

代码如下:

```
import pytest

@pytest.mark.parametrize("test_case", [1, 2, 3, 'orange', 'apple'])
def test_string(test_case):
    print("\n要测试的数据:{}".format(test_case))
```

一个测试用例,有多少条数据就自动执行多少遍。执行的结果如下:

```
test_mark_parametrize.py .
要测试的数据:1
.
我们的测试数据:2
.
我们的测试数据:3
.
我们的测试数据:orange
.
我们的测试数据:apple
                                              [100%]

======================= 5 passed in 0.03s =======================
Process finished with exit code 0
```

4.2.2 多参数应用

测试输入的数据可以是表达式,输入的参数可以是多个。多个数据可以通过元组方式组织。下面是一个测试计算器的简单例子,前面两个是变量,后面是对应的数据。3+5 对应的是 test_input 参数名,8 对应的是 expected 参数名,下面的数据以此类推。eval 将字符串 str 当成有效表达式来求值并返回计算结果。

代码如下:

```python
import pytest

#参数化,前面两个是变量,后面是对应的数据. 3+5 ---> test_input, 8 --> expected
@pytest.mark.parametrize("test_input,expected",[("3+5", 8),
                                                ("2+5", 7),
                                                ("7*5", 30),
                                                ])
def test_eval(test_input, expected):
    #eval 将字符串 str 当成有效表达式来求值并返回计算结果
    assert eval(test_input) == expected
```

将其中一组数据写成错误的形式,验证断言的详细情况。执行结果如下:

```
..F
test_mark_parametrize.py:13 (test_eval[7*5-30])
35 != 30

Expected :30
Actual   :35
<Click to see difference>

test_input = '7*5', expected = 30

    @pytest.mark.parametrize("test_input,expected",[("3+5", 8),
                                                    ("2+5", 7),
                                                    ("7*5", 30),
                                                    ])
    def test_eval(test_input, expected):
        #eval 将字符串 str 当成有效表达式来求值并返回计算结果
>       assert eval(test_input) == expected
E       AssertionError: assert 35 == 30
E        +  where 35 = eval('7*5')

../chapter-4/test_mark_parametrize.py:20: AssertionError
                    [100%]

==================== FAILURES ====================
_____ test_eval[7*5-30] _____
test_input = '7*5', expected = 30

    @pytest.mark.parametrize("test_input,expected",[("3+5", 8),
                                                    ("2+5", 7),
                                                    ("7*5", 30),
                                                    ])
```

```
            def test_eval(test_input, expected):
                # eval 将字符串 str 当成有效表达式来求值并返回计算结果
>               assert eval(test_input) == expected
E               AssertionError: assert 35 == 30
E                + where 35 = eval('7 * 5')
```

4.2.3 多个参数化

一个用例可以标记多个@pytest.mark.parametrize 标记。

代码如下：

```
#test_multi.py

@pytest.mark.parametrize('test_input', [1, 2, 3])
@pytest.mark.parametrize('test_output, expected', [(1, 2), (3, 4)])
def test_multi(test_input, test_output, expected):
    pass
```

实际收集到的用例是它们所有可能的组合。

代码如下：

```
collected 6 items
<Module test_multi.py>
  <Function test_multi[1-2-1]>
  <Function test_multi[1-2-2]>
  <Function test_multi[1-2-3]>
  <Function test_multi[3-4-1]>
  <Function test_multi[3-4-2]>
  <Function test_multi[3-4-3]>
```

4.2.4 参数化与 fixture 的结合

当一个测试方法既是注入依赖，也就是使用 fixture，同时又要参数化时，使用 parametrize 会有冲突，此时可以通过 fixture 自带的参数 params 实现参数化。这也是参数化的一种方法，实现方法参见 3.8 节，应用参见 8.2.5 节。

4.2.5 pytestmark 实现参数化

可以尝试通过对 pytestmark 赋值，参数化一个测试模块。

代码如下：

```
#test_module.py

import pytest

pytestmark = pytest.mark.parametrize('test_input, expected', [(1, 2), (3, 4)])


def test_module(test_input, expected):
    assert test_input + 1 == expected
```

4.3　parametrize 源码详细讲解

下面通过两个例子讲解参数化技术。我们先来看一下它在源码中的定义。此方法在 structures.py 文件中。

源码及部分翻译如下：

```
def parametrize(self, argnames, argvalues, indirect = False, ids = None, scope = None):
    """ Add new invocations to the underlying test function using the list
    of argvalues for the given argnames.  Parametrization is performed
    during the collection phase.  If you need to setup expensive resources
    see about setting indirect to do it rather at test setup time.
```

在收集阶段执行参数化。向底层测试函数 argnames(参数名)添加新调用 argValue(参数值)，参数值使用列表。如果你需要设置有价值的资源，则可参阅 indirect 这个参数的设置。

```
    :arg argnames: a comma - separated string denoting one or more argument
                   names, or a list/tuple of argument strings.
```

一个用逗号分隔的字符串，或者一个列表/元组，表明指定的参数名。

```
    :arg argvalues: The list of argvalues determines how often a
           test is invoked with different argument values.  If only one
           argname was specified argvalues is a list of values.  If N
           argnames were specified, argvalues must be a list of N - tuples,
           where each tuple - element specifies a value for its respective
           argname.
```

argvalues：一个可迭代对象，表明对 argnames 参数的赋值，具体有以下几种情况：
如果 argnames 包含多个参数，那么 argvalues 的迭代返回元素必须是可度量的(即支持 len()方法)，并且长度和 argnames 声明参数的个数相等，所以它可以是元组/列表/集合等，表明所有入参的实参。

```
    :arg indirect: The list of argnames or boolean. A list of arguments'
          names (self, subset of argnames).  If True the list contains all names from  the
argnames.  Each argvalue corresponding to an argname in this list will be passed as request.param
to its respective argname fixture
          function so that it can perform more expensive setups during the
          setup phase of a test rather than at collection time.
```

indirect：argnames 的子集或者一个布尔值。将指定参数的实参通过 request.param 重定向到和参数同名的 fixture 中，以此满足更复杂的场景。默认 indirect 为 False，使用 mark.parametrize 后的数据 indirect = True。

:arg ids: list of string ids, or a callable.
    If strings, each is corresponding to the argvalues so that they are
    part of the test id. If None is given as id of specific test, the
    automatically generated id for that argument will be used.
    If callable, it should take one argument (self, a single argvalue) and return a string or return None. If None, the automatically generated id for that argument will be used.
    If no ids are provided they will be generated automatically from
    the argvalues.

ids：一个可执行对象，用于生成测试 ID，或者一个列表/元组，指明所有新增用例的测试 ID。ids 参数就是 id，因为与关键字雷同，所以不能用，因此改成 ids。通常不写 ids 时每次不同数据直接显示，也就是数据本身，如果写 ids，则显示的就是这个 ids。大家可以通过在 ids 中写内容来标记我们的测试要点。例如第 1 个数据是数字，第 2 个数据是中文，第 3 个数据是特殊字符。这样在报告中看到结果就知道是否测试完整。

:arg scope: if specified it denotes the scope of the parameters.
    The scope is used for grouping tests by parameter instances.
    It will also override any fixture-function defined scope, allowing
    to set a dynamic scope using test context or configuration.

scope：声明 argnames 中参数的作用域，并通过对应的 argvalues 实例划分测试用例，进而影响测试用例的收集顺序。

"""

4.4 argnames 参数

parametrize 方法中的第一个参数 argnames 是一个用逗号分隔的字符串，或者一个列表/元组，表明指定的参数名。argnames 通常是与被标记测试方法入参的参数名对应的，但实际上有一些限制，它只能是被标记测试方法入参的子集。

4.4.1 argnames 与测试方法中的参数关系

1. 测试方法未声明，mark.parametrize 中声明

test_sample1 中并没有声明 expected 参数，如果在标记中强行声明，则会得到如下错误。代码如下：

```
@pytest.mark.parametrize('input, expected', [(1, 2)])
def test_sample1(input):
    assert input + 1 == 1
```

执行的结果会提示下面所示的错误信息：

```
In test_sample1: function uses no argument 'expected'
```

2. 测试方法参数声明的范围小于 mark.parametrize 中声明的范围

不能是被标记测试方法入参中定义了默认值的参数。

代码如下：

```
@pytest.mark.parametrize('input, expected', [(1, 2)])
def test_sample2(input, expected = 2):
    assert input + 1 == expected
```

虽然 test_sample2 声明了 expected 参数，但同时也为其赋予了一个默认值，如果非要在标记中强行声明，则会得到如下错误：

```
In test_sample: function already takes an argument 'expected' with a default value
```

4.4.2　argnames 调用覆盖同名的 fixture

通常在使用 fixture 和参数 parametrize 时，可以一个参数使用参数化，另一个参数使用 fixture 和参数化，而同时使用 fixture 和参数化时，参数化的参数值会覆盖原来 fixture 返回的值。

代码如下：

```
@pytest.fixture()
def expected():
    return 2

@pytest.fixture()
def input():
    return 0

@pytest.mark.parametrize('input',[(1)])
def test_sample(input, expected):
    assert input + 1 == expected
```

可以看到 expected 参数未使用参数化传入数据，而是直接调用 fixture 中的返回值 2，input 同时使用参数化和 fixture，参数化中参数值 1 覆盖了原来 fixture 的返回值 0，因此执行结果断言应该是成功的。

参数化的参数可以不是 fixture 的，因此可以通过参数值传入。

代码如下：

```python
@pytest.fixture()
def expected():
    return 1


@pytest.mark.parametrize('input, expected', [(1, 2)])
def test_sample(input, expected):
    assert input + 1 == expected
```

test_sample 标记的 input 参数的值是由后面的(1,2)传入的，expected 参数（参数值为2）覆盖了同名的 fixture expected（返回值 1），所以这条用例是可以测试成功的。

注意：可参考第 3.12 节内容在用例参数中覆写 fixture。

4.5　argvalues 参数

参数化中参数值 argvalues 是一个可迭代对象，表明对 argnames 参数的赋值，具体有以下几种情况：如果 argnames 包含多个参数，那么 argvalues 的迭代返回元素必须是可度量的值，即支持 len()方法，并且长度和 argnames 所声明参数的个数相等，所以它可以是元组/列表/集合等，表明所有入参的实参。

代码如下：

```python
@pytest.mark.parametrize('input, expected', [(1, 2), [2, 3], set([3, 4])])
def test_sample4(input, expected):
    print(expected)
    assert input + 1 == expected
```

执行结果如下：

```
test_mark_param_sub.py .2
.3
.4
```

注意：考虑集合的去重特性，本书并不建议使用它。

4.5.1　argvalues 来源于 Excel 文件

argvalues 是一个可迭代对象，所以可以应用在更复杂的场景中，这在实际应用中被特别广泛使用。公司一般会将测试数据保存在 Excel 表中，或 csv 文件中，或数据库中。可以先将数据读取到列表中，这样便可以在参数化的参数值中直接调用。例如：从 Excel 文件中读取实参。

代码如下：

```python
def read_excel():
    #从数据库或者 Excel 文件中读取设备的信息,这里简化为一个列表
    for dev in ['dev1', 'dev2', 'dev3']:
        yield dev


@pytest.mark.parametrize('dev', read_excel())
def test_sample(dev):
    assert dev
```

实现这个场景有多种方法,也可以直接在一个 fixture 中加载 Excel 文件中的数据,但是它们在测试报告中的表现会有所区别。

4.5.2 使用 pytest.param 为 argvalues 赋值

在结合 pytest.param 方法对 skip 和 xfail 标记中,可以使用 pytest.param 为 argvalues 参数赋值,让执行有更详细说明。

代码如下:

```python
@pytest.mark.parametrize(
    ('n', 'expected'),
    [(4, 2),
     pytest.param(6, 3, marks=pytest.mark.xfail(), id='XPASS')])
def test_params(n, expected):
    assert n / 2 == expected      #其执行结果显示为 XPASS
```

执行结果如图 4-1 所示。

把上面计算器的例子修改一下,如果将具体数据执行分成不同结果,则可以采用 pytest.param 实现。

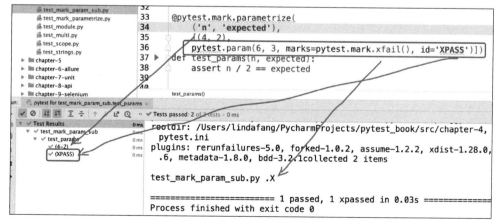

图 4-1　pytest.param 参数赋值后的结果

代码如下:

```
@pytest.mark.parametrize("test_input,expected",[("3+5", 8),
                                                ("2+5", 7),
                                                pytest.param("6*9", 42, marks=pytest.mark.xfail),
                                                ])
def test_eval(test_input, expected):
    #eval 将字符串 str 当成有效表达式来求值并返回计算结果
    assert eval(test_input) == expected
```

执行结果如下:

```
test_mark_param_sub.py ..x
test_input = '6 * 9', expected = 42

    @pytest.mark.parametrize("test_input,expected",[("3+5", 8),
                                                    ("2+5", 7),
                                                    pytest.param("6*9", 42, marks=pytest.mark.xfail),
                                                    ])
    def test_sample(test_input, expected):
        #eval 将字符串 str 当成有效表达式来求值并返回计算结果
>       assert eval(test_input) == expected
E       AssertionError: assert 54 == 42
E        +  where 54 = eval('6 * 9')

test_mark_param_sub.py:48: AssertionError
```

下面探究原理,实际使用者可以略看。

无论 argvalues 中传递的是可度量对象(列表、元组等)还是具体的值,在源码中都会将其封装成一个 ParameterSet 对象,它是一个具名元组(namedtuple),包含 values、marks、id 3 个元素,代码如下:

```
from _pytest.mark.structures import ParameterSet as PS
PS._make([(1, 2), [], None])
ParameterSet(values=(1, 2), marks=[], id=None)
```

如果直接传递一个 ParameterSet 对象会发生什么呢?下面去源码里找答案。
源码如下:

```
#_pytest/mark/structures.py

class ParameterSet(namedtuple("ParameterSet", "values, marks, id")):
    ...
```

```
@classmethod
def extract_from(cls, parameterset, force_tuple = False):
    """
    :param parameterset:
        a legacy style parameterset that may or may not be a tuple,
        and may or may not be wrapped into a mess of mark objects

    :param force_tuple:
        enforce tuple wrapping so single argument tuple values
        don't get decomposed and break tests
    """

    if isinstance(parameterset, cls):
        return parameterset
    if force_tuple:
        return cls.param(parameterset)
    else:
        return cls(parameterset, marks = [], id = None)
```

可以看到,如果直接传递一个 ParameterSet 对象,那么返回的就是它本身(return parameterset),所以下面例子中的两种写法是等价的。

```
import pytest

from _pytest.mark.structures import ParameterSet

@pytest.mark.parametrize(
    'input, expected',
    [(1, 2), ParameterSet(values = (3, 4), marks = [], id = None)])
def test_sample10(input, expected):
    assert input + 1 == expected
```

查看到这里,或许你已经猜到了,pytest.param 的作用就是封装一个 ParameterSet 对象。我们去源码里求证一下吧!
源码如下:

```
# _pytest/mark/__init__.py

def param( * values, ** kw):
    """Specify a parameter in `pytest.mark.parametrize`_ calls or
    :ref:`parametrized fixtures < fixture - parametrize - marks >`.

    .. code - block:: python
```

```
            @pytest.mark.parametrize("test_input,expected", [
                ("3+5", 8),
                pytest.param("6*9", 42, marks=pytest.mark.xfail),
            ])
            def test_eval(test_input, expected):
                assert eval(test_input) == expected

        :param values: variable args of the values of the parameter set, in order.
        :keyword marks: a single mark or a list of marks to be applied to this parameter set.
        :keyword str id: the id to attribute to this parameter set.
        """
        return ParameterSet.param(*values, **kw)
```

正如上面所料,现在你应该更明白怎么给 argvalues 传参了。

4.6 indirect 参数

indirect 是 argnames 的子集或者一个布尔值。将指定参数的实参通过 request.param 重定向到和参数同名的 fixture 中,以此满足更复杂的场景。默认 indirect 为 False,使用 mark.parametrize 后的数据。当 indirect 为 True 时,使用 fixture 中的数据。

代码如下:

```
import pytest

@pytest.fixture()
def max(request):
    return request.param - 1


@pytest.fixture()
def min(request):
    return request.param + 1


#默认 indirect 为 False,min 和 max 使用的是后面的数据。
@pytest.mark.parametrize('min, max', [(1, 2), (3, 4)])
def test_indirect(min, max):
    assert min <= max


#min 和 max 对应的实参重定向到同名的 fixture 中,min 和 max 使用的是 fixture 的数据
@pytest.mark.parametrize('min, max', [(1, 2), (3, 4)], indirect=True)
```

```
def test_indirect_indirect(min, max):
    assert min >= max


# 只将 max 对应的实参重定向到 fixture 中,min 使用的是后面的数据,max 使用的是 fixture 的
# 数据
@pytest.mark.parametrize('min, max', [(1, 2), (3, 4)], indirect = ['max'])
def test_indirect_part_indirect(min, max):
    assert min == max
```

indirect=True,min 和 max 对应的实参重定向到同名的 fixture 中,min 和 max 使用的是 fixture 的数据。

indirect=['max'],只将 max 对应的实参重定向到 fixture 中,min 使用的是后面的数据,max 使用的是 fixture 的数据。

其实这是一种间接参数化的方式,当 indirect=True 时,允许在将值传递给测试之前使用接收值的 fixture 对测试进行参数化。

4.7 ids 参数

ids 参数就是 id,因为与关键字雷同所以不能用,因此改成 ids。通常不写 ids 时每次不同数据直接显示,也就是数据本身,如果定义 ids 值,则显示的就是这个值。大家可以通过在 ids 中写内容来标记我们的测试要点。通常我们在测试时分测试数字、字母、边界值等,因此我们可以通过对这个参数的设置检查是不是覆盖全面。例如第 1 个数据是数字,第 2 个数据是中文,第 3 个数据是特殊字符。这样在报告中看到结果就知道是否测试完整。

ids 是一个可执行对象,用于生成测试 id,或者一个列表/元组,指明所有新增用例的测试 id。这些 id 可用于 -k 选择要运行的特定用例,当某个用例失败时,它们还将识别该特定用例。运行 pytest --collect-only 将显示生成的 id。

4.7.1 ids 的长度

如果使用列表/元组直接指明测试 id,那么它的长度等于 argvalues 的长度。
代码如下:

```
@pytest.mark.parametrize('input, expected', [(1, 2), (3, 4)],
                         ids = ['first', 'second'])
def test_ids_with_ids(input, expected):
    pass
```

input 参数的 id 是 first,第 1 次的值是 1,第 2 次的值是 3,expected 参数的 id 是 second,第 1 次的值是 2,第 2 次的值是 4。

4.7.2　ids 相同

如果测试 id 相同，pytest 则会在后面自动添加索引，例如[num0]和[num1]。

```
@pytest.mark.parametrize('input, expected', [(1, 2), (3, 4)],
                ids = ['num', 'num'])
def test_ids_with_ids(input, expected):
    pass
```

执行的结果如下：

```
test_ids.py::test_ids_with_ids1[first] PASSED
[ 11%]
test_ids.py::test_ids_with_ids1[second] PASSED
[ 22%]
test_ids.py::test_ids_with_ids2[num0] PASSED
[ 33%]
test_ids.py::test_ids_with_ids2[num1] PASSED
[ 44%]
test_ids.py::test_ids_with_ids3[Windows0] PASSED
[ 55%]
test_ids.py::test_ids_with_ids3[Windows1] PASSED
[ 66%]
test_ids.py::test_ids_with_ids3[Non-Windows] PASSED
```

4.7.3　ids 中使用中文

测试 ID 中可以使用中文，默认显示的是字节序列。

```
@pytest.mark.parametrize('input, expected', [(1, 2), (3, 4)],
                ids = ['num', '中文'])
def test_ids_with_ids(input, expected):
    pass
```

收集到的测试 ID 如下：

```
test_ids.py::test_ids_with_ids4[num] PASSED
[ 88%]
test_ids.py::test_ids_with_ids4[\u4e2d\u6587] PASSED
```

从上面的结果可以看出，期望显示"中文"，但实际上显示的是\u4e2d\u6587。如何解决此问题，笔者查了一下源码 python.py。

源码如下：

```python
# _pytest/python.py

def _ascii_escaped_by_config(val, config):
    if config is None:
        escape_option = False
    else:
        escape_option = config.getini(
            "disable_test_id_escaping_and_forfeit_all_rights_to_community_support"
        )
    return val if escape_option else ascii_escaped(val)
```

解决中文乱码,可以在 pytest.ini 中将 disable_test_id_escaping_and_forfeit_all_rights_to_community_support 选项设置为 True。

代码如下:

```
[pytest]
disable_test_id_escaping_and_forfeit_all_rights_to_community_support = True
```

再次收集到的测试 ID 如下:

```
test_mark_param_ids.py::test_ids_with_ids4[num] PASSED
[ 88%]
test_mark_param_ids.py::test_ids_with_ids4[中文] PASSED
```

4.7.4 通过函数生成 ids

```python
def idfn(val):
    # 将每个 val 都加 1
    return val + 1


@pytest.mark.parametrize('input, expected', [(1, 2), (3, 4)], ids=idfn)
def test_ids_with_ids(input, expected):
    pass
```

执行结果显示如下:

```
collected 2 items
<Module test_ids.py>
  <Function test_ids_with_ids[2-3]>
  <Function test_ids_with_ids[4-5]>
```

通过上面的例子不难看出,对于一个具体的 argvalues 参数(1,2)来讲,它被拆分为 1

和2分别传递给 idfn，并将返回值通过-符号连接在一起，以此作为一个测试 id 返回，而不是将(1，2)作为一个整体传入。

源码如下：

```
#_pytest/python.py

def _idvalset(idx, parameterset, argnames, idfn, ids, item, config):
    if parameterset.id is not None:
        return parameterset.id
    if ids is None or (idx >= len(ids) or ids[idx] is None):
        this_id = [
            _idval(val, argname, idx, idfn, item=item, config=config)
            for val, argname in zip(parameterset.values, argnames)
        ]
        return "-".join(this_id)
    else:
        return _ascii_escaped_by_config(ids[idx], config)
```

和猜想的一样，先通过 zip(parameterset.values，argnames)将 argnames 和 argvalues 的值一一对应，再将处理过的返回值通过"-".join(this_id)连接。

4.7.5　ids 的覆盖

从上面的源码还可以看出，假设已经通过 pytest.param 指定了 id 属性，那么将会覆盖 ids 中对应的测试 id。

代码如下：

```
@pytest.mark.parametrize(
    'input, expected',
    [(1, 2), pytest.param(3, 4, id='id_via_pytest_param')],
    ids=['first', 'second'])
def test_ids_with_ids(input, expected):
    pass
```

执行结果如下：

```
collected 2 items
<Module test_ids.py>
  <Function test_ids_with_ids[first]>
  <Function test_ids_with_ids[id_via_pytest_param]>
```

测试 id 是 id_via_pytest_param，而不是 second。

4.7.6　ids 的作用

ids 最主要的作用就是更进一步细化测试用例，区分不同的测试场景，为有针对性的执行测试提供了一种新方法。

例如，对于以下测试用例，可以通过 -k 'Non-Windows' 选项，只执行和 Non-Windows 相关的场景。

代码如下：

```python
# src/chapter4/test_ids.py

import pytest

@pytest.mark.parametrize('input, expected', [
    pytest.param(1, 2, id='Windows'),
    pytest.param(3, 4, id='Windows'),
    pytest.param(5, 6, id='Non-Windows')
])
def test_ids_with_ids(input, expected):
    pass
```

执行结果如图 4-2 所示。

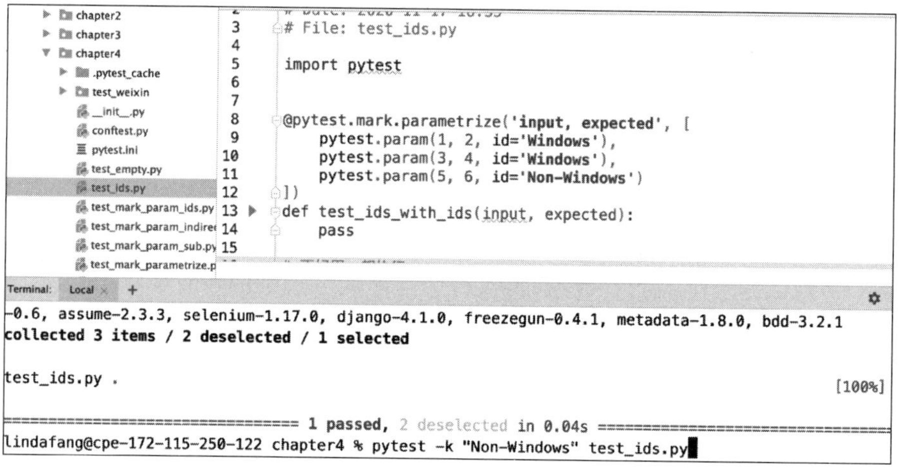

图 4-2　ids 的作用

4.8　scope 参数

scope 参数声明 argnames 中参数的作用域，并通过对应的 argvalues 实例划分测试用例，进而影响测试用例的收集顺序。

4.8.1　module 级别

如果我们显式地指明 scope 参数，例如，将参数作用域声明为模块级别，这样设置后测试方法会进行一起统筹，也就是执行的顺序是先执行所有测试方法的第一组数据，再整体执行第二组数据，直到执行完成。

代码如下：

```python
# test_scope.py
import pytest

@pytest.mark.parametrize('test_input, expected', [(1, 2), (3, 4)], scope = 'module')
def test_scope1(test_input, expected):
    pass


@pytest.mark.parametrize('test_input, expected', [(1, 2), (3, 4)], scope = 'module')
def test_scope2(test_input, expected):
    pass
```

执行结果如下：

```
collected 4 items
<Module test_scope.py>
  <Function test_scope1[1-2]>
  <Function test_scope2[1-2]>
  <Function test_scope1[3-4]>
  <Function test_scope2[3-4]>
```

当未将 scope 设置为 module 时，默认的收集顺序是按测试方法的先后执行的。也就是先执行第一个测试方法中的所有数据，再执行第二测试方法中的所有数据。

```
collected 4 items
<Module test_scope.py>
  <Function test_scope1[1-2]>
  <Function test_scope1[3-4]>
  <Function test_scope2[1-2]>
  <Function test_scope2[3-4]>
```

4.8.2　未指定 scope

在 scope 未指定的情况下（或者 scope＝None），当 indirect 被设置为 True 或者包含所有的 argnames 参数时，作用域为所有 fixture 作用域的最小范围，否则，其永远为 function。

代码如下：

```python
# test_scope.py

@pytest.fixture(scope = 'module')
def test_input(request):
    pass


@pytest.fixture(scope = 'module')
def expected(request):
    pass


@pytest.mark.parametrize('test_input, expected', [(1, 2), (3, 4)],
                        indirect = True)
def test_scope1(test_input, expected):
    pass


@pytest.mark.parametrize('test_input, expected', [(1, 2), (3, 4)],
                        indirect = True)
def test_scope2(test_input, expected):
    pass
```

test_input 和 expected 的作用域都是 module，所以参数的作用域也是 module，用例的收集顺序和 4.8.1 节相同。

代码如下：

```
collected 4 items
<Module test_scope.py>
  <Function test_scope1[1-2]>
  <Function test_scope2[1-2]>
  <Function test_scope1[3-4]>
  <Function test_scope2[3-4]>
```

4.9　pytest_generate_tests 钩子方法

pytest 实现参数化有 3 种方式：
- pytest.fixture() 使用 fixture 传 params 参数实现参数化；
- @pytest.mark.parametrize 允许在测试函数或类中定义多组参数；
- pytest_generate_tests 允许定义自定义参数化方案或扩展。

分别在 3.8 节和 4.2 节介绍了前面两种方式。本节简单介绍自定义参数化方案。

pytest_generate_tests 在测试用例参数化收集前调用此钩子函数，根据测试配置或定义测试函数的类或模块中指定的参数值生成测试用例，可以使用此钩子实现自定义参数化方案或扩展。

有时可能要实现自己的参数化方案或实现某种动态性来确定 fixture 的参数或范围，因此，可以使用 pytest_generate_tests 在收集测试函数时调用的钩子。通过传入的 metafunc 对象，可以检查请求的测试上下文，最重要的一点是，可以调用 metafunc.parametrize()引起参数化。

我们先看一看源码中是怎么使用这种方法的。

源码如下：

```
# _pytest/python.py

def pytest_generate_tests(metafunc):
    # those alternative spellings are common - raise a specific error to alert
    # the user
    alt_spellings = ["parameterize", "parametrise", "parameterise"]
    for mark_name in alt_spellings:
        if metafunc.definition.get_closest_marker(mark_name):
            msg = "{0} has '{1}' mark, spelling should be 'parametrize'"
            fail(msg.format(metafunc.function.__name__, mark_name), pytrace=False)
    for marker in metafunc.definition.iter_markers(name="parametrize"):
        metafunc.parametrize(*marker.args, **marker.kwargs)
```

首先，它检查了 parametrize 的拼写错误，如果不小心将 parametrize 写成了["parameterize"，"parametrise"，"parameterise"]中的一个，pytest 会返回一个异常，并提示正确的单词，然后循环遍历所有的 parametrize 的标记，并调用 metafunc.parametrize 方法。

例如，假设我们要运行一个测试，并接收通过新的 pytest 命令行选项设置的字符串输入。我们首先需要编写一个接收 stringinput 函数参数的简单测试。

我们检查给定的 stringinput 是否只由字母组成，但是我们并没有为其打上 parametrize 标记，所以 stringinput 被认为是一个 fixture。

代码如下：

```
# test_strings.py

def test_valid_string(stringinput):
    assert stringinput.isalpha()
```

现在，我们期望把 stringinput 当成一个普通的参数，并且从命令行赋值。

首先，我们定义一个命令行选项。

代码如下：

```
#conftest.py

def pytest_addoption(parser):
    parser.addoption(
        "--stringinput",
        action = "append",
        default = [],
        help = "list of stringinputs to pass to test functions",
    )
```

然后,我们通过 pytest_generate_tests 方法,将 stringinput 的行为由 fixture 改成 parametrize。代码如下:

```
#conftest.py

def pytest_generate_tests(metafunc):
    if "stringinput" in metafunc.fixturenames:
        metafunc.parametrize("stringinput", metafunc.config.getoption("stringinput"))
```

最后,我们可以通过--stringinput 命令行选项为 stringinput 参数赋值。代码如下:

```
pytest -q --stringinput = 'hello' --stringinput = 'world' src/chapter-11/test_strings.py
..                                                                          [100%]
2 passed in 0.02s
```

如果我们不加--stringinput 选项,相当于 parametrize 的 argnames 中的参数没有接收到任何的实参,那么测试用例的结果将会被置为 SKIPPED。

```
pytest -q test_strings.py
s                                                                           [100%]
1 skipped in 0.02s
```

不管是 metafunc.parametrize 方法还是@pytest.mark.parametrize 标记,它们的参数(argnames)不能是重复的,否则会产生一个错误:ValueError: duplicate 'stringinput'。

4.10 本章小结

本章主要讲解 pytest 数据驱动 paramtrize 中各种参数及情况的使用:
(1) 使用 pytest.mark 中的数据驱动。
(2) 多个参数与多个数据的结合。
(3) indirect、ids、scope 参数的使用。
(4) 其他几种情况(这几种应用不广泛)。

第 5 章 pytest 的相关插件及插件管理

5.1 pytest 的插件安装

pytest 不仅是一个功能强大的测试框架，同时还是一个插件化的测试平台。插件的使用方法与功效各有千秋，它们的共同特点是只需配置就可以直接使用，而不需要测试代码配合。

安装和卸载第三方插件的命令如下：

```
pip install pytest-NAME
pip uninstall pytest-NAME
```

如果安装了插件，pytest 则可以自动查找并集成它，而无须激活它。

5.2 常见插件介绍

pytest-xdist：将测试分发到 CPU 和远程主机，以盒装模式运行，该模式可以保留分段错误，以 looponfail 模式运行，并根据文件更改自动重新运行失败的测试。

pytest-instafail：在测试运行时报告失败。

pytest-bdd：使用行为驱动的测试编写测试。在第 11 章将详细介绍。

pytest-timeout：基于功能标记或全局定义的超时测试。

pytest-pep8：启用 PEP8 符合性检查的选项。

pytest-flakes：使用 pyflakes 检查源代码。

pytest-django：使用 pytest 集成为 Django 应用编写测试。在本书中未介绍，大家可自行学习。

pytest-cov：覆盖率报告，与分布式测试兼容。

要查看所有插件的完整列表，以及它们针对不同的 pytest 和 Python 版本的最新测试状态，可访问 http://plugincompat.herokuapp.com/，或搜索 https://pypi.org/search/?q=pytest。

5.3 常用插件的使用

按字母顺序简单介绍常用插件的使用。

5.3.1 pytest-assume 断言报错后依然执行

pytest 的断言失败后,后面的代码就不会执行了,通常在一个用例中我们会写多个断言,有时候我们希望第一个断言失败后,后面能继续断言。pytest-assume 插件可以解决断言失败后继续执行后面断言的问题。

环境准备：先安装 pytest-assume 依赖包,此插件存在与 pytest 和 Python 的兼容性问题。

```
pip install pytest-assume
```

有些断言是并行的,我们同样想知道其他断言执行结果。举例说明,输入的测试数据有 3 种,我们需要断言同时满足 3 种情况：x==y,x+y>1,x>1,这 3 个条件是并行的。

代码如下：

```python
import pytest

@pytest.mark.parametrize(('x', 'y'),
                         [(1, 1), (1, 0), (0, 1)])
def test_simple_assume(x, y):
    print("测试数据 x=%s, y=%s" % (x, y))
    assert x == y
    assert x + y > 1
    assert x > 1
```

未导入插件前运行的结果如下：遇到第一个错误就停下来,当 x=1,y=1 时,x==y 的断言通过了,x+y>1 的断言也通过了,由于错误便停在 x>1 的断言上。当 x=1,y=0 时,在第一个断言 x==y 时就停止了。执行结果如下：

```
F 测试数据 x=1, y=1

src/chapter5/test_assume.py: 12 (test_simple_assume[1-1])
x = 1, y = 1

    @pytest.mark.parametrize(('x', 'y'),
                             [(1, 1), (1, 0), (0, 1)])
    def test_simple_assume(x, y):
```

```
            print("测试数据 x = %s, y = %s" % (x, y))
            assert x == y
            assert x + y > 1
>           assert x > 1
E           assert 1 > 1

test_assume.py: 19: AssertionError
F 测试数据 x = 1, y = 0

src/chapter5/test_assume.py: 12 (test_simple_assume[1 - 0])
1 != 0

Expected : 0
Actual   : 1
 <Click to see difference>

x = 1, y = 0

    @pytest.mark.parametrize(('x', 'y'),
                             [(1, 1), (1, 0), (0, 1)])
    def test_simple_assume(x, y):
        print("测试数据 x = %s, y = %s" % (x, y))
>       assert x == y
E       assert 1 == 0

test_assume.py: 17: AssertionError
F 测试数据 x = 0, y = 1

src/chapter5/test_assume.py: 12 (test_simple_assume[0 - 1])
0 != 1

Expected : 1
Actual   : 0
 <Click to see difference>

x = 0, y = 1

    @pytest.mark.parametrize(('x', 'y'),
                             [(1, 1), (1, 0), (0, 1)])
    def test_simple_assume(x, y):
        print("测试数据 x = %s, y = %s" % (x, y))
>       assert x == y
E       assert 0 == 1

test_assume.py: 17: AssertionError
```

导入插件并修改代码如下：

```python
import pytest

@pytest.mark.parametrize(('x', 'y'),
                         [(1, 1), (1, 0), (0, 1)])
def test_simple_assume1(x, y):
    print("测试数据 x = %s, y = %s" % (x, y))
    pytest.assume(x == y)
    pytest.assume(x + y > 1)
    pytest.assume(x > 1)
    print("测试完成!")
```

导入插件后执行的结果如下，此时出现错误后不会停止执行，但会统计错误的次数，例如，Failed Assumptions：3，(1,1)的这组数据中的结果就是 Failed Assumptions：1。

```
F 测试数据 x = 1, y = 1
测试完成!

src/chapter5/test_assume.py: 21 (test_simple_assume1[1-1])
test_assume.py: 28: AssumptionFailure
    pytest.assume(x > 1)

--------------------------------------------------------------
Failed Assumptions: 1
F 测试数据 x = 1, y = 0
测试完成!

src/chapter5/test_assume.py: 21 (test_simple_assume1[1-0])
test_assume.py: 26: AssumptionFailure
    pytest.assume(x == y)


test_assume.py: 27: AssumptionFailure
    pytest.assume(x + y > 1)


test_assume.py: 28: AssumptionFailure
    pytest.assume(x > 1)

--------------------------------------------------------------
Failed Assumptions: 3
F 测试数据 x = 0, y = 1
测试完成!
```

```
src/chapter5/test_assume.py: 21 (test_simple_assume1[0-1])
test_assume.py: 26: AssumptionFailure
    pytest.assume(x == y)


test_assume.py: 27: AssumptionFailure
    pytest.assume(x + y > 1)


test_assume.py: 28: AssumptionFailure
    pytest.assume(x > 1)

--------------------------------------------------------------
Failed Assumptions: 3
```

还可以使用 with 上下文管理器编写，代码如下：

```
import pytest
# 笔者的 assume 版本是 2.3.3
from pytest_assume.plugin import assume

@pytest.mark.parametrize(('x', 'y'),
                         [(1, 1), (1, 0), (0, 1)])
def test_simple_assume_with(x, y):
    print("测试数据 x = %s, y = %s" % (x, y))
    with assume: assert x == y
    with assume: assert x + y > 1
    with assume: assert x > 1
    print("测试完成!")
```

执行结果如下：

```
F 测试数据 x = 1, y = 1
测试完成!

src/chapter5/test_assume.py: 33 (test_simple_assume_with[1-1])
x = 1, y = 1

    @pytest.mark.parametrize(('x', 'y'),
                             [(1, 1), (1, 0), (0, 1)])
    def test_simple_assume_with(x, y):
        print("测试数据 x = %s, y = %s" % (x, y))
        with assume: assert x == y
        with assume: assert x + y > 1
>       with assume: assert x > 1
```

```
E       pytest_assume.plugin.FailedAssumption:
E       1 Failed Assumptions:
E
E       test_assume.py:40: AssumptionFailure
E       >>     with assume: assert x > 1
E       AssertionError: assert 1 > 1

test_assume.py:40: FailedAssumption
F 测试数据 x = 1, y = 0
测试完成!

src/chapter5/test_assume.py:33 (test_simple_assume_with[1-0])
x = 1, y = 0

    @pytest.mark.parametrize(('x', 'y'),
                             [(1, 1), (1, 0), (0, 1)])
    def test_simple_assume_with(x, y):
        print("测试数据 x = %s, y = %s" % (x, y))
        with assume: assert x == y
        with assume: assert x + y > 1
>       with assume: assert x > 1
E       pytest_assume.plugin.FailedAssumption:
E       3 Failed Assumptions:
E
E       test_assume.py:38: AssumptionFailure
E       >>     with assume: assert x == y
E       AssertionError: assert 1 == 0
E
E       test_assume.py:39: AssumptionFailure
E       >>     with assume: assert x + y > 1
E       AssertionError: assert (1 + 0) > 1
E
E       test_assume.py:40: AssumptionFailure
E       >>     with assume: assert x > 1
E       AssertionError: assert 1 > 1

test_assume.py:40: FailedAssumption
F 测试数据 x = 0, y = 1
测试完成!

src/chapter5/test_assume.py:33 (test_simple_assume_with[0-1])
x = 0, y = 1

    @pytest.mark.parametrize(('x', 'y'),
                             [(1, 1), (1, 0), (0, 1)])
    def test_simple_assume_with(x, y):
```

```
                print("测试数据 x = % s, y = % s" % (x, y))
                with assume: assert x == y
                with assume: assert x + y > 1
>               with assume: assert x > 1
E               pytest_assume.plugin.FailedAssumption:
E               3 Failed Assumptions:
E
E               test_assume.py: 38: AssumptionFailure
E               >>    with assume: assert x == y
E               AssertionError: assert 0 == 1
E
E               test_assume.py: 39: AssumptionFailure
E               >>    with assume: assert x + y > 1
E               AssertionError: assert (0 + 1) > 1
E
E               test_assume.py: 40: AssumptionFailure
E               >>    with assume: assert x > 1
E               AssertionError: assert 0 > 1

test_assume.py: 40: FailedAssumption
```

整体运行结果如图 5-1 所示。

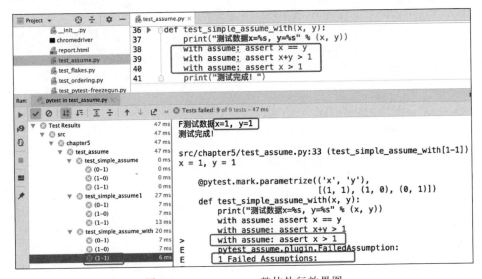

图 5-1 pytest-assume 整体执行效果图

5.3.2 pytest-cov 测试覆盖率

pytest-cov 是自动检测测试覆盖率的一个插件,在测试中被广泛应用。提到覆盖率,先介绍一下 Python 自带的代码覆盖率的命令行检测工具 coverage.py。它监视你的程序,并指出代码的哪些部分已执行,然后分析源代码以识别可能已执行但尚未执行的代码。要理解 pytest-cov 首先要了解 coverage 这个命令行工具。

1. coverage

coverage 在覆盖率中是语句覆盖的一种,是白盒测试中最低级的用例设计方法和要求,还有分支覆盖、条件判定覆盖、条件分支覆盖、路径覆盖等,语句覆盖不能对逻辑进行判断,逻辑的真实意义需要多结合项目本身,这个覆盖率数据不具有强大说服力,不要盲目追求。

一般来讲,覆盖率测试通常用于评估测试的有效性。有效性从高到低的顺序依次是"路径覆盖率">"判定覆盖">"语句覆盖"。coverage 可以显示测试正在执行代码的哪些部分,哪些没有被执行。目前最新版本是 2020 年 7 月 5 日发布的 coverage.py 5.2,参考文档网址:https://coverage.readthedocs.io/en/coverage-5.2/。实现语言覆盖的步骤如下:

第 1 步,安装 coverage.py。

```
pip install coverage
```

下面对 coverage 命令参数进行简单介绍。

coverage 命令共有 10 种参数形式,分别是:
- run:运行一个 Python 程序并收集运行数据;
- report:生成报告;
- html:把结果输出 html 格式;
- xml:把结果输出 xml 格式;
- annotate:运行一个 Python 程序并收集运行数据;
- erase:清楚之前 coverage 收集的数据;
- combine:合并 coverage 收集的数据;
- debug:获取调试信息;
- help:查看 coverage 帮助信息;
- coverage help 动作或者 coverage 动作 -help:查看指定动作的帮助信息。

第 2 步,运行命令。

通过 coverage run 命令运行 Python 程序,并收集信息,命令如下:

```
coverage run test.py
```

第 3 步,报告结果。

```
coverage report
```

提供 4 种风格的输出文件格式,分别对应 html 和 xml 命令。最简单的报告是 report

命令输出的概要信息,执行结果如下,report 包括执行的行数 stmts,没有执行的行数 miss,以及覆盖百分比 cover。

```
coverage report
Name                    Stmts    Miss    Cover
---------------------------------------------
my_program.py             20       4      80%
my_module.py              15       2      86%
my_other_module.py        56       6      89%
---------------------------------------------
TOTAL                     91      12      87%
```

2. pytest-cov

pytest-cov 是 pytest 的一个插件,其本质也是引用 Python 的 coverage 库,用来统计代码覆盖率。我们新建 3 个文件,my_program.py 是程序代码,test_my_program.py 是测试代码,在同一个目录 coverage-cov 下还建立一个 run.py 执行文件。

(1) pip 安装,命令如下:

```
pip install pytest-cover
```

(2) 建立 my_program.py 文件,代码如下:

```python
def cau(type, n1, n2):

    if type == 1:
        a = n1 + n2
    elif type == 2:
        a = n1 - n2
    else:
        a = n1 * n2
    return a
```

可以看出函数有 3 个参数,里面的逻辑由 3 条条件分支组成,即 type 等于 1 时为加法,type 等于 2 时为减法,type 为其他值时为乘法,最后返回结果。

(3) 新建 test_my_program.py 测试文件,代码如下:

```python
from my_program import cau
class Test_cover:
    def test_add(self):
        a = cau(1, 2, 3)
        assert a == 3
```

上面代码用于测试 type 等于 1 时这个语句的覆盖率。

（4）新建执行脚本 run.py 文件，代码如下：

```
import pytest

if __name__ == '__main__':
    pytest.main(["--cov=要测试的绝对路径","--cov-report=html","--cov-config=绝对路径/.coveragerc"])
    #执行某个目录下 case
```

上述代码说明：--cov 参数后面接的是测试的目录，程序代码跟测试脚本必须在同一个文件夹下。--cov-report=html 用于生成报告。

只需输入命令 python run.py 就可以运行。也可以在 run.py 上直接右击使 pytest 运行。这样执行情况如图 5-2 所示，HTML 报告如图 5-3 所示。

```
test_my_program.py:8: AssertionError

---------- coverage: platform darwin, python 3.6.8-final-0 -----------
Coverage HTML written to dir htmlcov

============================= 1 failed in 0.07s =============
lindafang            coverage-cov % python run.py
```

图 5-2 执行情况

Coverage report: 57%				
Module ↓	statements	missing	excluded	coverage
my_program.py	7	3	0	57%
Total	**7**	**3**	**0**	**57%**

图 5-3 HTML 覆盖率的报告

run.py 文件中 coveragerc 是配置文件，配置用于跳过 omit 某些脚本，这些脚本不用于覆盖率测试。例如：跳过所有非开发文件的统计，即 run.py、test_my_program.py 文件跟 init 文件。在 coveragerc 文件中增加以下内容：

```
[run]
omit =
    */__init__.py
    */run.py
    */test_my_program.py
```

再次执行 run.py 文件，HTML 报告如图 5-4 所示。
生成结果后进入 htmlcov 文件夹，可以直接单击 index.html 文件，如图 5-4 所示，跳过测试文件和初始化文件。单击进入 my_program.py 文件，my_program.py 的覆盖率详细说明，如图 5-5 所示。

```
Coverage report: 86%

Module ↓         statements    missing    excluded    coverage
my_program.py        7            1           0          86%
Total                7            1           0          86%
```

图 5-4　跳过非开发代码的 HTML 报告

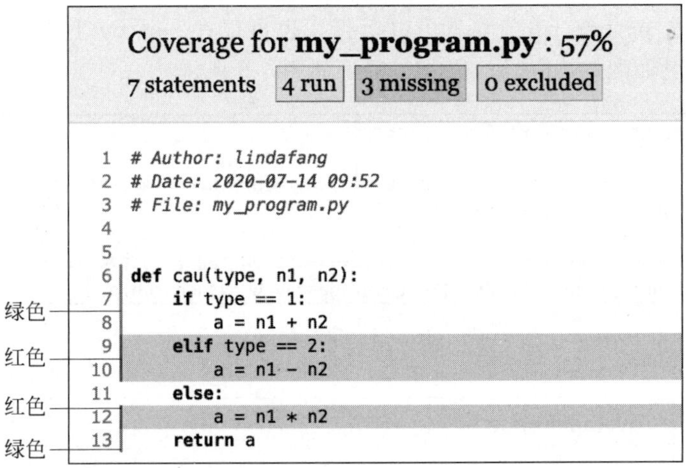

```
Coverage for my_program.py : 57%
7 statements  4 run  3 missing  0 excluded

 1   # Author: lindafang
 2   # Date: 2020-07-14 09:52
 3   # File: my_program.py
 4
 5
 6   def cau(type, n1, n2):
 7       if type == 1:
 8           a = n1 + n2         ← 绿色
 9       elif type == 2:         ← 红色
10           a = n1 - n2
11       else:                   ← 红色
12           a = n1 * n2
13       return a                ← 绿色
```

图 5-5　每个文件的覆盖率

从图 5-5 中可以看到以下的执行情况，绿色代表已运行的代码（6、7、9、13 行已覆盖），红色代表未被执行（9、10、12 行未被覆盖），自己检查下代码逻辑，可以得出该结果是正确的。在测试代码中增加其他分支的测试，再执行则覆盖率会提高，直到把所有分支都测试完成，覆盖率便为 100% 了。

在 test_my_program.py 文件中增加测试代码如下：

```
def test_sub(self):
    a = cau(2,3,2)
    assert a == 1
```

再次执行，单击进入 my_program.py 文件查看结果，如图 5-6 所示。

测试覆盖率，除了成功和失败以外，最重要的测试数据。上面的测试还差一个分支没有完成，所以增加测试把所有分支至少执行一次。这个是采用单元测试分支覆盖方法写出的测试用例。100% 测试覆盖率，只是完成 Python 项目单元测试的一个基本要求。因此，这个插件是十分重要的一个插件。

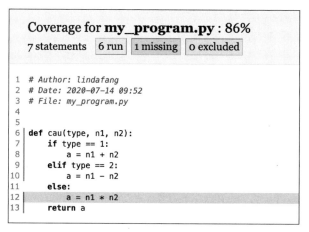

图 5-6 修改后再次执行的结果

5.3.3 pytest-freezegun 冰冻时间

这个插件随时可以变化当前系统时间,freezer 可以冰冻时间,freezer.move_to 可以改变时间,解决验证某一时间点的代码触发,或未来时间的代码变化问题。

下面的代码在 test_frozen_date 中是未冰冻的,即当前时间是相等的。test_moveing_date 通过 move_to 修改时间,再验证此时的时间不等于当前时间。还可以通过标识修改当前时间,在 test_current_date 方法上加 @pytest.mark.freeze_time('修改的时间,时间格式见下面')同样可以修改时间。此外,还可以与 fixture 结合实现修改时间,在 test_changing_date 测试上添加 @pytest.mark.freeze_time,使用 fixture 依赖注入传参的方式实现修改时间。

代码如下:

```
#File: test_pytest-freezegun.py

import time
from datetime import datetime, date
import pytest


def test_frozen_date(freezer):
    now = datetime.now()
    time.sleep(1)
    later = datetime.now()
    assert now == later
```

```python
def test_moving_date(freezer):
    now = datetime.now()
    freezer.move_to('2017-05-20')
    later = datetime.now()
    assert now != later


@pytest.mark.freeze_time('2017-05-21')
def test_current_date():
    assert date.today() == date(2017, 5, 21)


@pytest.fixture
def current_date():
    return date.today()


@pytest.mark.freeze_time
def test_changing_date(current_date, freezer):
    freezer.move_to('2017-05-20')
    assert current_date == date(2017, 5, 20)
    freezer.move_to('2017-05-21')
    assert current_date == date(2017, 5, 21)
```

5.3.4　pytest-flakes 静态代码检查

这是一个基于pyflakes的插件，对Python代码做一个快速的静态代码检查。使用方式和pytest-pep8类似，效果也十分显著。环境准备，输入命令如下：

```
pip install pytest-flakes
```

在test_flakes.py文件导入os模块，但后面的代码未用到这个导入，pyflakes这个插件就可以自动检查出来这是无用的导入（unused）。

代码如下：

```
import os

print('Hello world!')
```

执行pytest --flakes test_flakes.py命令，运行结果如下：

```
lindafang@cpe-172-115-247-185 chapter-5 % pytest --flakes test_flakes.py
=========================== test session starts ===========================
```

```
platform darwin -- Python 3.6.8, pytest-5.2.1, py-1.8.0, pluggy-0.13.1
rootdir: /Users/lindafang/PyCharmProjects/pytest_book
plugins: rerunfailures-5.0, forked-1.0.2, pep8-1.0.6, flakes-4.0.0, assume-1.2.2, cov
-2.10.0, xdist-1.28.0, ordering-0.6, metadata-1.8.0, bdd-3.2.1
collected 1 item


test_flakes.py F
    [100%]


====================== FAILURES ========================
_____ pyflakes-check _____
/PyCharmProjects/pytest_book/src/chapter-5/test_flakes.py:5: UnusedImport
'os' imported but unused
======================= 1 failed in 0.05s ==================
```

5.3.5 pytest-html 生成 HTML 报告

可以使用的两个 HTML 报告框架，pytest-html 和 allure，本节主要介绍 pytest-html，在测试的内部通用。

pytest-html 是个插件，此插件用于生成测试结果的 HTML 报告，兼容 Python 2.7 和 Python 3.8。GitHub 源码网址：https://github.com/pytest-dev/pytest-html。

环境准备，执行命令如下：

```
pip install pytest-html
```

执行时加入目标目录即可。

```
pytest --html=./report/html/report.html
```

执行完后会在当前目录生成一个 report.html 的报告文件。生成的报告如图 5-7 所示。css 是独立的，通过邮件分享报告的时候样式就会丢失，不好阅读，也无法筛选。

5.3.6 pytest-httpserver 模拟 HTTP 服务

在 Python 程序中，用 requests 发起网络请求是常见的操作，但如何测试是一个麻烦的问题。如果是单元测试，则可以用 pytest-mock，但如果是集成测试，用 Stub 的思路，则可以考虑 pytest-httpserver。

如何使用 pytest-httpserver 来对 requests 等涉及网络请求操作的代码进行集成测试呢？可以利用 pytest 的 fixture 机制为测试函数提供一个 httpserver。以下提供一个简单的代码样例，便于理解完整流程。

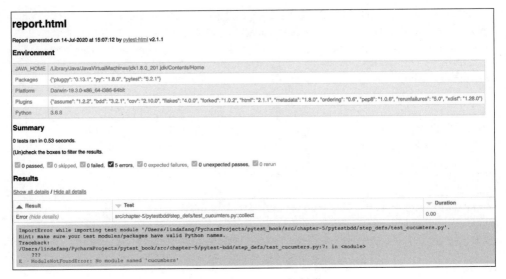

图 5-7　pytest-html 的报告

代码如下：

```
#File: test_httpserver.py

import requests
from pytest_httpserver import HTTPServer
from pytest_httpserver.httpserver import RequestHandler


def test_root(httpserver:HTTPServer):
    handler = httpserver.expect_request('/')
    assert isinstance(handler, RequestHandler)
    handler.respond_with_data('', status=200)

    response = requests.get(httpserver.url_for('/'))
assert response.status_code == 200
```

httpserver 需要设置两方面内容，输入（Request）和输出（Response）。先通过 expect_request 指定输入，再通过 respond_with_data 指定输出。最后，通过 url_for 获取随机生成 Server 的完整 URL。这里，仅对"/"的 Request 响应，返回 status=200 的 Response。

如果在一些不方便使用 fixtures 的场景，则可以通过 with 来使用相同功能。

代码如下：

```
def test_root():
    with HTTPServer() as httpserver:
        handler = httpserver.expect_request('/')
```

```python
        assert isinstance(handler, RequestHandler)
        handler.respond_with_data('', status = 200)

        response = requests.get(httpserver.url_for('/'))
        assert response.status_code == 200
```

上面的代码定义了"/"路径的响应，下面代码增加其他路径（'/status'、'/method'和'/data'）的响应，代码如下：

```python
#/status 这个路径返回的状态码为 302
def test_status(httpserver:HTTPServer):
    uri = '/status'
    handler = httpserver.expect_request(uri)
    handler.respond_with_data('', status = 302)

    response = requests.get(httpserver.url_for(uri))
    assert response.status_code == 302

#/method 这个路径请求方法为 get，返回的状态码为 200

def test_method(httpserver:HTTPServer):
    uri = '/method'
    handler = httpserver.expect_request(uri = uri, method = 'GET')
    handler.respond_with_data('', status = 200)

    response = requests.get(httpserver.url_for(uri))
    assert response.status_code == 200
    response = requests.post(httpserver.url_for(uri))
    assert response.status_code == 500

#/data 这个路径，方法 method 为 post，返回的状态码为 200

def test_respond_with_data(httpserver:HTTPServer):
    uri = '/data'
    handler = httpserver.expect_request(
        uri = uri,
        method = 'POST',
    )
    handler.respond_with_data('good')

    response = requests.post(httpserver.url_for(uri))
    assert response.status_code == 200
    assert response.content == 'good'

#/data 这个路径，方法 method 为 post，数据为 json 类型，返回的状态码为 200
```

```python
def test_respond_with_json(httpserver:HTTPServer):
    uri = '/data'
    expect = {'a': 1, 'b': 2}
    handler = httpserver.expect_request(
        uri = uri,
        method = 'POST',
    )
    handler.respond_with_json(expect)
    handler.respond_with_data

    response = requests.post(httpserver.url_for(uri))
    assert response.status_code == 200
    assert expect == response.json()
```

5.3.7　pytest-instafail 用于用例失败时立刻显示错误信息

用例失败时立刻显示错误的堆栈回溯信息，安装插件及执行如下：

```
pip install pytest-instafail
pytest --instafail
```

执行结果如图 5-8 所示。

```
=================== short test summary info ====================
FAILED coverage-cov/test_my_program.py::Test_cover::test_add - assert 5 == 3
FAILED test_assume.py::test_simple_assume[1-1] - assert 1 > 1
FAILED test_assume.py::test_simple_assume[1-0] - assert 1 == 0
FAILED test_assume.py::test_simple_assume[0-1] - assert 0 == 1
FAILED test_assume.py::test_simple_assume1[1-1] - pytest_assume.plugin.FailedAssumption:
FAILED test_assume.py::test_simple_assume1[1-0] - pytest_assume.plugin.FailedAssumption:
FAILED test_assume.py::test_simple_assume1[0-1] - pytest_assume.plugin.FailedAssumption:
FAILED test_assume.py::test_simple_assume_with[1-1] - pytest_assume.plugin.FailedAssumption:
FAILED test_assume.py::test_simple_assume_with[1-0] - pytest_assume.plugin.FailedAssumption:
FAILED test_assume.py::test_simple_assume_with[0-1] - pytest_assume.plugin.FailedAssumption:
FAILED test_pytest-freezegun.py::test_changing_date - assert datetime.date(2020, 11, 26) == Fake...
================= 11 failed, 16 passed in 2.34s ================
lindafang@cpe-172-115-242-99 chapter5 % pytest --instafail
```

图 5-8　pytest-instafail 的执行结果

5.3.8　pytest-mock 模拟未实现的部分

1. mock 的定义

mock 通常用于测试，mock 的意思是虚假的、模拟的。在 Python 的单元测试中，由于一切都是对象（object），而 mock 的技术就是在测试时不修改源码的前提下，替换某些对象，从而模拟测试环境。

2. mock 的源起

单元测试的条件有限，在测试过程中，有时会遇到难以准备的环境。例如，与服务器的网络交互、对数据库的读写等。

传统思路是利用 fixture 进行测试环境准备。这种做法的优点是，与真实环境非常相似，测试效果好，但缺点是，测试代码开发时间长，测试执行时间也很长。

另一种思路是，准备一个虚假的沙箱，对代码的执行效果进行模拟。这样虽然不能测试真正的最终效果，但是更容易保证 100% 测试覆盖率，并且避免重复测试，从而降低测试执行时间。

例如，在一个函数中调用了 3 个函数。只需测试这 3 个函数是否被依次调用，而无须测试真实的调用修改。

代码如下：

```python
def put_elepent_into_fridge(elepent, fridge):
    fridge.open()
    fridge.put(elepent)
    fridge.close()
```

假设 fridge 这个类已经完全被测试覆盖了。这里如果用传统的测试方法，只能让这 3 种方法再被测试一遍。而如果把 fridge 换成一个 mock，那么就可以避免重复测试，并且达到测试目的。

在 Python 标准库中，有 unittest 这个库。在 Python 3.3 以后，其中包含一个 unittest.mock，就是 Python 最常用的 mock 库。此外，PyPI 上还有一个 mock 库，是进入标准库前的 mock，可以在旧的版本使用。

虽然可以直接在 pytest 的测试中，直接使用 mock，但是并不方便。所以，在此直接推荐 pytest-mock。

3. pytest-mock 插件

pytest-mock 是一个 pytest 的插件，安装即可使用。它提供了一个名为 mocker 的 fixture，仅在当前测试 function 或 method 生效，而不用自行包装。模拟一个 object，是最常见的需求。由于 function 也是一个 object，所以以 function 举例。

代码如下：

```python
import os


def rm(filename):
    os.remove(filename)


def test_rm(mocker):
    filename = 'test.file'
    mocker.patch('os.remove')
    rm(filename)
```

这里在给 os.remove 打了一个补丁,让它变成了一个 MagicMock,然后利用 assert_called_once_with 查看它是否被调用一次,并且参数为 filename。

注意:只能对已经存在的东西使用 mock。

有时,仅仅需要模拟一个 object 里的 method,而无须模拟整个 object。例如,在对当前 object 的某个 method 进行测试时可以用 patch.object。

代码如下:

```
class ForTest:
    field = 'origin'

    def method():
        pass


def test_for_test(mocker):
    test = ForTest()
    mock_method = mocker.patch.object(test, 'method')
    test.method()
    assert mock_method.called

    assert 'origin' == test.field
    mocker.patch.object(test, 'field', 'mocked')
    assert 'mocked' == test.field
# 上例中,分别对 field 和 method 进行了模拟。当然,对一个给定 module 的 function,也能使用

def test_patch_object_listdir(mocker):
    mock_listdir = mocker.patch.object(os, 'listdir')
    os.listdir()
    assert mock_listdir.called
# 用 spy 包装
# 如果只是想用 MagicMock 包装一个东西,而又不想改变其功能,则可以用 spy

def test_spy_listdir(mocker):
    mock_listdir = mocker.spy(os, 'listdir')
    os.listdir()
    assert mock_listdir.called
```

与上例中的 patch.object 不同的是,上例的 os.listdir() 不会真的执行,而本例中则会真的执行。

4. MagicMock

即使使用 pytest-mock 简化使用过程,对 mock 本身还是要有基本的了解,尤其是 MagicMock。

MagicMock 属于 unittest.mock 中的一个类,是 mock 这个类的一个默认实现。在构

造时，还常用 return_value、side_effect 和 wraps 这 3 个参数。当然，还有其他不常用参数，详见 mock。

代码如下：

```python
import os
import pytest


def name_length(filename):
    if not os.path.isfile(filename):
        raise ValueError('{} is not a file!'.format(filename))
    print(filename)
    return len(filename)


def test_name_length0(mocker):
    isfile = mocker.patch('os.path.isfile', return_value = True)
    assert 4 == name_length('test')
    isfile.assert_called_once()

    isfile.return_value = False
    with pytest.raises(ValueError):
        name_length('test')
    assert 2 == isfile.call_count


def test_name_length1(mocker):
    mocker.patch('os.path.isfile', side_effect = TypeError)
    with pytest.raises(TypeError):
        name_length('test')


def test_name_length2(mocker):
    mocker.patch('os.path.isfile', return_value = True)
    mock_print = mocker.patch('builtins.print', wraps = print)
    mock_len = mocker.patch(__name__ + '.len', wraps = len)
    assert 4 == name_length('test')
    assert mock_print.called
    assert mock_len.called
```

以上展示了 return_value、side_effect 和 wraps 的用法。不仅可以在构造 MagicMock 时作为参数传入，还可以在传入参数之后调整。return_value 修改了 os.path.isfile 的返回值，控制程序执行流，而无须在文件系统中生成文件。side_effect 可以令某些函数抛出指定的异常。wraps 可以既把某些函数包装成 MagicMock，又不改变它的执行效果（这一点类似 spy）。当然，也完全可以替换成另一个函数。

在 Python 3 中,内置函数可以通过 builtins.* 进行模拟,然而某些内置函数牵涉甚广,例如 len,不适合在 Builtin 作用域进行模拟,可以在被测试的函数所在的 Global 作用域进行模拟。如本例中,就对当前 module 的 Global 作用域里的 len 进行了模拟。

此外,上例中还展示了 MagicMock 中的一些属性,如 assert_called_once、call_count、called 等,详见 mock。

5. 总结

无论是 pytest-mock 这层薄薄的封装,还是 unittest.mock 本身,都还有很多未介绍的细节,但以上介绍的内容,应该已经可以满足绝大部分使用场景。

在弄懂了 mock 之后,Python 的单元测试功能终于算是大成了。验证函数返回值是否相等,断言你的函数返回了某个值。如果此断言失败,将看到函数调用的返回值。

5.3.9 pytest-ordering 调整执行顺序

用例执行顺序的基本原则根据名称的字母逐一进行 ASCII 码比较,其值越大越先执行。当含有多个测试模块(.py 文件)时,根据基本原则执行。在一个测试模块(.py 文件)中,先执行测试函数,然后执行测试类。多个测试类则遵循基本原则,类中的测试方法遵循代码编写顺序。

如果想调整这个顺序,则可以通过插件进行,可以在测试方法上加 @pytest.mark.run(order=1),其值 1 表示最先执行。

代码如下:

```python
#File: test_ordering.py
import pytest


def test_03():
    print("\ntest_03")



def test_04():
    print("test_04")



class TestA(object):
    def test_05(self):
        print("test_05")

    #@pytest.mark.last
    def test_06(self):
        print("test_06")



class TestC(object):
```

```
#@pytest.mark.run(order = 1)
def test_01(self):
    print("\ntest_01")

def test_02(self):
    print("test_02")
```

未修改执行顺序前执行结果如图 5-9 所示。test_01 的执行顺序是倒数第二个执行。

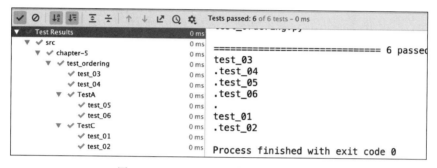

图 5-9　pytest-ordering 调整执行顺序前

通过 pytest-ordering 改变执行顺序，执行结果如图 5-10 所示。test_01 的执行顺序是第 1 个。

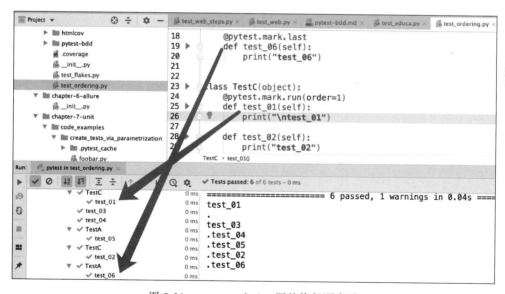

图 5-10　pytest-ordering 调整执行顺序后

5.3.10　pytest-pep8 自动检测代码规范

PEP 8 是 Python 中一个通用的代码规范。Python 是一门优雅的语言，然而，如果连这个规范都不遵守,则 Python 代码根本谈不上优雅,而 pytest-pep8 就是在进行 pytest 测试时自动检测代码是否符合 PEP 8 规范的插件,安装命令如下:

```
pip install pytest-pep8
```

安装后,增加--pep8 参数,即可执行测试。

```
pytest -- pep8
```

只要有一行代码不符合规范,就会让整个测试失败,如图 5-11 所示。

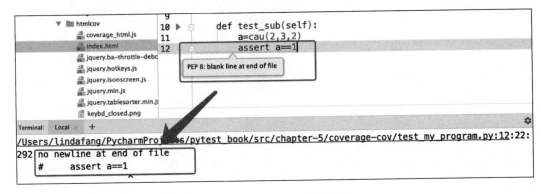

图 5-11　pytest-pep8 执行结果

5.3.11　pytest-picked 运行未提交 git 的用例

我们每天写完自动化用例后都会将代码提交到 git 仓库,随着用例的增多,为了保证仓库代码的干净,当有新增用例的时候,我们希望只运行新增的且尚未提交到 git 仓库的用例。

pytest-picked 插件可以实现只运行尚未提交到 git 仓库的代码。

环境准备,安装插件命令如下:

```
pip install pytest-picked
```

在 git 中文件从新建到暂存库期间有 4 种状态,可以通过不同参数执行不同状态的文件,如图 5-12 所示。

运行尚未提交 git 的测试用例的步骤如下:

第 1 步,在已提交过 git 仓库的用例中新增两个文件 test_new.py 和 test_new_2.py,如图 5-13 所示。

第 2 步,使用 git status 查看当前分支状态。

第5章 pytest的相关插件及插件管理

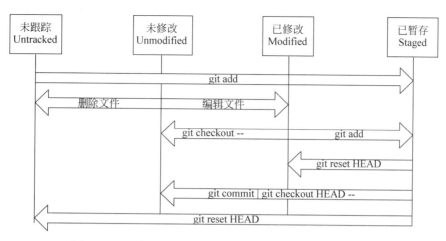

图 5-12　git 中文件从新建到提交到暂存库期间的 4 种状态

图 5-13　pytest-picked 的测试方法

执行结果如下，有两个新文件。

```
> git status
On branch master
Your branch is up-to-date with 'origin/master'.
Changes to be committed:
  (use "git reset HEAD <file>..." to unstage)

        new file:   pytest_demo/test_new.py
        new file:   pytest_demo/test_new_2.py

Changes not staged for commit:
  (use "git add <file>..." to update what will be committed)
  (use "git checkout -- <file>..." to discard changes in working directory)

        modified:   pytest_demo/test_new.py
        modified:   pytest_demo/test_new_2.py
```

第 3 步，使用 pytest --picked 运行用例。

执行结果如下，所有测试都将从已修改但尚未提交的文件和文件夹中运行。

```
> pytest -- picked

Changed test files... 2. ['pytest_demo/test_new.py', 'pytest_demo/test_new_2.py']
Changed test folders... 0. []
=========================== test session starts ===========================
platform win32 -- Python 3.6.6, pytest-6.0.2, py-1.9.0, pluggy-0.13.1
Test order randomisation NOT enabled. Enable with --random-order or --random-order-bucket=<bucket_type>
rootdir: ...
collected 4 items

pytest_demo\test_new.py ..                                           [ 50%]
pytest_demo\test_new_2.py ..                                         [100%]

=========================== 4 passed in 0.20s ===========================
```

不同参数具有不同的执行效果，下面举例分析。

1. 参数--picked=first

首先运行修改后的测试文件中的测试，然后运行所有未修改的测试。

代码如下：

```
> pytest -- picked=first
=========================== test session starts ===========================
platform win32 -- Python 3.6.6, pytest-6.0.2, py-1.9.0, pluggy-0.13.1
rootdir: ...
collected 11 items

pytest_demo\test_new.py ..                                           [ 18%]
pytest_demo\test_new_2.py ..                                         [ 36%]
pytest_demo\test_b.py ......                                         [ 90%]
pytest_demo\test_c.py .                                              [100%]

=========================== 11 passed in 0.10s ===========================
```

2. 参数--mode=unstaged 执行未提交的所有文件

--mode 有 2 个参数可选 unstaged 和 branch，默认为--mode=unstaged。当 git 文件的状态为 untrack 时，执行没添加到 git 中的新文件。unstaged 表示未暂存状态，也就是没有被 git add 过的文件。staged 表示已暂存状态，执行 git add 后文件状态。

为更好地理解什么是 untrack 状态，举例说明。当我们用 PyCharm 打开 git 项目，并且新增一个文件时，会弹出询问框：是否将文件添加到 git，如图 5-14 所示。

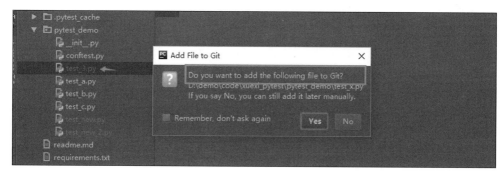

图 5-14　pytest-picked 添加 git 文件

如果选择是，文件会变为绿色，也就是 unstaged 状态（没 git add 过）。选择否，表示此文件是一个新文件，未被加到当前分支的 git 目录中，文件颜色是棕色。

当使用 git status 查看当前分支的状态时，会看到 pytest_demo/test_3.py 是 Untracked files。Test_new.py 和 test_new_2.py 文件状态是 unstage。

执行结果如下：

```
> git status
On branch master
Your branch is up-to-date with 'origin/master'.
Changes to be committed:
    (use "git reset HEAD <file>..." to unstage)

        new file:    pytest_demo/test_new.py
        new file:    pytest_demo/test_new_2.py

Changes not staged for commit:
    (use "git add <file>..." to update what will be committed)
    (use "git checkout -- <file>..." to discard changes in working directory)

        modified:    pytest_demo/test_new.py
        modified:    pytest_demo/test_new_2.py

Untracked files:
    (use "git add <file>..." to include in what will be committed)

        .idea/
        pytest_demo/__pycache__/
        pytest_demo/test_3.py
```

运行 pytest --picked，执行所有的 Untracked 文件和 not staged 文件，此时参数默认为 --mode=unstaged。

```
> pytest -- picked

Changed test files... 3. ['pytest_demo/test_new.py',
'pytest_demo/test_new_2.py', 'pytest_demo/test_3.py']
Changed test folders... 2. ['.idea/', 'pytest_demo/__pycache__/']
========================= test session starts =========================
platform win32 -- Python 3.6.6, pytest-6.0.2, py-1.9.0, pluggy-0.13.1
collected 5 items

pytest_demo\test_new.py ..
[ 40%]
pytest_demo\test_new_2.py ..
[ 80%]
pytest_demo\test_3.py .
[100%]

========================= 5 passed in 0.06s =========================
```

3. 参数--mode＝branch 运行分支上已经被暂存但尚未提交的代码

如果只需运行当前分支上已经被暂存，但尚未提交的文件（不包含 Untracked files），则可以使用 git diff 查看分支代码的差异。

执行命令及结果如下：

```
> git diff -- name-only master
pytest_demo/test_new.py
pytest_demo/test_new_2.py
```

运行 pytest --picked --mode＝branch，即运行分支上已经被暂存但尚未提交的代码。

执行命令及结果如下：

```
> pytest -- picked -- mode = branch

Changed test files... 2. ['pytest_demo/test_new.py', 'pytest_demo/test_new_2.py']
Changed test folders... 0. []
========================= test session starts =========================
platform win32 -- Python 3.6.6, pytest-6.0.2, py-1.9.0, pluggy-0.13.1
collected 4 items

pytest_demo\test_new.py ..
[ 50%]
pytest_demo\test_new_2.py ..
[100%]

========================= 4 passed in 0.04s =========================
```

5.3.12 pytest-rerunfailures 失败重试

测试过程中经常在执行测试用例时会有失败的情况出现,这种失败可能是断言失败,可能是代码问题,可能是环境问题,还可能是未知问题。我们为了排除部分原因会在失败时重试这些用例。失败重试依赖 pytest-rerunfailures 插件实现。

```
pip install pytest-rerunfailures
```

用例失败再重新执行一次,需要在命令行加参数--reruns。参数 reruns 有两种用法:
--reruns=RERUNS RERUNS 是失败重执行的次数,默认为 0;
--reruns-delay=RERUNS_DELAY RERUNS_DELAY 是失败后间隔多少秒重新执行。

```
pytest -- reruns 1 - html = report.html -- self - contained - html
```

5.3.13 pytest-repeat 重复运行测试

重复运行测试:pytest-repeat。环境准备的代码如下:

```
pip install pytest-repeat;
```

pytest test_x.py --count=n(重复运行的次数)。

pytest-repeat 允许用户重复执行单个用例或多个测试用例,并指定重复次数。提供 marker 功能,允许单独指定某些测试用例的执行次数。

GitHub 网址 https://github.com/pytest-dev/pytest-repeat。

源码解析:

```
# pytest-repeat.py:
# pytest_addoption(parser):添加一个 command line 的 option

def pytest_addoption(parser):
    parser.addoption(
        '--count',
        action='store',
        default=1,
        type=int,
        help='Number of times to repeat each test')
```

pytest_configure(config):一般用来注册 marker,这样当用户使用 pytest --markers 时便可了解有哪些可用的 marker 了,如果不加这个 hook,则功能上没什么影响,建议使用这种规范的写法。

代码如下:

```python
def pytest_configure(config):
    config.addinivalue_line(
        'markers',
        'repeat(n): run the given test function `n` times.')
# pytest_generate_tests(metafunc): 参数化生成 test case 的 hook

@pytest.fixture(autouse=True)
def __pytest_repeat_step_number(request):
    if request.config.option.count > 1:
        try:
            return request.param
        except AttributeError:
            if issubclass(request.cls, TestCase):
                warnings.warn(
                    "Repeating unittest class tests not supported")
            else:
                raise UnexpectedError(
                    "This call couldn't work with pytest-repeat. "
                    "Please consider raising an issue with your usage.")


@pytest.hookimpl(trylast=True)
def pytest_generate_tests(metafunc):
    count = metafunc.config.option.count
    m = metafunc.definition.get_closest_marker('repeat')
    if m is not None:
        count = int(m.args[0])
    if count > 1:
        def make_progress_id(i, n=count):
            return '{0}-{1}'.format(i + 1, n)

        scope = metafunc.config.option.repeat_scope
        metafunc.parametrize(
            '__pytest_repeat_step_number',
            range(count),
            indirect=True,
            ids=make_progress_id,
            scope=scope
        )
# config.option.xx 和 marker.args[n]或者 marker.kwargs(xx)优先级处理代码

count = metafunc.config.option.count
m = metafunc.definition.get_closest_marker('repeat')
if m is not None:
```

```
        count = int(m.args[0])
if count > 1:
    # do something here
# 利用 parametrize fixture 的 hook function 生成 repeat 的 test case items


    metafunc.parametrize(
        '__pytest_repeat_step_number',
        range(count),
        indirect = True,
        ids = make_progress_id,
        scope = scope
    )
```

5.3.14 pytest-random-order 随机顺序执行

pytest-random-order 插件允许用户按随机顺序执行测试，它提供包括 module、class、package 及 global 等不同粒度的随机性，并且允许用户使用 mark 标记特定粒度的测试集，从而保证部分 test cases 的执行顺序不被更改，具有高度灵活性。

源码解析：

主要介绍 plugin 这个 module，直接与 pytest 插件开发相关。

random_order/plugin.py：

pytest_addoption：Hook function，这里创建了一个 argparser 的 group，通过 addoption 方法添加 option，使得显示 help 信息时相关 option 显示在同一个 group 下面，更加友好。

代码如下：

```
def pytest_addoption(parser):
    group = parser.getgroup('pytest-random-order options')
    group.addoption(
        '--random-order',
        action = 'store_true',
        dest = 'random_order_enabled',
        help = 'Randomise test order (by default, it is disabled) with default configuration.',
    )
    group.addoption(
        '--random-order-bucket',
        action = 'store',
        dest = 'random_order_bucket',
        default = Config.default_value('module'),
        choices = bucket_types,
        help = 'Randomise test order within specified test buckets.',
```

```python
    )
    group.addoption(
        '--random-order-seed',
        action = 'store',
        dest = 'random_order_seed',
        default = Config.default_value(str(random.randint(1, 1000000))),
        help = 'Randomise test order using a specific seed.',
    )
# pytest_report_header: Hook function,在这里给出插件运行的相关信息,
# 方便出现问题时定位和复现问题

def pytest_report_header(config):
    plugin = Config(config)
    if not plugin.is_enabled:
        return "Test order randomisation NOT enabled. Enable with --random-order or --random-order-bucket=<bucket_type>"
    return (
        'Using --random-order-bucket={plugin.bucket_type}\n'
        'Using --random-order-seed={plugin.seed}\n'
    ).format(plugin = plugin)
# pytest_collection_modifyitems: Hook function,在测试项收集完以后执行,
# 用于过滤或者重排测试项


def pytest_collection_modifyitems(session, config, items):
    failure = None

    session.random_order_bucket_type_key_handlers = []
    process_failed_first_last_failed(session, config, items)

    item_ids = _get_set_of_item_ids(items)

    plugin = Config(config)

    try:
        seed = plugin.seed
        bucket_type = plugin.bucket_type
        if bucket_type != 'none':
            _shuffle_items(
                items,
                bucket_key = bucket_type_keys[bucket_type],
                disable = _disable,
                seed = seed,
                session = session,
            )
```

```
        except Exception as e:
            # See the finally block -- we only fail if we have lost user's tests
            _, _, exc_tb = sys.exc_info()
            failure = 'pytest-random-order plugin has failed with {0!r}: \n{1}'.format(
                e, ''.join(traceback.format_tb(exc_tb, 10)))
            )
            if not hasattr(pytest, "PytestWarning"):
                config.warn(0, failure, None)
            else:
                warnings.warn(pytest.PytestWarning(failure))

        finally:
            # Fail only if we have lost user's tests
            if item_ids != _get_set_of_item_ids(items):
                if not failure:
                    failure = 'pytest-random-order plugin has failed miserably'
                raise RunTimeError(failure)
```

pytest-random-order 是一个 pytest 插件,用于随机化测试顺序。这对于按顺序检测通过的测试可能是有用的,因为该测试恰好在不相关的测试之后运行,从而使系统处于良好状态。

该插件允许用户控制他们想要引入的随机性级别,并禁止对测试子集进行重新排序。通过传递先前测试运行中报告的种子值,可以按特定顺序重新运行测试,如图 5-15 所示。

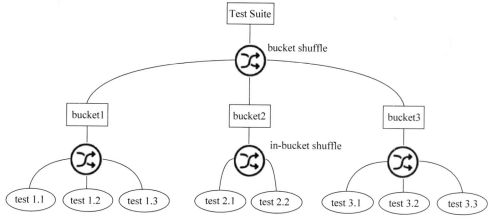

图 5-15　pytest-random-order 的测试用例结构

使用 pip 安装插件,命令如下:

```
pip install pytest-random-order
```

使用命令 pytest -h 查看,命令行有 3 个参数供选择。

代码如下：

```
pytest-random-order options:
  --random-order        Randomise test order (by default, it is disabled) with default
configuration.
  --random-order-bucket={global,package,module,class,parent,grandparent,none}
                        Randomise test order within specified test buckets.
  --random-order-seed=RANDOM_ORDER_SEED
                        Randomise test order using a specific seed.
```

从版本 v1.0.0 开始，默认情况下，此插件不再将测试随机化。要启用随机化，必须以下列方式之一运行 pytest：

```
pytest --random-order
pytest --random-order-bucket=<bucket_type>
pytest --random-order-seed=<seed>
```

如果要始终随机化测试顺序，需配置 pytest。有很多种方法可以做到这一点，笔者最喜欢的一种方法是 addopts =--random-order，即在 pytest 选项（通常是[pytest]或[tool: pytest]部分）下添加特定用于项目的配置文件。

```
# pytest.ini 文件内容
[pytest]
addopts = --random-order
# --random-order 随机测试
```

先写几个简单的用例，如图 5-16 所示，目录结构如下：

```
# module1/test_order1.py

class TestRandom():

    def test_01(self):
        print("用例1")

    def test_02(self):
        print("用例2")

    def test_03(self):
        print("用例3")

# module2/test_order2.py
```

```
class TestDemo():

    def test_04(self):
        print("用例 4")

    def test_05(self):
        print("用例 5")

    def test_06(self):
        print("用例 6")

> pytest -- random - order - v
================ test session starts ======================
Using -- random - order - bucket = module
Using -- random - order - seed = 357703
collected 6 items

module2/test_order2.py: : TestDemo: : test_04 PASSED
[ 16 % ]
module2/test_order2.py: : TestDemo: : test_05 PASSED
[ 33 % ]
module2/test_order2.py: : TestDemo: : test_06 PASSED
[ 50 % ]
module1/test_order1.py: : TestRandom: : test_03 PASSED
[ 66 % ]
module1/test_order1.py: : TestRandom: : test_02 PASSED
[ 83 % ]
module1/test_order1.py: : TestRandom: : test_01 PASSED
[100 % ]

=========================== 6 passed in 0.05s =======
```

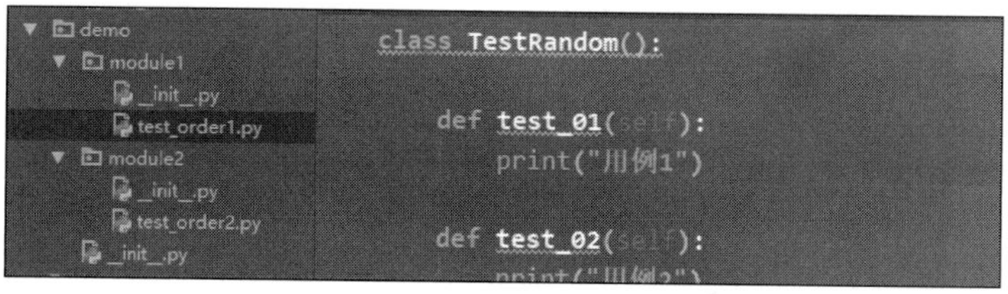

图 5-16 pytest-random-order 的测试用例

5.3.15　pytest-sugar 显示彩色进度条

很多程序员喜欢将执行结果用不同颜色进行显示，即显示色彩和进度条(也能显示错误的堆栈信息)。如果有更好的报告模板，则此插件就没什么用了，而且有些时候跟某些插件或版本有冲突。插件安装后立即生效：pip install pytest-sugar。

执行结果如图 5-17 所示。

```
plugins: rerunfailures-5.0, forked-1.0.2, pep8-1.0.6, flakes-4.0.0, sugar-0.9.4
28.0, ordering-0.6, metadata-1.8.0, bdd-3.2.1
collecting ...
 src/chapter-1/test_first.py ✓✓
 src/chapter-1/test_frame.py ✓✓✓✓✓
 src/chapter-1/test_frame_1.py ✓✓
 src/chapter-1/test_frame_2.py ✓✓

Results (0.07s):
    11 passed
```

图 5-17　pytest-sugar 的效果

5.3.16　pytest-selenium 浏览器兼容性测试

在兼容性测试中测试网站在不同浏览器中各种功能是否正常。通常使用自动化的方式实现。pytest-selenium 可以将浏览器的名字通过参数传入，这样就可以通过命令行方式进行兼容性测试了。

插件安装：

```
pip install pytest-selenium
```

举例说明：自动化实现启动一个浏览器、打开网址、运行 Web 应用、填充表单等。代码如下：

```python
# Author: lindafang
# Date: 2020-07-14 16:29
# File: test_pytest_selenium_browser.py
import sys
import pytest_selenium
import pytest


def test_baidu_title(selenium):
    selenium.get('http://www.baidu.com/')
    assert selenium.title == '百度一下,你就知道'


def test_baidu_current_URL(selenium):
    selenium.get('http://www.baidu.com/')
    assert selenium.current_URL == 'https://www.baidu.com/'
```

```python
def test_baidu_so_getValue(selenium):
    selenium.get('http://www.baidu.com/')
    so = selenium.find_element_by_id('kw')
    so.send_keys('linda')
    assert so.get_attribute('value') == 'linda'
```

执行命令如下：

```
pytest --driver Chrome --driver-path /path/to/Chromedriver
```

注意：有时执行会有问题，说明与其他插件有冲突，逐步找到冲突的插件。

5.3.17 pytest-timeout 设置超时时间

为测试设置时间限制：pytest-timeout。

安装插件：pip install pytest-timeout。

pytest test_x.py --timeout=n（时间限制，单位：秒）

5.3.18 pytest-xdist 测试并发执行

pytest-xdist 这款插件允许用户将测试并发执行（进程级并发）。主要开发者是 pytest 目前的核心开发人员 Bruno Oliveira，截至笔者写作此文时，该项目已有 711 个 star，应用于 7850 个项目。需要注意的是，由于插件动态决定测试用例执行的顺序，为了保证各个测试能在各自独立线程中正确地执行，用例的作者应该保证测试用例的独立性（这也符合测试用例设计的最佳实践）。

具体的执行流程如下：

第 1 步，收集测试项。

第 2 步，测试收集检查。

第 3 步，测试分发。

第 4 步，测试执行。

第 5 步，测试结束。

把本章的所有测试用例使用并发的形式执行一下，命令为 pytest -n 3，这里的数字 3 是并发 3 个线程执行，结果如下：

```
pytest -n 3
========================= test session starts =========================
platform darwin -- Python 3.6.8, pytest-6.0.1, py-1.9.0, pluggy-0.13.1
sensitiveURL: .*
rootdir: /Users/lindafang/PyCharmProjects/pytest_book
```

```
plugins: xdist-2.1.0,
gw0 [36] / gw1 [36] / gw2 [36]
.FFFFFFFF......FF...F.....FEEE.....
[100%]
========================= ERRORS ==================================
#......略过一些执行结果
test_pytest-freezegun.py:37: AssertionError
==================== short test summary info =====================
FAILED test_assume.py::test_simple_assume[0-1] - assert 0 == 1
#......略过一些执行结果

================== 13 failed, 20 passed, 3 errors in 3.59s ==================
```

注意：测试用例执行时间短，并发的效果可能会有相反的效果，因为多建立一个线程也需要时间。

5.4 插件管理

5.4.1 在测试模块或 conftest 文件中加载插件

可以在测试模块或 conftest 文件中加载插件，代码如下：

```
pytest_plugins = ("myapp.testsupport.myplugin",)
```

加载测试模块或 conftest 插件时，也会加载指定的插件。

注意：不推荐使用 pytest_plugins 非根 conftest.py 文件中的变量来加载插件。参阅插件部分中的完整说明。该名称 pytest_plugins 是保留名称，不应用作自定义插件模块的名称。

5.4.2 找出哪些插件处于活动状态

如果要找出环境中哪些插件处于活动状态，则可以输入如下命令：

```
pytest --trace-config
```

此命令可获得扩展的测试标头，其中显示了已激活的插件及其名称。

5.4.3 通过名称停用/注销插件

可以阻止插件加载或注销它们，命令如下：

```
pytest -p no:NAME
```

这意味着任何随后的激活/加载命名插件的尝试都将不起作用。

如果要无条件禁用项目插件,则可以将此选项添加到 pytest.ini 文件中,代码如下:

```
[pytest]
addopts = -p no: NAME
```

或者,仅在某些环境中(例如,在 CI 服务器中)禁用它,可以将 PYTEST_ADDOPTS 环境变量设置为-p no:NAME。

5.5 本章小结

本章主要讲解 pytest 的插件及管理:
(1) 常用插件的介绍和使用。
(2) 常用插件管理。

第 6 章 与 Allure 框架结合定制测试报告

6.1 Allure 框架介绍

Allure 的全名：Allure Test Report，对于不同的编程语言，有很多很酷的测试框架。Allure Framework 是一种灵活的轻量级多语言测试报告工具，它不仅可以以简洁的 Web 报告形式显示已测试的内容，而且还允许参与开发过程的每个人从日常执行中最大限度地提取有用测试信息。

从开发及质量保证的角度来看，Allure 报告可以缩短常见缺陷的生命周期：可以将测试失败划分为 Bug 和破坏性测试，还可以配置日志、步骤、初始化和销毁、附件、时间、历史记录与测试管理系统及 Bug 跟踪系统的集成，因为负责任的开发人员和测试人员需要掌握所有信息。

从管理人员的角度来看，Allure 提供了一个清晰的"全局"，涵盖了所需要的功能、缺陷聚集的位置、执行时间表的外观及许多其他方面的事情。Allure 的模块化和可扩展性确保开发人员始终能够微调某些东西，以使 Allure 更适合项目开发。官方网址 http://allure.qatools.ru/，参考文档网址 https://docs.qameta.io/allure/。

6.2 Allure 如何生成测试报告

Allure 基于标准 xUnit 结果输出，但添加了一些补充数据。任何报告都通过两个步骤生成。在测试执行期间（第一步），通过与不同语言的客户端（Python 语言是 allure-pytest 库）将有关已执行测试的信息保存到 xml 文件中。在报告生成（第二步）期间，xml 文件将被转换为 HTML 报告。这可以通过命令行工具、CI 插件或构建工具来完成。

6.3 Allure 报告组成

Allure 报告主要由六部分组成，总览（Overview）、缺陷（Defects）、测试执行结果（xUnit）、数据驱动测试（Behaviors）、图表（Graph）和时间轴（Timeline）。从不同方面展示报告的各种测试信息。

6.3.1 总览

总览包括测试套件、环境、趋势、类别及具体场景。可以单击链接进入不同的具体展示页面详细查看,如图6-1所示。单击左下角En按钮可选择更换为中文。

图6-1 Allure报告中的总览

图片有点大,此处只展示部分内容,即图的下半部分,如图6-2所示。

图6-2 Allure报告中总览的下面部分

6.3.2 类别

按测试用例状态进行过滤,可以选择通过、跳过、失败等进行选择并查看,如图6-3所示。

6.3.3 测试套件

缺陷页面提供了测试执行过程中所发现的缺陷的详细清单,可以区分与失败测试相对应的产品缺陷(Product Defects)及与破坏测试相对应的测试缺陷(Test Defects)。大家可以查看每个测试套件的测试统计信息及每个测试用例的详细信息,如图6-4所示。

图 6-3　Allure 报告中的类别

图 6-4　Allure 报告中的测试套件

6.3.4　功能

通过下面的报告可以轻松查看哪些功能存在问题，如图 6-5 所示。

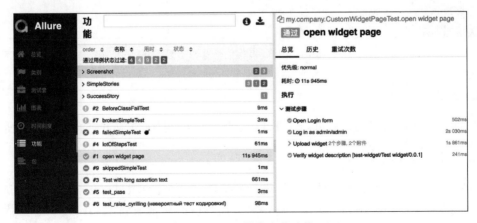

图 6-5　Allure 报告中的功能

6.3.5 图形

图形可以帮助开发人员直观地评估测试结果,如图 6-6 所示。

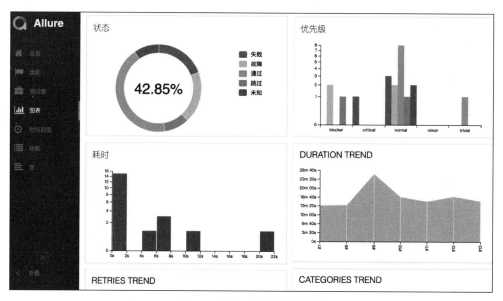

图 6-6　Allure 报告中的图形

6.3.6 时间轴

在时间轴页面显示每个测试用例在哪个时间点开始执行,并且它运行了多长时间,如图 6-7 所示。

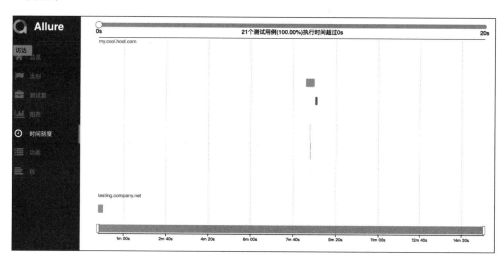

图 6-7　Allure 报告中的时间轴

6.3.7 包

展示不同文件夹（包 Packages）执行情况，如图 6-8 所示。

图 6-8　Allure 报告中的包

6.4　Allure 的初体验

要生成第一份报告，只需执行以下几个简单的步骤：
（1）下载并安装所需的 Allure 命令行应用程序（客户端）。
（2）查找要生成报告的测试执行数据。
（3）生成在线报告。

6.4.1　Allure 在 Windows 系统下安装

官网提供的安装方式有点复杂，下面提供另一种安装方式。

在官网进入 Download 页面，自动跳转到 GitHub 上，下载安装包，如图 6-9 所示的标注。

将下载的文件夹及其内容存放到一个目录下，配置系统环境变量：在 PATH 路径中加入文件夹下的 bin 目录，如图 6-10 所示。

客户端安装完成后，使用下面命令安装与 pytest 配合的客户端：

```
pip install allure - pytest
```

这将安装 allure-pytest 和 allure-python-commons 程序包，以生成与 Allure 2 兼容的报告数据。如果安装了用于第一代 Allure 报告的早期版本，则首先需要将其卸载，这一点很重要。

验证安装完成，输入 allure --version 会显示版本信息：

```
allure -- version
```

图 6-9　Allure 下载安装的具体位置

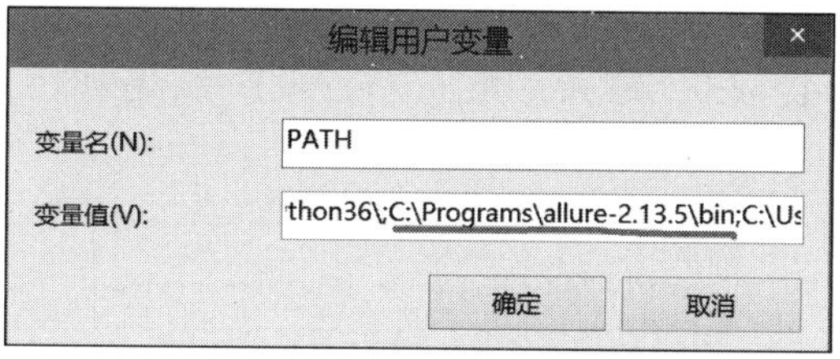

图 6-10　Allure 配置系统环境变量

6.4.2　Allure 在 Linux 环境下安装

Linux 对于基于 Debian 的存储库，提供了 PPA，命令如下：

```
sudo apt-add-repository ppa:qameta/allure
sudo apt-get update
sudo apt-get install allure
```

如果使用的是 CentOS 系统，可以先安装 Node.js，再使用 npm 进行安装，如图 6-11 所示。

执行 Allure 需要 Java 1.8 及以上版本。如果没有安装，则应按下述步骤安装。

```
[root@VM-0-7-centos ~]# yum install nodejs
已安装：
  nodejs.x86_64 1:6.17.1-1.el7
作为依赖被安装：
  libicu.x86_64 0:50.2-4.el7_7            libuv.x86_64 1:1.38.0-2.el7
  npm.x86_64 1:3.10.10-1.6.17.1.1.el7
完毕！
[root@VM-0-7-centos ~]# npm install -g allure-commandline --save-dev
/usr/bin/allure -> /usr/lib/node_modules/allure-commandline/bin/allure
/usr/lib
└── allure-commandline@2.13.0
```

图 6-11　Linux 下安装 Allure 客户端

yum -y list java * 命令可以查看有哪些版本可供安装。选择一个正确的版本进行安装，如图 6-12 所示。

```
java-1.7.0-openjdk-src.x86_64              1:1.7.0.261-2.6.22.2.el7_8    updates
java-1.8.0-openjdk.i686                    1:1.8.0.252.b09-2.el7_8       updates
java-1.8.0-openjdk.x86_64                  1:1.8.0.252.b09-2.el7_8       updates
java-1.8.0-openjdk-accessibility.i686      1:1.8.0.252.b09-2.el7_8       updates
java-1.8.0-openjdk-accessibility.x86_64    1:1.8.0.252.b09-2.el7_8       updates
java-1.8.0-openjdk-demo.i686               1:1.8.0.252.b09-2.el7_8       updates
java-1.8.0-openjdk-demo.x86_64             1:1.8.0.252.b09-2.el7_8       updates
java-1.8.0-openjdk-devel.i686              1:1.8.0.252.b09-2.el7_8       updates

[root@VM-0-7-centos ~]# yum install java-1.8.0-openjdk.x86_64
```

图 6-12　Linux 下安装所需的 Java 1.8

验证 Allure 是否安装成功，输入 allure --version，结果如图 6-13 所示，则表明安装成功。

```
完毕！
[root@VM-0-7-centos ~]# allure --version
2.13.0
```

图 6-13　Linux 下验证 Allure 是否安装成功

6.4.3　Allure 在 Mac OS 系统下安装

对于 Mac OS 系统，可通过 Homebrew 进行自动安装，命令如下：

```
brew install allure
```

安装前确认已经安装 homebrew。

```
lindafang@linda ~ % /usr/local/Homebrew/bin/brew -v
Homebrew 2.1.6
Homebrew/homebrew-core (git revision 860e; last commit 2020-05-20)
```

如果没安装，则需要把下面命令粘贴到终端执行安装，在中国执行的方式如下所示：

```
/bin/zsh -c "$(cURL -fsSL https://gitee.com/cunkai/HomebrewCN/raw/master/Homebrew.sh)"
```

之后会显示具体的安装步骤，如图 6-14 所示。

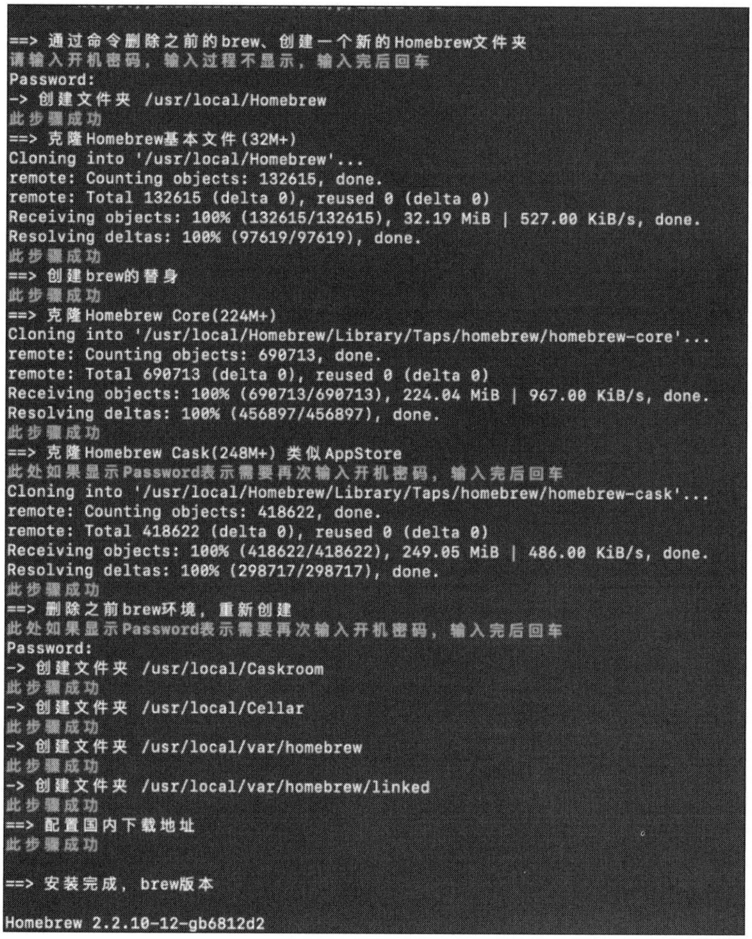

图 6-14 Mac 系统下使用 brew 安装

6.4.4 Allure 的简单用法

1. 收集结果

要使 Allure 侦听器能够在测试执行过程中收集结果，只需添加 --alluredir 选项并提供路径便可以存储结果。示例代码如下：

```
pytest --alluredir=/tmp/my_allure_results
```

2. 查看报告

要在测试完成后查看报告,需要使用 Allure 命令行实用程序从结果生成报告,命令如下:

```
allure serve /tmp/my_allure_results
```

此命令将在默认浏览器中显示所生成的报告。

6.4.5 Allure 的帮助说明

通过帮助查看具体的用法。

```
allure -- help
```

结果及常用参数的解释,执行结果如下:

```
Usage: allure [options] [command] [command options]
  Options:
    -- help
      Print commandline help.
    #命令行输出帮助信息
    -q, -- quiet
    #切换成静默模式,执行后只显示简短的结果
      Switch on the quiet mode.
      Default: false
    -v, -- verbose
      Switch on the verbose mode.
    #切换为显示详细信息模式,多个 v 显示的信息更详细
      Default: false
    -- version
      Print commandline version.
    #命令行输出版本
      Default: false
  Commands:
    generate      Generate the report    #生成报告
      # Usage: generate [options] The directories with allure results
          #报告的路径默认为当前路径,可以编写输出路径
      Options:
        -c, -- clean
          Clean Allure report directory before generating a new one.
            #清除以前在此路径下生成的报告信息
          Default: false
        -- config
          Allure commandline config path. If specified overrides values from
```

```
            --profile and --configDirectory.
         --configDirectory
            Allure commandline configurations directory. By default uses
            ALLURE_HOME directory.
         --profile
            Allure commandline configuration profile.
         -o, --report-dir, --output
            The directory to generate Allure report into.
                 #这个路径是生成报告信息的目录,默认为 allure-report
            Default: allure-report

 serve       Serve the report        #通过结果生成报告
   Usage: serve [options] The directories with allure results
                 #测试结果所在的路径,与上面参数的路径一致才能找到结果
      Options:
         --config
            Allure commandline config path. If specified overrides values from
            --profile and --configDirectory.
         --configDirectory
            Allure commandline configurations directory. By default uses
            ALLURE_HOME directory.
         -h, --host
            This host will be used to start web server for the report.
                 #这个主机将被应用启动报告的 Web 服务器
         -p, --port
            This port will be used to start web server for the report.
                 #这个端口将被应用启动报告的 Web 服务器
            Default: 0
         --profile
            Allure commandline configuration profile.

 open        Open generated report    #打开生成的报告
   Usage: open [options] The report directory
      Options:
         -h, --host
            This host will be used to start web server for the report.
         -p, --port
            This port will be used to start web server for the report.
            Default: 0

 plugin      Generate the report
   Usage: plugin [options]
      Options:
         --config
            Allure commandline config path. If specified overrides values from
            --profile and --configDirectory.
```

```
--configDirectory
    Allure commandline configurations directory. By default uses
    ALLURE_HOME directory.
--profile
    Allure commandline configuration profile.
```

6.5 定制测试报告

除了具有 pytest 的环境外,Allure 当前支持绝大多数可用功能。可以在代码中加入报告需要的信息,让报告更丰富且更易理解。

6.5.1 定制详细的步骤说明

Allure 报告最显著的特点是,它可以使每个测试得到非常详细的说明。这可以通过@allure.step 装饰器实现,该装饰器将带注释的方法或函数的调用与提供的参数一起添加到报表中。

1. 步骤(step)可外部调用也可嵌套

带有注释的方法@step 可以存储在测试之外,并在需要的时候导入。步骤方法可以具有任意深度的嵌套结构。在本章根目录下建立 steps.py 文件。

代码如下:

```
#File: steps.py

def imported_step():
    print("非常重要的步骤!")
```

同级建立 test_step_nested.py 文件,步骤可以直接加在测试方法上,也可以调用其他文件。如:在 test_with_imported_step 方法中调用 steps 文件中的 imported_step 方法。

同样可以嵌套使用,调用的方法参数随意。step_with_nested_steps 方法嵌套调用 nested_step 方法再嵌套调用 nested_step_with_arguments 方法。

代码如下:

```
import allure

from .steps import imported_step


@allure.step
def passing_step():
```

```python
        print("通过的步骤")
        pass


@allure.step
def step_with_nested_steps():
    print("带有嵌套的步骤 nested")
    nested_step()


@allure.step
def nested_step():
    print("调用带有参数 arg 的步骤")
    nested_step_with_arguments(1, 'abc')


@allure.step
def nested_step_with_arguments(arg1, arg2):
    print("带有两个参数：",arg1,arg2)
    pass


def test_with_imported_step():
    passing_step()
    print("外部导入")
    imported_step()


def test_with_nested_steps():
    passing_step()
    step_with_nested_steps()
```

在生成 Allure 报告时，每个步骤的状态都显示在名称右侧的"执行"中。嵌套步骤以树状可折叠结构组织，如图 6-15 所示。

2. 步骤(step)可以有带参数的描述行

步骤可以具有一个描述行，该行支持传递位置和关键字参数的占位符。关键字参数的默认参数也将被捕获，这样在 Allure 报告的描述中就可以看到描述信息和具体的参数。

代码如下：

```python
# Author: lindafang
# Date: 2020 - 10 - 05 14:19
# File: test_step_placeholders.py
import allure
```

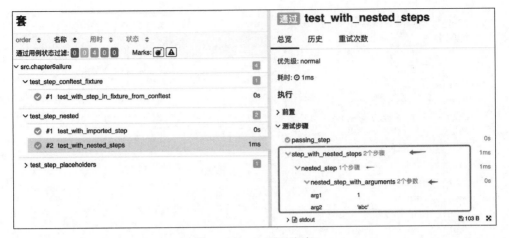

图 6-15　step 可以嵌套

```
@allure.step('步骤可以有描述行, 位置参数显示输入参数: "{0}", 关键字显示输入参数: "{key}"')
def step_with_title_placeholders(arg1, key = None):
    pass


def test_steps_with_placeholders():
    step_with_title_placeholders(1, key = '这是关键字参数')
    step_with_title_placeholders(2)
    step_with_title_placeholders(3, '这是位置参数')
```

执行结果如图 6-16 所示。

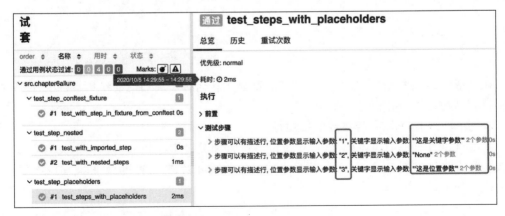

图 6-16　step 可以有带参数的描述行

3. fixture 和 conftest 同样支持步骤 step

fixture 也支持步骤，下面是使用 conftest.py 模块中定义的 fixture 进行测试的示例（即使不直接导入，这些 fixture 也将由 pytest 解析）。

在 Allure 的报告中初始化的步骤（在 conftest.py 中）显示在单独"前置"的树中，用于配置初始化和销毁。

代码如下：

```
conftest.py

import allure
import pytest

@allure.step('step in conftest.py')
def conftest_step():
    pass

@pytest.fixture
def fixture_with_conftest_step():
    conftest_step()


# 初始化中的步骤显示在单独前置的树中，用于配置和销毁

test_step_conftest_fixture.py

import allure

@allure.step
def passing_step():
    pass
def test_with_step_in_fixture_from_conftest(fixture_with_conftest_step):
    passing_step()
```

执行结果如图 6-17 所示。

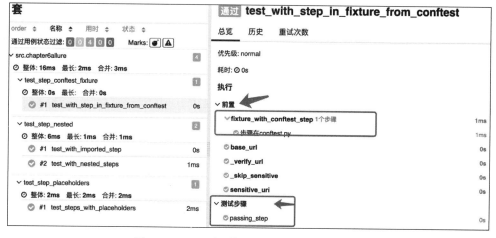

图 6-17　step 可以使用 fixture 和 conftest.py

6.5.2 不同类型的附件补充测试说明

报告可以显示许多不同类型的附件,这些附件可以补充测试、步骤或 fixture 的结果。可以通过以下方式创建附件。

allure.attach(body, name, attachment_type, extension):

(1) body:要写入文件的原始内容。

(2) name:带有文件名的字符串。

(3) attachment_type:allure.attachment_type 值之一。

(4) extension:用作创建文件的扩展名。

allure.attach.file(source, name, attachment_type, extension):source 表示包含文件路径的字符串,其他参数与 allure.attach 的参数相同。

代码如下:

```python
#File: test_attach_fixture_finalizer.py
import allure
import pytest


@pytest.fixture
def attach_file_in_module_scope_fixture_with_finalizer(request):
    allure.attach('这个文本的附件是文件级的', '这可以是文件名字', allure.attachment_type.TEXT)
    def finalizer_module_scope_fixture():
        allure.attach('这个文本的附件是文件级的结束部分', '下面调用finalizer',
                      allure.attachment_type.TEXT)
    request.addfinalizer(finalizer_module_scope_fixture)


def test_with_attacments_in_fixture_and_finalizer(attach_file_in_module_scope_fixture_with_finalizer):
    print("这是通过fixture调用上面的步骤,步骤中包含结束finalizer调用")
    pass
```

执行结果如图 6-18 所示。

Allure 报告中可以添加附件,附件的形式有多种,可以是图片、文本或 HTML 网页等。

代码如下:

```python
#File: test_attach_html_png.py
import allure
```

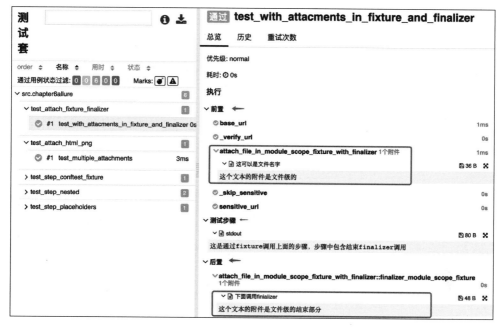

图 6-18 attach 可以使用 fixture 和 finalizer

```
def test_multiple_attachments():
    '''
    attachment_type =
    TEXT = ("text/plain", "txt")
    CSV = ("text/csv", "csv")
    TSV = ("text/tab-separated-values", "tsv")
    URI_LIST = ("text/uri-list", "uri")
    HTML = ("text/html", "html")
    XML = ("application/xml", "xml")
    JSON = ("application/json", "json")
    YAML = ("application/yaml", "yaml")
    PCAP = ("application/vnd.tcpdump.pcap", "pcap")

    PNG = ("image/png", "png")
    JPG = ("image/jpg", "jpg")
    SVG = ("image/svg-xml", "svg")
    GIF = ("image/gif", "gif")
    BMP = ("image/bmp", "bmp")
    TIFF = ("image/tiff", "tiff")
    MP4 = ("video/mp4", "mp4")
    OGG = ("video/ogg", "ogg")
    WEBM = ("video/webm", "webm")
    PDF = ("application/pdf", "pdf")
```

```
        '''
        allure.attach.file('1.jpg', attachment_type = allure.attachment_type.JPG)
        allure.attach.file('2.yaml', attachment_type = allure.attachment_type.YAML)
        allure.attach('< head ></head>< body > 这是个网页说明：下面可有多个 </body>', '附加一个
网页的具体说明', allure.attachment_type.HTML)
```

附件显示在它们所属的测试实体的上下文中。HTML 类型的附件将呈现并显示在报告页面。这是一种方便的方式,可以为自己的测试结果提供一些自定义内容。

代码注释中所列的是可接受的所有类型,执行结果如图 6-19 所示。

图 6-19　attach.file 可以附加各种类型文件

6.5.3　定制各种类型内容描述

可以添加测试的详细说明,也可以根据需要为报告提供尽可能多的上下文。这些内容可以通过几种方法来完成：可以添加 @allure.description 提供描述字符串的装饰器,也可以用 @allure.description_html 提供一些 HTML,以便在测试用例的"描述"部分呈现,或者仅从测试方法的文档字符串中获取描述。

1. 支持 unicode 字符串

可以使用不同国家的语言,以便使不同语言都可以正确显示。

代码如下：

```
#File: test_describe_html.py
import allure
def test_unicode_in_docstring_description():
    """unicode 描述使用不同国家的语言
    Этот тест проверяет юникод
    こんにちは、友達
```

```
    你好伙计
    """
    assert 42 == int(6 * 7)
```

执行结果如图 6-20 所示。

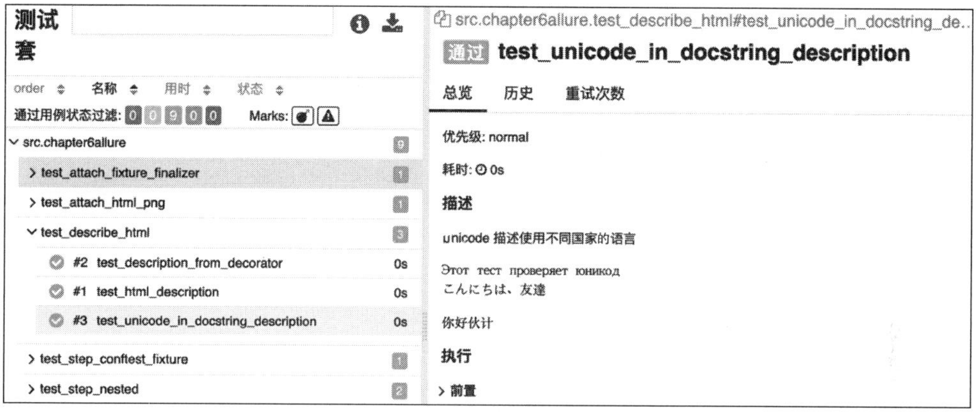

图 6-20　description 中使用多国文字描述

2．来自 HTML 的渲染 description_html

可以添加 html 格式的描述，在报告中显示，没有一点违和感。

代码如下：

```
import allure

@allure.description_html("""
<h1>测试带有一些复杂的描述</h1>
<table style = "width:100 %">
  <tr>
    <th>名</th>
    <th>姓</th>
    <th>年龄</th>
  </tr>
  <tr align = "center">
    <td>小明</td>
    <td>李</td>
    <td>50</td>
  </tr>
  <tr align = "center">
    <td>小三</td>
    <td>唐</td>
    <td>94</td>
  </tr>
```

```
</table>
""")
def test_html_description():
    print("测试通过网页描述")
    assert True



@allure.description("""
这是一个多功能描述,计算
""")
def test_description_from_decorator():
    assert 42 == int(6 * 7)
```

执行结果如图 6-21 所示。

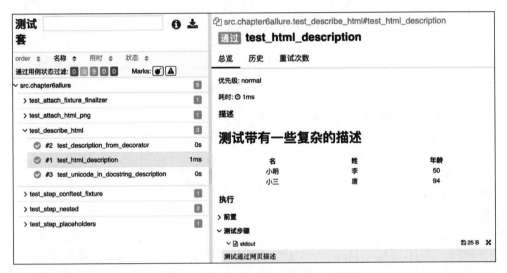

图 6-21　description 中使用 HTML 描述

3．动态描述,可替换开始时的描述

还可以从测试内动态更新最初的描述 allure.dynamic.description,这是非常灵活的方式,代码如下：

```
@allure.description("""
这段描述将在测试结束后被替换
""")
def test_dynamic_description():
    assert 42 == int(6 * 7)
    allure.dynamic.description('这是最后的描述,用于替换前面的描述!')
```

执行结果如图 6-22 所示。

图 6-22 description 中使用动态描述

6.5.4 定制测试标题

可以使用特殊的@allure.title 装饰器使测试标题更具可读性。标题支持参数的占位符并支持动态替换，也可以在"功能"中查看。

代码如下：

```python
# Author: lindafang
# Date: 2020 - 10 - 05 16:11
# File: test_title.py

import allure
import pytest


@allure.title("这是一个正常的标题")
def test_with_a_title():
    assert 2 + 2 == 4


@allure.title("这个标题支持其他国家文字：Привет!こんにちは!")
def test_with_unicode_title():
    assert 3 + 3 == 6


@allure.title("参数化标题: adding {param1} with {param2}")
@pytest.mark.parametrize('param1,param2,expected', [
    (2, 2, 4),
    (1, 2, 5),
])
def test_with_parameterized_title(param1, param2, expected):
    assert param1 + param2 == expected
```

```
@allure.title("动态标题")
def test_with_dynamic_title():
    assert 2 + 2 == 4
    allure.dynamic.title('替换动态标题为最终标题')
```

执行结果如图 6-23 所示。

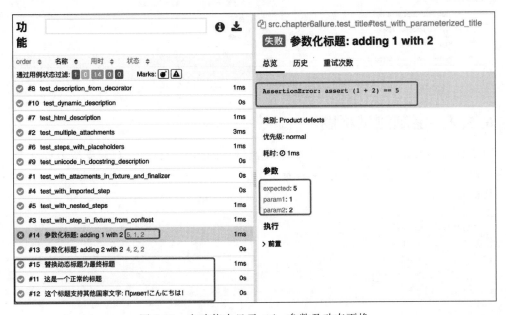

图 6-23　在功能中显示 title、参数及动态更换

6.5.5　各种链接

报告可展示缺陷系统或测试管理系统链接，Allure 具有整合 @allure.link、@allure.issue 和 @allure.testcase 描述的功能。通过这些功能可以直接链接到测试用例和缺陷管理系统及其他地方。

代码如下：

```
import allure

TEST_CASE_LINK = 'http://81.70.24.116:8088/zentao/bug-view-26.html'

@allure.link('http://81.70.24.116:8088/zentao/testcase-browse-1.html')
def test_with_link():
    pass

@allure.link('http://81.70.24.116:8088/zentao/product-index-no.html', name='单击进入项目')
```

```
def test_with_named_link():
    pass


@allure.issue('140', '测试出现问题的结果')
def test_with_issue_link():
    pass


@allure.testcase(TEST_CASE_LINK, '测试登录健壮性')
def test_with_testcase_link():
    pass
```

@allure.link 将在"链接"部分提供指向所提供的网址的可单击链接,如图 6-24 所示。

图 6-24 @link 在链接中显示

@allure.issue 提供带有小错误图标的链接,如图 6-25 所示。

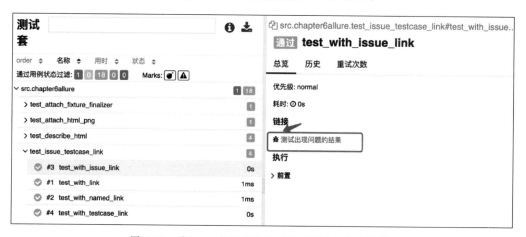

图 6-25 @issue 在链接中显示出错误"小虫子"图标

6.5.6 自定义各种标签

有时,需要灵活地执行要执行的测试。pytest 允许使用标记装饰器 @pytest.mark (pytest docs)。

Allure 允许使用 3 种类型的标记修饰符以类似的方式标记测试,这些修饰符可以构造报告的结构形式:BDD 样式标记,此标记可以表示模块、功能、故事、严重性标签、自定义标签。

代码如下:

```
包括以下:
    EPIC = 'epic'
    FEATURE = 'feature'
    STORY = 'story'
    PARENT_SUITE = 'parentSuite'
    SUITE = 'suite'
    SUB_SUITE = 'subSuite'
    SEVERITY = 'severity'
    THREAD = 'thread'
    HOST = 'host'
    TAG = 'tag'
    ID = 'as_id'
    FRAMEWORK = 'framework'
    LANGUAGE = 'language'
```

BDD 样式标记有两个装饰器:@allure.feature 和 @allure.story。用于根据特定项目的功能和故事来标记测试(有关背景,可参阅 Wikipedia 上的 BDD 文章)。要标记某些功能或故事属于子系统,可使用以 epic_前缀开头的名称。

代码如下:

```
#File: test_feature_story_epic.py


import allure


def test_without_any_annotations_that_wont_be_executed():
    pass


@allure.story('epic_1')
def test_with_epic_1():
    pass
```

```python
@allure.feature('feature_1')
def test_with_feature_1():
    pass

@allure.story('story_1')
def test_with_story_1():
    pass

@allure.feature('feature_2')
@allure.story('story_1')
def test_with_story_1_and_feature_2():
    pass


@allure.feature('feature_2')
@allure.story('story_2')
def test_with_story_2_and_feature_2():
    pass
```

可以使用以下命令行选项指定不同的测试集，以执行以逗号分隔的值的列表的操作：

```
--allure-epics
--allure-features
--allure-stories
```

在控制台输入不同的命令和参数会有不同的执行结果。pytest test_feature_story_epic.py --allure-stories story_1,story_2 执行这个代码中@allure.story(story_1)和@allure.story(story_2)的测试方法。pytest test_feature_story_epic.py --allure-features feature_2 --allure-stories story_2 中的两个参数是或的关系，也就是只要满足一个条件就执行，所以此处执行2个测试。

执行结果如下：

```
pytest test_feature_story_epic.py --allure-stories story_1,story_2 --alluredir=./result

========================= test session starts =========================
collected 6 items

test_feature_story_epic.py ...
[100%]

========================= 3 passed in 0.06s =========================
pytest -s -v test_feature_story_epic.py --allure-features feature_2 --allure-stories story_2
```

```
collected 6 items

test_feature_story_epic.py::test_with_story_1_and_feature_2 PASSED
test_feature_story_epic.py::test_with_story_2_and_feature_2 PASSED

============== 2 passed in 0.01 seconds ==============================

 pytest test_feature_story_epic.py -- allure-epics epic_1

collected 6 items

test_feature_story_epic.py .
[100%]


======================= 1 passed in 0.04s ===================
```

执行的结果如图 6-26 所示。

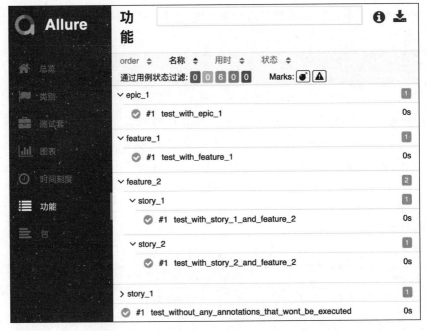

图 6-26 @feature@epic@story 在功能中显示

6.5.7 严重性标记

要按严重性级别标记测试级别,可以使用 @allure.severity 装饰器。它以 allure.severity_level 枚举值作为参数。在这个枚举值中 Bug 的严重级别通常包括 4～5 级。在下面的代码注释中有相关解释。

代码如下：

```
#File: test_severity.py
import allure
'''
    BLOCKER = 'blocker'
    CRITICAL = 'critical'
    NORMAL = 'normal'
    MINOR = 'minor'
    TRIVIAL = 'trivial'

Bug 的严重程度(Severity)
1. Blocker
   即系统无法执行、崩溃或严重资源不足、应用模块无法启动或异常退出、无法测试、造成系统不稳定
严重花屏
内存泄漏
用户数据丢失或破坏
系统崩溃/死机/冻结
模块无法启动或异常退出
严重的数值计算错误
功能设计与需求严重不符
其他导致无法测试的错误，如服务器 500 错误
2. Critical
即影响系统功能或操作，主要功能存在严重缺陷，但不会影响系统稳定性
功能未实现
功能错误
系统刷新错误
数据通信错误
轻微的数值计算错误
影响功能及界面的错误字或拼写错误
安全性问题
3. Major
即界面、性能缺陷、兼容性缺陷
操作界面错误(包括数据窗口内列名定义、含义是否一致)
边界条件下错误
提示信息错误(包括未给出信息、信息提示错误等)
长时间操作无进度提示
系统未优化(性能问题)
光标跳转设置不好，鼠标(光标)定位错误
兼容性问题
4. Minor/Trivial
即易用性及建议性问题
界面格式等不规范
辅助说明描述不清楚
操作时未给用户提示
可输入区域和只读区域没有明显的区分标志
```

```
    个别不影响产品理解的错别字
    文字排列不整齐等一些小问题
    '''

    def test_with_no_severity_label():
        pass


    @allure.severity(allure.severity_level.TRIVIAL)
    def test_with_trivial_severity():
        pass


    @allure.severity(allure.severity_level.NORMAL)
    def test_with_normal_severity():
        pass


    @allure.severity(allure.severity_level.NORMAL)
    class TestClassWithNormalSeverity(object):

        def test_inside_the_normal_severity_test_class(self):
            pass

        @allure.severity(allure.severity_level.CRITICAL)
        def test_inside_the_normal_severity_test_class_with_overriding_critical_severity(self):
            pass
```

严重性修饰符可以应用于函数、方法或整个类。

通过将--allure-severities命令行选项与逗号分隔的严重性级别结合使用，仅运行具有相应严重性的测试。这个级别的关系也是或的关系，只要有一条符合条件就执行。

执行结果如下：

```
pytest -s -v test_severity.py --allure-severities normal,critical

collected 5 items

test_severity.py::test_with_normal_severity PASSED
test_severity.py::TestClassWithNormalSeverity::test_inside_the_normal_severity_test_class PASSED
test_severity.py::TestClassWithNormalSeverity::test_inside_the_normal_severity_test_class_with_overriding_critical_severity PASSED

========================= 3 passed in 0.04s =============================
```

在 Allure 报告中的 severity 显示具体的级别，如图 6-27 所示。

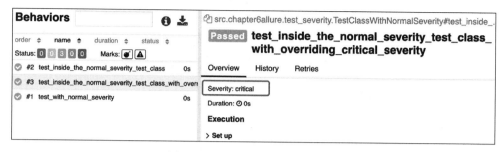

图 6-27　@severity 在报告中显示

6.5.8　重试信息展示

Allure 允许汇总有关在一次测试运行中重新执行的测试信息，以及一段时间内测试执行的历史记录。对于重试，可以使用 pytest 重新运行失败插件。

例如，如果我们有一个非常不可靠的步骤方法，该方法经常失败，那么可以添加参数 --reruns=5，pytest 启动选项中指定参数后，我们会在 Retries 选项卡上看到所有未成功运行此测试的尝试。

代码如下：

```python
import allure
import random
import time


@allure.step
def passing_step():
    pass


@allure.step
def flaky_broken_step():
    if random.randint(1, 5) != 1:
        raise Exception('Broken!')


def test_broken_with_randomized_time():
    passing_step()
    time.sleep(random.randint(1, 3))
    flaky_broken_step()
```

在控制台执行如下命令，执行结果如图 6-28 所示。

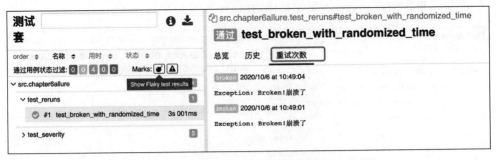

图 6-28　重试的插件执行后可以在重试次数中查看

```
pytest test_reruns.py -- reruns = 5 -- alluredir = ./result
```

6.6　本章小结

pytest 与 Allure 框架结合定制测试报告，可以通过添加不同的装饰器增加报告的可读性。

（1）@allure.step：定制详细步骤。

（2）@allure.attach：添加附件信息。

（3）@allure.description：添加各种类型的内容描述。

（4）@allure.title：添加测试标题。

（5）@allure.issue、@allure.link、@allure.test_case：添加 Bug、链接和测试用例。

（6）@allure.feature、@allure.story、@allure.epic：添加项目的功能和模块。

（7）@allure.severity：添加测试严重级别。

（8）与插件 reruns 结合，在报告重试次数中显示 reruns 执行时重试的次数。

第 7 章 单元自动化测试实践

7.1 什么是单元测试

单元测试的英文是 Unit Testing，特指在一个分隔的代码单元中的测试。一个单元可以是整个模块、一个单独的类或者函数，或者这两者间的任何代码。然而，重要的是，测试代码要与没有测试到的其他代码相互隔离，因为如果其他代码本身有错误会因此混淆测试结果。

7.2 pytest 测试框架是单元测试的框架

Python 单元测试有自带的框架 unittest，同样能进行单元测试，但很多时候需要固定的写法，因而不是特别灵活。pytest 框架的写法相对随意，并且是一个单元、接口和 UI 通用的自动化测试框架。既然通用，应用一种框架就能完成测试流程活动中的全部测试项目，当然是我们喜闻乐见的。

7.3 单元测试与质量

单元测试大部分是由高级程序员自己测试自己编写的代码而产生的。编写测试代码这是一个优秀的程序员所必备的技能，单元测试的作用非常大，可以在早期找到漏洞（Bug），这时修改成本较小。对于一个对质量要求较高的公司，通常对质量的要求是深入人心的，也就是每个人对于产品的质量都负有责任与义务，所以有些公司没有专门的测试部门，而是全员测试，而且有很多开源的工具同样可以看到单元测试代码。

有很多公司并没有编写单元测试用例，这是由中国软件测试的历史决定的。

7.4 单元测试一个函数

先编写一段代码，再使用 pytest 进行测试。

编写一个函数，判断输入数据是否是质（素）数。首先开发人员需要知道定义，质数定义为在大于 1 的自然数中，除了 1 和它本身以外不再有其他因数，这样的数称为质数。

编写第 1 个版本，代码如下：

```python
# Author: lindafang
# File: is_prime.py


def is_prime(number):
    """如果是质数，返回值为 True"""
    for element in range(number):
        if number % element == 0:
            return False
    return True


def print_next_prime(number):
    """打印输入 number 数字最近的质数"""
    index = number
    while True:
        index += 1
        if is_prime(index):
            print(index)
```

开发人员使用测试驱动开发（TDD）方式，编写的最简单的测试代码如下：

```python
# Author: lindafang
# File: test_is_prime.py

import pytest
from src.chapter7unit.method_examples.is_prime import is_prime


class TestPrimes(object):
    """测试 is_prime"""

    def test_is_five_prime(self):
        """5 是不是质数？"""
        assert is_prime(5) is True


if __name__ == '__main__':
    pytest.main()
```

执行结果如下：

```
    number = 5

    def is_prime(number):
        """如果是质数,返回值为 True"""
        for element in range(number):
>           if number % element == 0:
E           ZeroDivisionError: integer division or modulo by zero

method_examples/is_prime.py:9: ZeroDivisionError
```

这时发现代码未考虑到除 0 的问题,于是修改代码把 for 循环换成 for element in range(2,number):也就是不从 0 开始,也跳过 1,这是由质数的概念决定的。修改代码后从 2 开始执行。修改后的执行结果通过测试。

再添加一个测试用例,测试 4 不是质数,这是第 2 个版本,代码如下:

```
def test_is_four_non_prime(self):
    """4 不是质数"""
    assert is_prime(4) is False
```

运行结果如图 7-1 所示。

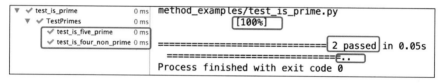

图 7-1　测试一个函数的运行结果

再添加一个测试用例,测试边界值 0 是不是质数,这是第 3 个版本,代码如下:

```
def test_is_zero_not_prime(self):
    """测试边界值 0 是不是质数"""
    assert is_prime(0) is False
```

执行结果如下:

```
    def test_is_zero_not_prime(self):
        """测试边界值 0 是不是质数"""
>       assert is_prime(0) is False
E       assert True is False
E        +  where True = is_prime(0)

method_examples/test_is_prime.py:23: AssertionError
```

7.5 单元测试一个类

现在面向对象的思想已经深入人心,类也是大工程项目的基石。针对类的测试通过了,就能确信对类所做的改进没有意外地破坏其原有的行为。

类的测试与函数的测试相似,所做的大部分工作是测试类中方法的行为,但存在一些不同之处,下面来编写一段代码对类进行测试。下面来看一个在帮助管理中匿名调查问卷的类。

开发的代码如下:

```python
#File: survey.py


class AnonymousSurvey():
    """收集匿名调查问卷的答案"""

    def __init__(self, question):
        """存储一个问题,并为存储答案做准备"""
        self.question = question
        self.responses = []

    def show_question(self):
        """显示调查问卷"""
        print(self.question)

    def store_respond(self, new_response):
        """存储单份调查答卷"""
        self.responses.append(new_response)

    def show_result(self):
        """显示收集到的所有答卷"""
        print("Survey results:")
        for response in self.responses:
            print('- ' + response)
```

7.5.1 类的说明

这个类首先存储了一个指定的调查问卷,并创建了一个空列表,用于存储答案。这个类包含打印调查问卷问题的方法、在答案列表汇总添加新答案的方法,以及将存储在列表中的答案都打印出来的方法。

要创建这个类的实例,只需提供一个问题。有了表示调查的实例后,就可以使用 show_question()来显示其中的问题,可使用 store_reponse()来存储答案,并使用 show_result()来显示调查结果。

7.5.2 开发的调用

为证明 AnonymousSurvey 类能够正确地工作,通常开发人员编写一个调用它的程序来简单测试。这种调用仅是调试,而非测试。

代码如下:

```python
#File: girlfriend_survey.py

from survey import AnonymousSurvey

#定义一个问题,并创建一个表示调查的 AnonymousSurvey 对象
question = "你喜欢的女朋友有什么样的气质?"
my_survey = AnonymousSurvey(question)

#显示问题并存储答案
my_survey.show_question()
print("输入'退出',可退出问卷。\n")
while True:
    response = input("气质: ")
    if response == '退出':
        break
    my_survey.store_respond(response)

#显示调查结果
print("\n感谢每位进行问卷调查的人员!")
my_survey.show_result()
```

这个程序定义了一个问题("你喜欢的女朋友有什么样的气质?"),并使用这个问题创建了一个 AnonymousSurvey 对象。接下来,这个程序调用 show_question() 来显示问题,并提示用户输入答案。收到每个答案的同时将其存储起来。用户输入所有答案(输入'退出',可退出问卷)后,调用 show_result() 来打印调查结果:

```
你喜欢的女朋友有什么样的气质?
输入'退出',可退出问卷。

气质: 温柔
气质: 聪明
气质: 美丽
气质: 退出

感谢每位进行问卷调查的人员!
Survey results:
- 温柔
```

```
- 聪明
- 美丽

Process finished with exit code 0
```

7.5.3　类持续开发：功能的增加及修改

AnonymousSurvey 类可用于进行简单的匿名调查。假设将它放在了 survey 中，并想对其进行改进：让每位用户都可输入多个答案。编写一种方法，它只列出不同的答案，并指出每个答案出现了多少次，再编写一个类，用于管理非匿名调查。

进行上述修改存在风险，可能会影响 AnonymousSurvey 类的当前行为。例如，当允许每位用户输入多个答案时，可能会出现不小心写出单个答案的情况。要确认在这个模块没有破坏既有行为，可以编写针对这个类的测试。

这种思路源于测试驱动开发，也就是先根据需求编写测试代码，验证在多种情况下程序都可用。

7.5.4　类的单元测试

我们利用 pytest 来编写一个测试，对 AnonymousSurvey 类的行为进行验证。例如，提供一个答案，多个答案，公用的数据的实例化在初始化方法中实现。

1. 提供一个答案的测试用例编写

如果用户面对调查问题时只提供了一个答案，这个答案也能被存储，使用方法 assert 来核实它是否包含在答案列表中。

代码如下：

```
#File: test_survey.py
#将 survey.py 上级包设置为 Sources   root,在文件夹上右击选择 Mark as Sources root,
#这样使用 pytest 框架执行和使用命令行执行都不会报错
from .survey import AnonymousSurvey


class TestAnonymousSurvey(object):
    """针对 AnonymousSurvey 类的测试"""

    def test_single_response(self):
        """测试单个答案是否会被妥善地存储"""
        question = "你喜欢的女朋友有什么样的气质?"
        my_survey = AnonymousSurvey(question)
        my_survey.store_respond('温柔')

        assert '温柔', my_survey.responses
```

首先导入要测试的类 AnonymousSurvey。将测试用例命名为 TestAnonymousSurvey。第一个测试方法验证调查问题的单个答案被存储后,答案包含在调查列表中。对于这种方法,一个不错的描述性名称是 test_store_single_response()。如果这个测试未通过,就能通过输出的方法名得知,在存储单个调查答案方面存在问题。

要测试类的行为,需要创建实例。最后,使用问题"你喜欢的女朋友有什么样的气质?"创建了一个名为 my_survey 的实例,然后使用方法 store_response()存储了单个答案'温柔'。接下来检查'温柔'是否包含在列表 my_survey.responses 中,以核实这个答案是否被妥善地存储了。

在控制台运行 test_survey 时,如果测试通过了,则只显示一个点:

```
test_survey.py                                              [100%]

======================== 1 passed in 0.04s ========================
.
Process finished with exit code 0
```

2. 提供多个答案的测试用例编写

测试需要考虑多种情况是否正确,只收集一个答案的调查用途不大。下面来核实用户提供 3 个答案的情况,它们也将被妥善地存储。为此,在 TestAnonymousSurvey 中再添加一种方法。

代码如下:

```
#File: test_survey.py
from .survey import AnonymousSurvey


class TestAnonymousSurvey(object):
    """针对 AnonymousSurvey 类的测试"""

    def test_single_response(self):
        """测试单个答案是否会被妥善地存储"""
        question = "你喜欢的女朋友有什么样的气质?"
        my_survey = AnonymousSurvey(question)
        my_survey.store_respond('温柔')

        assert '温柔', my_survey.responses

    def test_store_three_response(self):
        """测试 3 个答案是否会被妥善地存储"""
        question = "你喜欢的女朋友有什么样的气质?"
        my_survey = AnonymousSurvey(question)
        responses = ['温柔', '聪明', '美丽']
```

```
    for response in responses:
        my_survey.store_respond(response)
        # print(my_survey.responses)
        assert(response in my_survey.responses)
```

将这种方法命名为 test_store_responses(),并像 test_store_single_response()一样,在其中创建一个调查对象。定义一个包含 3 个不同答案的列表,再对其中的每个答案都调用 store_reponses()。存储这些答案后,使用断言确认每个答案都包含在 my_survey.response 中。

再次运行 test_survey,两个测试(针对单个答案的测试和针对 3 个答案的测试)都通过了(选择 PyCharm-pytest 执行):

```
collected 2 items

test_survey.py::TestAnonymousSurvey::test_single_response PASSED
test_survey.py::TestAnonymousSurvey::test_store_three_response PASSED

========================= 2 passed in 0.04s =========================
```

3. 使用公共的初始化方法

使用 pytest 及 fixture 功能实现每个测试方法公用一组实例化对象和数据。

代码如下:

```python
# File: test_survey.py
from .survey import AnonymousSurvey
import pytest


class TestAnonymousSurvey(object):
    """针对 AnonymousSurvey 类的测试"""
    """使用 fixture 自动应用的方式加载初始化数据和对象"""

    @pytest.fixture(autouse=True)
    def setUp_object_data(self):
        """
        创建一个调查对象和一组答案,供测试方法使用
        """
        question = "你喜欢的女朋友有什么样的气质?"
        self.my_survey = AnonymousSurvey(question)
        self.responses = ['温柔', '聪明', '美丽']

    def test_store_single_response(self):
        """测试单个答案是否会被妥善地存储"""
        self.my_survey.store_respond(self.responses[0])
```

```
            assert self.responses[0] in self.my_survey.responses

    def test_store_three_responses(self):
        """测试 3 个答案是否会被妥善地存储"""
        for response in self.responses:
            self.my_survey.store_respond(response)
        for response in self.responses:
            assert response in self.my_survey.responses
```

首先导入 pytest,创建 setUp_object_data()方法,方法名可以随意命名,这种方法做了两件事情:创建一个调查对象和创建一个答案列表。存储这两种数据的变量名包含前缀 self(即存储在属性中),因此可在这个类的任何地方使用。在这种方法上加 @pytest.fixture(autouse=True),下面的两个测试方法都共用这种方法中的变量及实例。这让两个测试方法都更简单,因为它们都不用创建调查对象和答案。方法 store_respond()核实 self.responses 中的第 1 个答案和 self.responses 中的全部 3 个答案是否都被妥善地存储。

执行的结果如下:

```
class_examples/test_survey2.py                                      [100%]

========================= 2 passed in 0.04s =========================
..
Process finished with exit code 0
```

这个测试类将在原有的 test_survey.py 文件中进行修改,然后执行测试,以确保 survey.py 编写及执行正确,修改的代码并未影响原来的功能。

7.6 本章小结

本章主要介绍如何使用 pytest 进行单元测试,测试一个函数,以及测试一个类。以下是本章的要点:

(1) 单元测试是成本最低的有效测试。
(2) 单元测试一个函数。
(3) 单元测试一个类。

第 8 章 API 自动化测试实践

8.1 测试微信公众号接口

以微信公众号接口为例,微信提供公开的 API,微信提供的接口文档相当标准。其测试接口网址 https://mp.weixin.qq.com/debug/cgi-bin/sandbox?t＝sandbox/login,可以通过微信扫描二维码登录,即使没有公众号也可以使用。

下面按通用的接口测试流程逐步推进。
- 需求分析(熟悉接口文档并分析);
- 测试用例设计;
- 测试执行(编写代码或使用工具执行);
- 测试报告。

8.1.1 熟悉接口文档以便获取信息

先熟悉接口文档,接口文档重点应熟悉输入参数和输出参数。熟悉接口需求的原则是每句话都要细读,每句都可写出多个测试用例。

登录后可看到接口文档,获取本人的测试号,此测试号是唯一的,如图 8-1 所示。

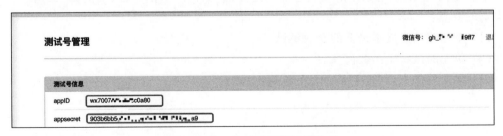

图 8-1 接口文档登录测试号信息

1. 获取本人的测试号信息

此处选择两个接口为大家展示接口 API 测试。一个是用户管理下的用户标签管理中的增、删、改、查 4 个接口,这个体现了接口的主要功能。另一个是消息管理下的群发文字的接口。

所有的接口发送请求都需要先执行 access_token 接口,以便获取它返回的 access_token。

2. 获取 token

进入 API 文档→开始开发→获取 Access token 的接口文档。

```
接口调用请求说明

HTTPS 请求方式：GET
https://api.weixin.qq.com/cgi-bin/token?grant_type=client_credential&appid=APPID&secret=APPSECRET

参数说明

参数            是否必须      说明
grant_type      是           获取 access_token 填写 client_credential
appid           是           第三方用户唯一凭证
secret          是           第三方用户唯一凭证密钥,即 appsecret
返回说明

正常情况下,微信会返回下述 JSON 数据包给公众号：

{"access_token":"ACCESS_TOKEN","expires_in":7200}
参数说明

参数            说明
access_token    获取的凭证
expires_in      凭证有效时间,单位：秒
错误时微信会返回错误码等信息,JSON 数据包示例如下(该示例为 AppID 无效错误)：

{"errcode":40013,"errmsg":"invalid appid"}
返回码说明

返回码          说明
-1             系统繁忙,此时请开发者稍候再试
0              请求成功
40001          AppSecret 错误或者 AppSecret 不属于这个公众号,请开发者确认 AppSecret 的正确性
40002          请确保 grant_type 字段值为 client_credential
```

进入 API 文档→开始开发→用户标签管理→创建标签、获取公众号已创建的标签、编辑标签、删除标签、根据标签进行群发的接口调用请求说明。这个模块包括多个接口,增、删、改、查功能是对数据的基本功能,以这个几个接口为例基本包括 API 的精髓。

```
标签管理

1. 创建标签

一个公众号最多可以创建 100 个标签。
```

接口调用请求说明

HTTP 请求方式：POST(请使用 HTTPS 协议)
https://api.weixin.qq.com/cgi-bin/tags/create?access_token = ACCESS_TOKEN

POST 数据格式：JSON POST 数据示例：

{ "tag" : { "name" : "广东" } }
参数说明

参数	说明
access_token	调用接口凭据
name	标签名(30 个字符以内)

返回说明(正常时返回的 json 数据包示例)

{ "tag":{ "id":134, "name":"广东" } }
返回参数说明

参数	说明
id	标签 id,由微信分配
name	标签名,UTF-8 编码

错误码说明

错误码	说明
-1	系统繁忙
45157	标签名非法,需要注意不能和其他标签重名
45158	标签名长度超过 30 字节
45056	创建的标签数过多,需要注意不能超过 100 个

2. 获取公众号已创建的标签

接口调用请求说明

HTTP 请求方式：GET(请使用 HTTPS 协议)
https://api.weixin.qq.com/cgi-bin/tags/get?access_token = ACCESS_TOKEN

返回说明

{
"tags":[{
    "id":1,
    "name":"每天一罐可乐星人",
    "count":0 //此标签下粉丝数

```
},
{
    "id":2,
    "name":"星标组",
    "count":0
},
{
    "id":127,
    "name":"广东",
    "count":5
}
] }
```

3. 编辑标签

接口调用请求说明

HTTP 请求方式：POST(请使用 HTTPS 协议)
https://api.weixin.qq.com/cgi-bin/tags/update?access_token=ACCESS_TOKEN

POST 数据格式：JSON POST 数据例子：

{ "tag" : { "id" : 134, "name" : "广东人" } }

返回说明

{ "errcode":0, "errmsg":"ok" }

错误码说明

错误码	说明
-1	系统繁忙
45157	标签名非法，需要注意不能和其他标签重名
45158	标签名长度超过 30 字节
45058	不能修改 0/1/2 这 3 个系统默认保留的标签

4. 删除标签

需要注意，当某个标签下的粉丝超过 10 万人时，后台不可直接删除标签。此时，开发者可以对该标签下的 openid 列表，先进行取消标签的操作，直到粉丝数不超过 10 万人后，才可直接删除该标签。

接口调用请求说明

HTTP 请求方式：POST(请使用 HTTPS 协议)
https://api.weixin.qq.com/cgi-bin/tags/delete?access_token=ACCESS_TOKEN

POST 数据格式：JSON POST 数据例子：

{ "tag":{ "id" : 134 } }

返回说明

{ "errcode":0, "errmsg":"ok" }

错误码说明

错误码	说明
-1	系统繁忙
45058	不能修改 0/1/2 这 3 个系统默认保留的标签
45057	该标签下粉丝数超过 10 万人，不允许直接删除

群发文本接口需求文档
5.根据标签进行群发
接口调用请求说明

HTTP 请求方式：POST https://api.weixin.qq.com/cgi-bin/message/mass/sendall?access_token=ACCESS_TOKEN

POST 数据说明

POST 数据示例如下：

文本：

```
{
    "filter":{
        "is_to_all":false,
        "tag_id":2
    },
    "text":{
        "content":"CONTENT"
    },
    "msgtype":"text"
}
```

参数说明

参数	是否必须	说明
filter	是	用于设定图文消息的接收者
is_to_all	否	用于设定是否向全部用户发送，值为 true 或 false,选择 true 则该消息群发给所有用户,选择 false 则可根据 tag_id 发送给指定群组的用户
tag_id	否	群发的标签的 tag_id,参见用户管理中用户分组接口,若 is_to_all 值为 true,则可不填写 tag_id

返回说明

返回数据示例(正确时的 JSON 返回结果)：

```
{
    "errcode":0,
    "errmsg":"send job submission success",
    "msg_id":34182,
    "msg_data_id": 206227730
}
```

参数	说明
errcode	错误码
errmsg	错误信息

8.1.2 接口测试用例设计

下面以创建标签接口为例编写部分接口测试用例。

（1）正常提交，标签名为英文字母，预期结果为创建成功，返回正确响应信息。

（2）创建 100 个标签是否能成功创建，预期结果为创建成功，返回正确响应信息。

（3）创建 101 个标签是否能成功创建，预期结果为创建失败，有返回错误消息。

（4）协议改为 HTTP，其他正确，预期结果为创建成功。

（5）请求方法修改 GET、PUT、DELETE，其他正确，预期结果为创建失败，返回对应的响应信息。

（6）请求地址错误 create 改为其他接口的 get、update 等，预期结果为创建失败，返回对应的响应信息。

（7）access_token 的值不正确，预期结果为创建失败，返回对应的响应信息。

（8）access_token 参数名不正确，预期结果为创建失败，返回对应的响应信息。

（9）access_token 参数不提交，预期结果为创建失败，返回对应的响应信息。

（10）POST 提交的内容是 json 类型，改为其他格式，预期结果未知，返回对应的响应信息（参考文档中全局状态码）。

（11）提交的内容为不完整的 json 类型，预期结果为创建失败。

（12）标签名为汉字与其他字母数据组合，其他正确，预期结果为创建成功，返回正确响应信息。

（13）标签名改为 31 字符，预期结果为创建失败，有返回错误消息。

（14）标签名为空，预期结果为创建失败，有返回错误消息。

（15）标签名重复，预期结果为创建失败，有返回错误消息。

（16）标签名为"' or 0＝0 --"，安全测试用例一，预期结果未知。

（17）多条安全测试用例：略。

8.2 执行测试

使用编写的代码发请求调用接口，输入数据，验证预期结果与实际结果是否一致。

一般接口层我们习惯叫服务层。服务层的测试是介于单元测试和 UI 测试中间的测试，每个层次测试都有其重点，单元测试重在代码逻辑，服务接口层测试重在传递，UI 层重点是质量特性及业务逻辑。

本书不是专门介绍测试的，但使用 pytest 框架编写测试代码时要考虑上述情况，所以接口服务层的测试一般不关联，只要发请求验证响应正确就可以，但有时公司为在 UI 建完之前测试服务器端的业务逻辑的正确性，需要在服务接口层将几个接口关联测试，以达到测试 UI 业务逻辑的目的。通常我们会把相关接口存放在一个文件夹中，一个功能存放在一个文件夹中。一个接口可以是一个文件，也可以将相关接口测试方法都写在一个文件中。具体根据项目规模、数据相关性、响应关联性、依从性来确定。下面是具体步骤，细节大家可以自己体会。

在终端或 cmd 中导入 pip install requests，在 PyCharm 建立工程和文件。

8.2.1 使用 get()、post() 方法发送请求，返回响应

使用 get() 方法编写获取 access_token 接口返回响应，使用 post() 方法编写 create tag 接口方法。编写上述 8.1.2 节测试用例 1 和 7 的测试方法，并添加断言。

access_token 接口使用 requests 中的 get() 方法，参数是 URL，URL 的组成包含用户的信息，appid 和 secret 使用自己的个人信息。返回的响应进行了 3 个层次的断言。由于断言遇到失败就停止，因此第一层断言协议层，也就是验证 HTTP 协议的响应状态码是否正确。第二层验证功能业务数据是否正确。第三层验证性能方面，也就是响应时间是否满足需求。access_token 接口返回 access_token 的值。

在下面建立两个测试用例的例子，一个是正确的用例，测试方法名 test_createtag，使用 requests 中的 post() 方法，参数是这个接口的 URL 再加上 access_token 接口返回的数据和 data 数据，这个数据要求是 json 参数发送，再验证数据是否正确。另一个错误的用例选择测试用例 7，错误的 access_token。验证预期结果是当出现错误的 access_token 时返回的状态码。

代码如下：

```
#File: test_create_tag.py
import requests
import pytest

def access_token():
    URL = 'https://api.weixin.qq.com/cgi-bin/token?' \
          'grant_type=client_credential&appid=wx7002cc0a80&secret=903b6b342003a5a9'
    #发送 get 请求
    rep = requests.get(URL)
    #print(rep.text)
    #断言响应状态码为 200,协议层断言
    assert 200 == rep.status_code
    rep_json = rep.json()
    #断言返回具体业务信息,数据业务层断言
    assert 7200 == rep_json['expires_in']
```

```python
    #断言请示响应总时间小于 3s,性能断言
    assert rep.elapsed.total_seconds()< 3

    return rep_json['access_token']


#正确地创建标签用例 1
def test_createtag():
    URL = 'https://api.weixin.qq.com/cgi-bin/tags/create?access_token = ' + access_token()
    data = {"tag": {"name": "000"}}
    rep = requests.post(URL = URL, json = data)
    assert 200 == rep.status_code
    json_rep = rep.json()
    assert '000' == json_rep['tag']['name']


#创建错误 token 用例 7
def test_createtag_F_errortoken():
    URL = 'https://api.weixin.qq.com/cgi-bin/tags/create?access_token = 1111'
    rep = requests.get(URL)
    assert 200 == rep.status_code
    #编写断言,40001 的作用可查看接口文档中全局状态码一章
    assert 40001 == rep.json()['errcode']


if __name__ == '__main__':
    pytest.main(['-s','test_create_tag.py'])
```

8.2.2 使用 conftest 共享数据

确定共享数据及共享的级别,如果大型项目要在不同层级文件夹中建立共享数据,则可以通过 conftest 在一个文件夹下共享数据。在所在包下创建 conftest.py 文件,把 access_token 接口调用返回写在这里,方法名为 get_access_token(),该接口依赖注入 fixture,范围 scope 为 session 级别,也就是整个会话只调用一次这个接口,将接口返回值写入内存,其他方法直接读取。

代码如下:

```python
#File: conftest.py
import pytest
import requests


@pytest.fixture(scope = 'session')
```

```python
def get_access_token():
    URL = 'https://api.weixin.qq.com/cgi-bin/token?' \
          'grant_type=client_credential&appid=wx7007dc5642c80&secret=903b6bb50b6e22db74342003a5a9'
    rep = requests.get(URL)
    return rep.json()['access_token']
```

8.2.3 读取 yaml 数据文件进行 parametrize

对于测试用例和预期结果可以通过文件传入，实现数据与代码分离，我们以建立 tag 接口为例，添加不同标签 tag，写在 yaml、csv 或 txt 文件中，本次以 yaml 格式为例，显示如下：

```
- beautiful
- linda
- 995511
- 77 <>.;,005511
```

创建标签接口实现参数化，使用 yaml.safe_load() 方法安全加载 yaml 文件，打开文件使用 open() 方法。使用显式调用 fixture，也就是在 conftest 中写的名为 get_access_token() 的方法，通过参数传入，并使用其返回值组成创建标签接口的 URL。添加协议层断言和业务层断言。

代码如下：

```python
#File: test_tag_method.py

import pytest
import requests
import yaml

#使用参数化方式执行多次
@pytest.mark.parametrize('value',yaml.safe_load(open(yaml_file, 'r', encoding="UTF-8")))
def test_create_tag(get_access_token,value):
    URL = 'https://api.weixin.qq.com/cgi-bin/tags/create?access_token=' \
          + get_access_token

    data = {"tag": {"name":value}}
    rep = requests.post(URL=URL, json=data)
    assert rep.status_code == 200
    assert rep.json()['tag']['name'] == value
```

8.2.4 关联接口数据传递及更新删除接口测试

在上面的代码中加上更新和删除接口测试方法，从获取数据 get 接口中获取要更新和删除的数据中的 id 值,获取的方式是获得最后加入的两个 id 值,并保存到列表中。使用 pytest.fixture() 显式调用 get_tag 传递数据。

代码如下:

```python
# File: test_tag_method.py

import pytest
import requests
import yaml


# 使用参数化方式执行多次
@pytest.mark.parametrize('value', yaml.safe_load(open(yaml_file, 'r', encoding="UTF-8")))
def test_create_tag(get_access_token, value):
    URL = 'https://api.weixin.qq.com/cgi-bin/tags/create?access_token=' \
          + get_access_token

    data = {"tag": {"name": value}}
    rep = requests.post(URL=URL, json=data)
    assert rep.status_code == 200
    assert rep.json()['tag']['name'] == value


@pytest.fixture()
def get_tag(get_access_token):
    URL = 'https://api.weixin.qq.com/cgi-bin/tags/get?access_token=' \
          + get_access_token
    rep = requests.get(URL=URL)
    # 从返回中获取最后 1 个 id 号
    ids = rep.json()['tags'][-1]['id']
    # 获取倒数第 2 个 id
    ids1 = rep.json()['tags'][-2]['id']
    ides = []
    ides.append(ids)
    ides.append(ids1)
    assert rep.status_code == 200
    assert 'tags' in rep.text
    return ides

def test_update_tag(get_access_token, get_tag):
    URL = 'https://api.weixin.qq.com/cgi-bin/tags/update?access_token=' \
          + get_access_token
    data = {"tag": {"id": get_tag[0], "name": "只觉388"}}
    data1 = {"tag": {"id": get_tag[1], "name": "zenm338"}}
    rep = requests.post(URL=URL, json=data)
```

```python
        assert rep.status_code == 200
        assert rep.json()['errcode'] == 0
        rep1 = requests.post(URL=URL, json=data1)
        assert rep1.status_code == 200
        assert rep1.json()['errcode'] == 0


def test_udelete_tag(get_access_token, get_tag):
    URL = 'https://api.weixin.qq.com/cgi-bin/tags/delete?access_token=' \
          + get_access_token
    print(get_tag)
    data = {"tag": {"id": get_tag[0]}}
    data1 = {"tag": {"id": get_tag[1]}}
    rep = requests.post(URL=URL, json=data)
    assert rep.status_code == 200
    assert rep.json()['errcode'] == 0
    rep1 = requests.post(URL=URL, json=data1)
    assert rep1.status_code == 200
    assert rep1.json()['errcode'] == 0


if __name__ == '__main__':
    pytest.main()
```

8.2.5 fixture 的依赖接口需要测试，也需要参数化

本次实例是一个综合实例，将微信公众号接口统一实现。使用的技术及解决的问题如下：

（1）将 access_token 接口的调用放在 conftest.py 文件中，并使用 @pytest.fixture（scope='session'），让其在全局中有效，所有的测试开发工程师都可使用其返回的 access_token。

（2）使用 get 接口返回的 id 数据动态实现修改 update 和删除 delete 接口所需要的 id 数据。get 接口作为 fixture 依赖注入，但同时也要测试 get 接口，此外还需加上断言和实现参数化。这样使用 fixture 和 parametrize 会有冲突，因此使用 fixture 中的参数进行参数化数据驱动。

（3）测试用例最后使用 yaml 文件编写规定格式的数据，本例只写了一个流程正确的用例。其他用例大家可以尝试编写代码，可按 8.1.2 节中的用例编写。

（4）这里并未使用封装，也未做过度的封装，这样便于理解测试用例执行的精髓。大家可根据自己公司的具体情况，使用 requests.request 封装。

图 8-2 所示是整个实现的文件结构。

实现的步骤如下：

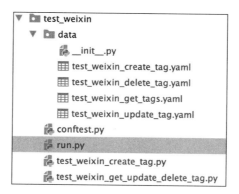

图 8-2　文件结构

（1）创建 conftest.py 文件，用于共享数据。

所有接口都使用 access_token 接口的返回值。

代码如下：

```python
# File: conftest.py
import pytest
import requests
# scope = session,整个 session 使用一个 fixture、token
# scope = function,每个函数使用一个 fixture、token
# scope class,整个类使用一个 fixture、token
@pytest.fixture(scope = 'session')
def get_access_token1():
    get_URL = 'https://api.weixin.qq.com/cgi-bin/token'
    params = {'grant_type': 'client_credential',
    # 需修改为自己的 appid 和 secret
              'appid': 'wxd8e5ab62d902',
              'secret': '2dbfc49b5b001646bee7cba28'
              }
    res = requests.get(URL = get_URL, params = params)
    res_json = res.json()
    return res_json['access_token']
```

（2）创建用于创建标签的数据文件 Test_weixin_create_tag.yaml。

所有动态数据都可以写在这里，读入数据后是一个列表中嵌套字典类型。

```yaml
- URL: https://api.weixin.qq.com/cgi-bin/tags/create?access_token=
  data:
    tag:
      name: linda780099
  result:
    name: linda780099
```

(3) 创建 test_weixin_create_tag.py 文件。

使用参数化读(2)所创建的创建标签 yaml 文件。

```
# File: test_weixin_create_tag.py
import requests
import pytest
import yaml
import allure


@pytest.mark.parametrize('create', yaml.safe_load(open('data/test_weixin_create_tag.yaml',
'r', encoding = 'utf8')))
def test_create_tag(get_access_token1, create):
    URL = create['URL'] + get_access_token1
    json_data = create['data']
    result = create['result']['name']
    res = requests.post(URL, json = json_data)
    assert res.status_code == 200
    assert res.json()['tag']['name'] == result
```

(4) 创建获得标签接口的数据文件 test_weixin_get_tags.yaml。

```
- URL: https://api.weixin.qq.com/cgi-bin/tags/get?access_token =
  result:
    name: 星标组
```

(5) 创建更新接口数据文件 test_weixin_update_tag.yaml。

```
- URL: https://api.weixin.qq.com/cgi-bin/tags/update?access_token =
  data:
    tag:
      id: 102
      name: linda102
  result:
    errcode: 0
```

(6) 创建删除接口数据文件 test_weixin_delete_tag.yaml。

```
- URL: https://api.weixin.qq.com/cgi-bin/tags/delete?access_token =
  data:
    tag:
      id: 101
  result:
    errcode: 0
    errmsg: ok
```

（7）创建 test_weixin_update_tag.py。

把 get 接口作为 fixture，提取最后一个添加的 id 值，使用参数化实现更新接口和删除接口，在调用前更新测试数据中 id 值，代码如下：

```python
#File: test_weixin_update_tag.py
import requests
import pytest
import yaml
import allure

@pytest.fixture(scope = 'session', params = yaml.safe_load(open('data/test_weixin_get_tags.yaml','r',encoding = 'utf8')))
def test_get_tags(get_access_token1,request):
    get_URL = request.param['URL'] + get_access_token1
    name = request.param['result']['name']
    res = requests.get(get_URL)
    res_json = res.json()
    assert res.status_code == 200
    assert name  ==  res_json['tags'][0]['name']
    return res_json['tags'][-1]['id']

@pytest.mark.parametrize('update',yaml.safe_load(open('data/test_weixin_update_tag.yaml','r',encoding = 'utf8')))
def test_update_tag(get_access_token1,test_get_tags,update):
    URL = update['URL'] + get_access_token1
    data = update['data']
    code = update['result']['errcode']
    data['tag']['id'] = test_get_tags
    res = requests.post(URL,json = data)
    assert res.status_code == 200
    assert res.json()['errcode'] == code


@pytest.mark.parametrize('delete',
          yaml.safe_load(open('data/test_weixin_delete_tag.yaml','r',encoding = 'utf8')))
def test_delete_tag(get_access_token1,delete,test_get_tags):
    URL = delete['URL'] + get_access_token1
    json_data = delete['data']
    #使用从 get 接口中获得的 id 号替换从数据文件中获得的 id
    json_data['tag']['id'] = test_get_tags
    result = delete['result']
    res = requests.post(URL,json = json_data)
    assert res.status_code == 200
    assert res.json() == result
```

(8) 创建 run.py，主要用于编写执行各类测试。

这个 run.py 可以是多个文件，用于执行不同的测试用例，可以把 p0 测试需要的测试用例写在一个文件中，把正确的用例写在一个文件中做验收测试用例，可以把错误和异常用例放在一个文件中作为稳健性测试用例，以下只是举例说明。

代码如下：

```
# File: run.py

import pytest
if __name__ == '__main__':
    pytest.main(['-s','test_weixin_create_tag.py'])
    pytest.main(['-s','test_weixin_get_update_delete_tag.py'])
    pytest.main(['-s','test_weixin_get_update_delete_tag.py::test_delete_tag'])
```

(9) 执行的结果如下，因为没有打印输出，所以看到点号表示通过。

```
/usr/local/bin/python3.6
/Users//PyCharmProjects/pytest_book/src/chapter4/test_weixin/run.py
=============== test session starts =======================
platform darwin -- Python 3.6.8, pytest-6.0.1, py-1.9.0, pluggy-0.13.1
sensitiveURL: .*
rootdir: /Users//PyCharmProjects/pytest_book/src/chapter4, configfile: pytest.ini
plugins: rerunfailures-9.0, allure-pytest-2.8.18, cov-2.10.0, xdist-1.34.0, metadata-1.8.0, bdd-3.2.1
collected 1 item

test_weixin_create_tag.py .

=============== 1 passed in 0.67s ================
=============== test session starts ==============
platform darwin -- Python 3.6.8, pytest-6.0.1, py-1.9.0, pluggy-0.13.1
sensitiveURL: .*
rootdir: /Users//PyCharmProjects/pytest_book/src/chapter4, configfile: pytest.ini
plugins: rerunfailures-9.0, xdist-1.34.0, ordering-0.6, assume-2.3.3, selenium-1.17.0, bdd-3.2.1
collected 2 items

test_weixin_get_update_delete_tag.py ..

================ 2 passed in 1.30s =====================
=========== test session starts ====================
platform darwin -- Python 3.6.8, pytest-6.0.1, py-1.9.0, pluggy-0.13.1
sensitiveURL: .*
rootdir: /Users//PyCharmProjects/pytest_book/src/chapter4, configfile: pytest.ini
```

```
plugins: rerunfailures-9.0, forked-1.0.2, pep8-1.0.6, apiritif-0.9.4, variables-1.9.
0, emoji-0.2.0, flakes-4.0.0, base-URL-1.4.2, allure-pytest-2.8.18, cov-2.10.0,
html-2.1.1, xdist-1.34.0, ordering-0.6, assume-2.3.3, selenium-1.17.0, freezegun-
0.4.1, metadata-1.8.0, bdd-3.2.1
collected 1 item

test_weixin_get_update_delete_tag.py .

==================== 1 passed in 0.96s ==================

Process finished with exit code 0
```

8.3 使用 Allure 定制报告

在 8.2.3 节和 8.2.4 节的例子基础上进行修改，定制需要的报告，在 test_create_tag.py 的文件中进行修改。

代码如下：

```python
#File: test_create_tag.py
import requests
import pytest
import allure

@allure.feature("获得 token")
@allure.step("获得所有接口的入门证书 token")
def access_token():
    URL = 'https://api.weixin.qq.com/cgi-bin/token?' \
          'grant_type=client_credential&appid=wx7007d2cc0a80&secret=903b6bb501713b6e22d5a9'

    allure.attach("发送 get 请求")
    rep = requests.get(URL)
    allure.attach("断言响应状态码为 200")
    assert 200 == rep.status_code
    rep_json = rep.json()
    allure.attach("断言返回具体业务信息")
    assert 7200 == rep_json['expires_in']
    allure.attach("断言请求响应全部时间小于 3s")
    assert rep.elapsed.total_seconds() < 3
    return rep_json['access_token']


# 正确地创建标签用例 1
@allure.feature("正确地创建标签")
@allure.story("全正确创建")
@allure.testcase("用例 1")
```

```python
def test_createtag(get_access_token):
    URL = 'https://api.weixin.qq.com/cgi-bin/tags/create?access_token=' + get_access_token
    allure.attach("数据准备")
    data = {"tag": {"name": "linda8090"}}
    allure.attach("发送 post 请求")
    rep = requests.post(URL=URL, json=data)
    allure.attach("断言响应状态码为 200")
    assert 200 == rep.status_code
    json_rep = rep.json()
    allure.attach("断言返回具体业务信息")
    assert 'linda8090' == json_rep['tag']['name']


#创建错误 token 用例 7
@allure.feature("错误地创建标签")
@allure.story("错误 token")
@allure.testcase("http://jira.com/weixin/tag/testcase/7","用例 7")
def test_createtag_F_errortoken():
    URL = 'https://api.weixin.qq.com/cgi-bin/tags/create?access_token=1111'
    allure.attach("发送 get 请求")
    rep = requests.get(URL)
    allure.attach("断言响应状态码为 200")
    assert 200 == rep.status_code
    allure.attach("断言返回具体业务码")
    assert 40001 == rep.json()['errcode']


if __name__ == '__main__':
    pytest.main(['-s','test_create_tag.py'])
```

执行结果如图 8-3 和图 8-4 所示。

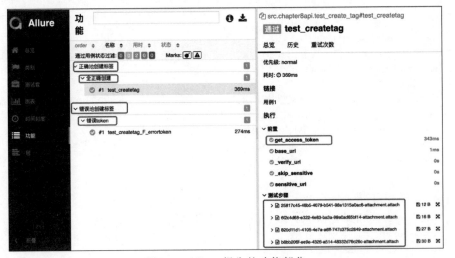

图 8-3　Allure 报告的功能部分

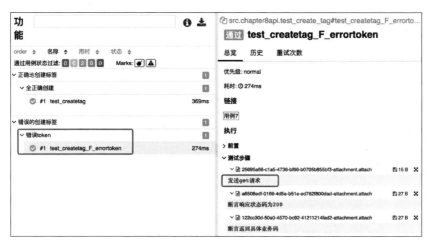

图 8-4　Allure 报告中附件

8.4　使用 pytest 进行各种执行

可以通过主函数自定义执行不同的测试用例。例如，p0 测试可以将所有测试用例中正确的用例都添加到 main()函数中。健壮性测试也可以单独建立一个文件，在此文件中添加负面的测试用例和异常的用例。

代码如下：

```
#File: run_p0.py
import pytest

if __name__ == '__main__':
    #p0 测试可以将所有测试用例中正确的用例执行一遍,可以添加多个文件
    pytest.main(['-s','test_create_tag.py','--allure-features','正确地创建标签'])
    #健壮性测试添加负面的测试用例和异常的用例.
    #pytest.main(['-s','test_create_tag.py','--allure-features','错误地创建标签'])
```

8.5　本章小结

这一章主要讲解 pytest 在 API 测试中的应用。
（1）了解接口的文档和信息。
（2）设计测试用例，包括功能、接口、安全、性能方面。
（3）编写单接口测试用例，添加协议层、业务层、性能方面断言。
（4）编写关联接口测试用例。
（5）使用 fixture 和 parametrize 实现数据传递（关联接口）和数据驱动。
（6）使用 Allure 定制报告。

第 9 章 Web 自动化测试持续集成实践

9.1 Web 自动化测试及持续集成源起

2010 年以来随着移动互联网的兴盛,带动了 IT 行业开发周期的缩短,需求的快速变化带动了版本的快速迭代。在整个项目产品流程中最大瓶颈就是测试自动化和环境部署自动化的速度和效率严重滞后。2018 年以来由于 Docker 等技术的出现形成了 DevOps 模型的逐步落地。各层次的自动化测试也是企业必须解决的重点问题。

pytest 框架的优势是整合各层次自动化,学习成本相对较低,Python 语言的开放和 pytest 的第三方插件及它与 Jenkins 持续集成平台的整合都为整个 DevOps 方案的落地提供了技术支撑。

下面以一个 Web 项目为例,使用 pytest 将 Web 自动化测试框架的整合编写过程及在 jenkins 中执行的过程展示给大家。我们从一个测试人员的角度进行编写,有些技术选择由于笔者经验限制或历史原因并未说明,框架整合只是雏形,希望起到抛砖引玉的作用,并且由于公司的项目需要保密,我们只能以一个公共系统为例,即以一个模块为例,以小见大,望能给大家探索一条路径。

9.2 被测试系统的安装和介绍

9.2.1 人力资源管理系统安装

Apache 公司提供 xampp 集成部署环境及插件系统。具体网址:https://www.apachefriends.org/download.html。可以下载需要的 xampp/lampp 等,按照说明安装在计算机中。安装简单说明的网址:https://www.apachefriends.org/bitnami_for_xampp.html♯。

在 https://bitnami.com/stack/xampp♯orangehrm 中下载人力资源管理系统。如图 9-1 所示,安装方法很简单,在 Linux 中修改权限直接运行,设置管理员密码便可完成安装。在 Windows 下可以直接安装即可。

在 Mac 系统的远程云服务器安装所需环境,简单的安装步骤仅供参考。

第9章 Web自动化测试持续集成实践

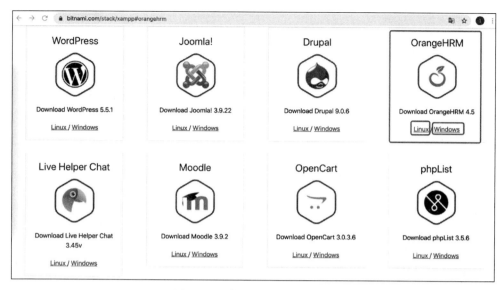

图 9-1 安装人力资源管理系统

代码如下：

```
#1.将下载的文件上传到云服务器/opt
scp 下载要上传到服务器的安装文件路径及文件名  root@云服务 IP 地址:/opt
#2.更改安装程序的权限
chmod 755 xampp-Linux-*-installer.run
#3.运行安装程序
./xampp-Linux-*-installer.run
#4.启动 lampp 服务(stop 停止、restart 重启服务)
/opt/lampp/lampp start
#5.上传下载的 orangehrm 系统文件
#6.更改权限后运行
./bitnami-orangehrm-4.4-0-module-Linux-x64-installer.run
#7.安装过程
----------------------------------------------------------------
Welcome to the Bitnami OrangeHrm Module Setup Wizard.

----------------------------------------------------------------
Installation folder

Please choose a folder that contains an installation of Bitnami or XAMPP.

Select a folder [/opt/lampp]:

Note: This module requires a pre-existing installation of Bitnami or a
Bitnami-compatible stack like XAMPP. Please select the previous platform
installation. For example: /opt/bitnami or /opt/lampp
```

```
Create Admin account
#密码输入时不显示,并且有要求
Enter the application password :
Retype password :
Do you want to configure mail support? [y/N]: N        #选择N可以不配置邮件

------------------------------------------------------------

Setup is now ready to begin installing Bitnami OrangeHrm Module on your
computer.

Do you want to continue? [Y/n]: Y

------------------------------------------------------------

Please wait while Setup installs Bitnami OrangeHrm Module on your computer.

  Installing
  0%             50%            100%
  #########################################

------------------------------------------------------------

Setup has finished installing Bitnami OrangeHrm Module on your computer.

Launch Bitnami OrangeHrm Module [Y/n]: Y
```

9.2.2 人力资源管理系统介绍

访问 http://云服务器 IP 地址/orangehrm/，进入首页。使用安装时设置的 admin 所对应的密码进入系统，在管理员→设置→本地化中将编辑语言设置为汉语，整个网站本地化语言为汉语，如图 9-2 所示。

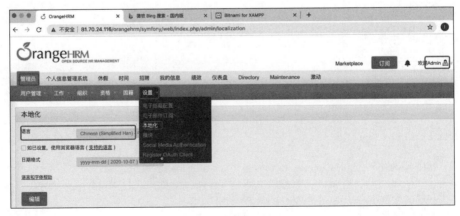

图 9-2　人力资源管理系统

很多系统功能需要添加基础配置后才会有后续的功能,大家可以在系统设置中进行控制。我们以这个系统为 Web 测试的示例项目,大家安装后可以自行熟悉。

9.3 Web 项目自动化原理及 Web 测试框架

9.3.1 自动化测试要达到的目标和涉及的技术

- 脚本生成和维护容易
  - ◆ 应对复杂变化的前端
  - ◆ 应对需求经常变化
  - ◆ 降低脚本对实现技术和语言的依赖
- 数据(输入数据及预期结果)代码分离
  - ◆ 应对测试数据及数据多样性变化
- 细粒度多层次断言
  - ◆ 页面级、控件级断言
  - ◆ 需求及业务逻辑断言
  - ◆ 自定义断言
- 不同浏览器的兼容性共用一脚本
  - ◆ 应对环境的变化
- 公司层面的持续集成及交付

要达到上述目标,需要的技术:脚本生成可以使用工具录制 Katalon Recorder(Selenium Tests),为了维护容易可以使用 POM(Page Object Module)将页面与测试分离,操作浏览器及浏览器上的元素可以使用 Selenium 框架,这个框架降低脚本对技术和语言的依赖,测试自动化运行管理和断言管理及数据参数化技术等使用 pytest 框架,报告使用 Allure 框架,持续集成使用 Jenkins 工具。

本书主要介绍 pytest,所以重点讲解 pytest 在 Web 自动化测试框架中的应用。Selenium 的技术并未从零开始进行,这方面不懂的地方可以自行学习。

9.3.2 Web 自动化测试框架 Selenium 介绍

Selenium 是一个涵盖了一系列工具和库的总体项目,这些工具和库支持 Web 浏览器的自动化,并且支持多平台、多浏览器,以及多语言。

Selenium 的核心是 WebDriver,它是一个编写命令集的接口,可以在许多浏览器中互换运行。

WebDriver 规范定义一组与平台、语言无关的接口,包括发现和操作页面上的元素及控制浏览器行为,主要用于支持 Web 应用的自动化测试。WebDriver 的核心通过 findElement 方法返回 DOM 对象(WebElement),通过 WebElement 可以对 DOM 对象进行操作(获取属性、触发事件等)。其中 findElement 方法需要的元素定位器(Locator)支持 ID、XPath、

CSS、超链接文本等多种方式。

简单来讲，Selenium 框架使用 WebDriver 协议调用浏览器的驱动程序 ChromeDriver 启动浏览器 Browser，并对浏览器页面元素进行操作，如图 9-3 所示。

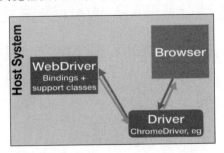

图 9-3　WebDriver 与浏览器关系

9.3.3　Selenium 框架技术简述

1. 浏览器操作

对浏览器进行操作，如打开浏览器、进入具体网页、设置等待时间、页面刷新、前进后退、设置窗口大小、截屏、退出网页等。

2. 定位页面元素技术

单个元素的定位包括 id、name、link_text（页面超链接文字）、class（类名）、tag_name（标签名）、partialLinkText（部分超链接文字）、cssSelector（样式表选择器）、xpath（路径位置）。使用 findElement_by_XX("具体值")方法来定位元素。多个元素的定位使用 findElements_by_XX("具体值")方法来定位元素，返回多个值后通过索引或循环获得具体需要的值。

3. 操作元素技术

对元素进行单击、输入、清除、提交、获得文本、获得属性值等常用操作。还有判断元素是否显示、元素是否被选择、元素是否被启用（按钮是否能被单击）等操作判断元素状态。此外还有处理下拉菜单、弹框、模拟鼠标操作等。

4. 执行中等待

执行中等待分为显式等待和隐式等待两种等待形式。这是 Web 自动化测试所需的一种技术，需要等待元素出现才能进行操作。

9.4　整合 Web 自动化测试框架

9.4.1　自动化测试准备

如果读者是领导、架构师或自己负责开发，则应遵循自动化测试要达到的目标设计自己的框架。为了更好讲解 pytest 在 Web 自动化测试的应用，以测试项目中一个登录模块和一个职称模块为例编写自动化测试代码，以此熟悉技术及流程。

(1) 使用 Katalon Recorder 录制脚本并导出。

这步可以略过,一般初学或元素不好定位时可参考此方法。

(2) 下载浏览器驱动程序(使用 ChromeDriver)。

谷歌浏览器驱动程序 ChromeDriver 下载网址 http://npm.taobao.org/mirrors/Chromedriver/,选择与自己 Chrome 浏览器版本相对应并适合自己的操作系统的版本。解压后放到工程中,如图 9-4 和图 9-5 所示。

图 9-4 ChromeDriver 的下载与 Chrome 版本匹配

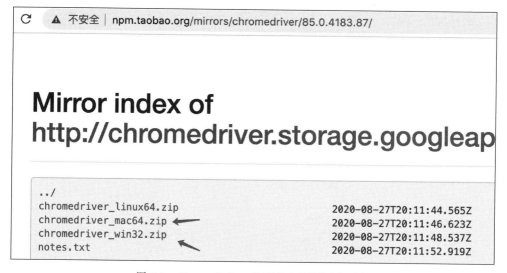

图 9-5 ChromeDriver 的下载与不同平台匹配

(3) 安装 Selenium。

在控制台、终端或 cmd 下输入 pip install selenium 命令,安装 Selenium。目前 Selenium 3 的最高版本是 3.141。

（4）在 PyCharm 中建立工程并编写测试系统的登录脚本。

大家仔细品味代码，未解释明白的代码需要多看注释。业务逻辑参见注释中的顺序。

代码如下：

```python
#File: baseoperation.py

#在 selenium 中导入 webdriver
from selenium import webdriver
import time

#1.打开浏览器(建立一个浏览器驱动程序),每个浏览器有不同驱动程序,驱动程序由浏览器厂商
#提供,这个驱动程序版本要跟浏览器版本配套(有方法可实现自动下载驱动程序)
driver = webdriver.Chrome(executable_path = '../driver/Chromedriver')

#2.进入被测试系统,输入网址并打开网页
driver.get("http://81.70.24.116/orangehrm/symfony/web/index.php/auth/login")

#等待2s
time.sleep(2)

#3.输入用户名及密码,(定位 id、name、xpath、css 等 8 种操作.sendkey 输入,click 单击)
driver.find_element_by_name("txtUsername").send_keys("admin")
driver.find_element_by_id("txtPassword").send_keys("root1234")

#4.单击登录
driver.find_element_by_id("btnLogin").click()

#5.断言登录成功

time.sleep(2)
assert '欢迎 Admin' == driver.find_element_by_id("welcome").text
#assert 'Admin' in driver.find_element_by_id("welcome").text
time.sleep(2)
#6.关闭浏览器
driver.close()
```

（5）确定浏览器驱动的相对位置。

executable_path='../driver/Chromedriver' 中的路径可以是 ChromeDriver 所在的绝对路径。

（6）对操作的元素定位。

元素的定位可以通过查看 Chrome-devtools 中元素的属性值来定位。可对所选择的元素右击"检查"进行查看,如图 9-6 所示。

Chrome 提供定位技术参考,可以对选中的元素右击→Copy→选择合适的定位方式的值,如图 9-7 所示。

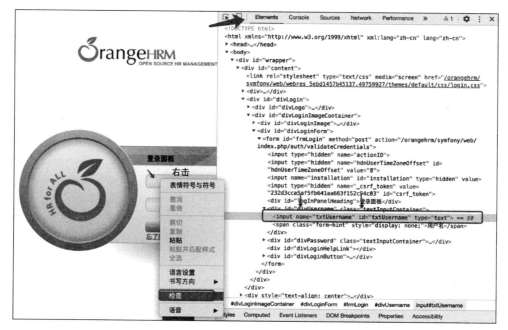

图 9-6　Chrome 页面元素属性

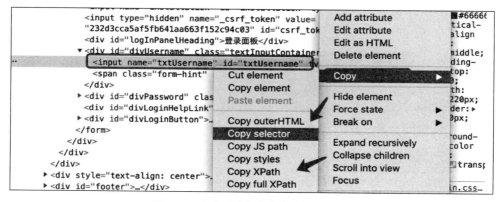

图 9-7　对页面元素定位方式进行复制

（7）使用 pytest 运行脚本以便查看结果。

```
pytest baseoperation.py
========================= test session starts =========================
platform darwin -- Python 3.6.8, pytest-6.0.1, py-1.9.0, pluggy-0.13.1
sensitiveURL: . *
rootdir: /Users/PyCharmProjects/pytest_book
plugins: rerunfailures-9.0, allure-pytest-2.8.18, assume-1.2.2, cov-2.10.0, html-2.1.1,
xdist-1.34.0, ordering-0.6, selenium-1.17.0, freezegun-0.4.1, metadata-1.8.0, bdd-3.2.1
```

```
collected 0 items

====================== no tests ran in 18.12s =========
```

由于编写的代码的文件名要以 test-开头,所以此次执行后会显示 no tests ran。修改文件名再次执行,通过以上步骤便可以熟悉自动化过程和技术。

9.4.2　创建工程目录

在 PyCharm 中创建一个工程(project),建立相关的文件夹(保存非代码)或包(保存代码文件),具体如图 9-8 所示。

(1) config:放置配置文件。
(2) datafile:放置数据文件。
(3) driver:放置浏览器驱动。
(4) log:执行的日志保存在这里。
(5) pages:页面代码。
(6) shootpicture:截图。
(7) test:测试代码。
(8) utils:工具代码。

图 9-8　工程目录

9.4.3　页面元素定位

使用 POM 技术将页面代码与测试代码分离,如何实现呢?在 pages 文件夹中建立每个页面的元素定位。

登录页的元素属性信息如图 9-9 所示。

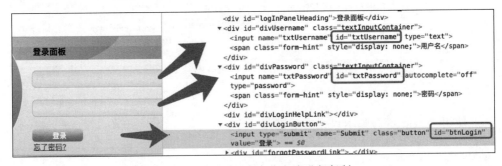

图 9-9　对页面元素定位方式进行复制

1. 编写登录页的元素定位

使用元组结构定义元素位置,定位方式为通过 id 定位。id 值来自图 9-9 页面源码中。登录页中元素包括用户名、密码这两个文本框和登录按钮,共 3 个元素。

代码如下:

```python
#File: login_locators.py

#登录页元素及定位方式
from selenium.webdriver.common.by import By

class LoginPageLocators(object):
    #登录用户名的元素定位,采用元组形式,通过id定位
    username_text = (By.ID, "txtUsername")
    password_text = (By.ID, "txtPassword")
    login_submit  = (By.ID, "btnLogin")
```

进入系统的首页元素信息,包括页面导航及欢迎等元素定位,查看信息如图9-10所示。

图 9-10 admin 页面元素定位

2. 编写首页元素定位

首页元素很多,可以先编写用到的元素,公共的元素可以是单独文件,也可以采用继承的方式。本次用到的是登录后要验证登录成功的元素,如图 9-10 右上角所示的元素,还有 admin 模块元素、工作模块元素、职称模块元素。如图 9-11 所示可以根据具体值和定位方式编写代码。

代码如下:

```python
#File: login_locators.py
#主页元素及定位方式
from selenium.webdriver.common.by import By

class MainPageLocators(object):
    welcome_link = (By.ID,"welcome")
    menu_admin_viewAdminModule_btn = (By.ID, "menu_admin_viewAdminModule")
    menu_admin_Job = (By.ID, "menu_admin_Job")
    menu_admin_viewJobTitleList = (By.ID, "menu_admin_viewJobTitleList")
```

职称页面添加职称功能的元素定位信息,如图 9-11 所示。

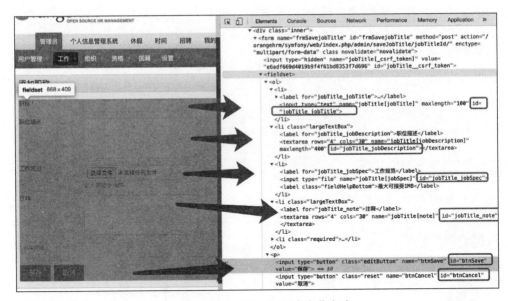

图 9-11　添加功能页面元素定位方式

3．职称添加功能页面元素定位

添加职称的添加按钮元素信息未截图，大家可以自行查找，其他添加职称信息如图 9-11 所示的信息。这个系统的定位比较简单，几乎是通过 id 定位的。其实我们也可以通过其他方式定位，下面代码将添加按钮改成以 XPATH 方式定位。具体值可以通过 Chrome 浏览器中的 devtools 获得信息，如图 9-7 所示。如果已掌握 XPATH 技术，可以自行编写。原则是有 id 则不用其他方式，因为定位元素的原则至少满足唯一性和高性能。

代码如下：

```
#File: job_title_locators.py
#职称添加页元素定位
from selenium.webdriver.common.by import By


class JobTitlePageLocators(object):
    #可以使用 id 定位，也可以换成 XPATH 定位
    btnAdd_btn = (By.XPATH, "//*[@id='btnAdd']")
    #btnAdd_btn = (By.ID, "btnAdd")
    jobTitle_jobTitle_text = (By.ID, "jobTitle_jobTitle")
    jobTitle_jobDescription_text = (By.ID, "jobTitle_jobDescription")
    jobTitle_jobSpec_upload = (By.ID, "jobTitle_jobSpec")
    jobTitle_note_text = (By.ID, "jobTitle_note")
    btnSave_submit = (By.ID, "btnSave")
```

其他页面大家可自行定位，通常按工作分配情况或模块来规划定位文件。通常采用一

个模块建立一个定位文件的方式。

9.4.4 页面元素操作

这里可以建立在 9.4.3 节所讲解的文件中，也可以独立创建一个文件。这类文件主要编写每个模块的小功能。公共的功能可以通过父类的方式让所有子类继承。

1. 公共功能的操作方法

把所有页面公共方法写在 Basepage 类中，例如打开浏览器的方法、截图方法等。浏览器驱动程序通过参数传入放在初始化的方法中，可以在类加载时加载。

代码如下：

```python
#File: base_public_method.py
#把所有页面公共方法写在里


class BasePage(object):
    #初始化是在类加载时加载的方法,在浏览器打开的基础上初始化浏览器参数
    #初始化打开浏览器
    def __init__(self, driver):
        self.driver = driver

    #截图
    def save_picture(self, filepath):
        self.driver.save_screenshot(filepath)
```

2. 登录页的操作方法

登录页的具体操作方法，输入用户名、输入密码、单击登录、验证登录成功。其中输入用户名方法采用封装了等待用户名文本框出现的智能等待方法，此外还包括单击用户名文本框、清空用户名文本框、输入用户名等几种方法。

代码如下：

```python
#File: login_method.py
#封装 login 页面的方法
from selenium.webdriver.support.wait import WebDriverWait
#绝对地址调用,默认路径为工程路径
from pages.login_locators import LoginPageLocators
from selenium.webdriver.support import expected_conditions as EC
#相对位置调用
#from .login_locators import LoginPageLocators
from pages.base_public_method import BasePage
from pages.main_locators import MainPageLocators


class LoginPage(BasePage):
```

```python
# 1. 输入用户名
def enter_username(self, username):
    # 智能等待加载这个元素
    # 等待加载用户名这个文本框
    WebDriverWait(self.driver, 10) \
        .until(lambda driver: driver.find_element( * LoginPageLocators.username_text))
    # 用户名元素定位,加" * "是将元组格式拆成两个参数
    element = self.driver.find_element( * LoginPageLocators.username_text)
    # 单击
    element.click()
    # 清空文本框
    element.clear()
    # 输入传入的内容
    element.send_keys(username)

# 2. 输入密码
def enter_password(self, password):
    WebDriverWait(self.driver, 10) \
        .until(lambda driver: driver.find_element( * LoginPageLocators.password_text))
    element = self.driver.find_element( * LoginPageLocators.password_text)
    element.clear()
    element.send_keys(password)

# 3. 单击登录
def click_login(self):
    element_click = self.driver.find_element( * LoginPageLocators.login_submit)
    element_click.click()

# 4. 返回要验证的文本,登录成功
def result_login_success(self):
    # 验证 welcome Admin(text)出现
    # presene_of_element_located(locator)    # locator 需要元组格式
    WebDriverWait(self.driver, 10)\
        .until(EC.presence_of_element_located(MainPageLocators.welcome_link))
    return self.driver.find_element( * MainPageLocators.welcome_link).text
```

3. 添加职称页的操作方法

进入职称页有两种方式,第 1 种可以通过光标悬停方式进入添加职称页,第 2 种是一步步单击进入职称页。光标悬停的方式通过链式方式 ActionChains 解决。首先执行 move_to_element 移动元素方法,然后执行 click()单击方法,最后不要忘记执行 perform()方法。单击及输入方法与上面代码类似。

代码如下:

```python
# File: add_jobtitle_method.py

from selenium.webdriver import ActionChains
```

```python
from selenium.webdriver.support.wait import WebDriverWait
# 绝对地址调用, 默认路径为工程路径
from pages.login_locators import LoginPageLocators
from selenium.webdriver.support import expected_conditions as EC
# 相对位置调用
# from .login_locators import LoginPageLocators
from pages.base_public_method import BasePage
from pages.job_title_locators import JobTitlePageLocators
from pages.main_locators import MainPageLocators


class AddJobtitlePage(BasePage):
    # 进入job页
    def click_movetoelement_job(self):
        WebDriverWait(self.driver, 10) \
            .until(lambda driver:
driver.find_element(*MainPageLocators.menu_admin_viewAdminModule_btn))
        # 可以通过链式方式解决光标悬停问题
        # ActionChains(self.driver)\
#         .move_to_element(self.driver.find_element(*MainPageLocators.menu_admin_viewAdminModule_btn))\
#         .move_to_element(self.driver.find_element(*MainPageLocators.menu_admin_viewJobTitleList))\
#         .click(self.driver.find_element(*MainPageLocators.menu_admin_Job)).perform()
        # 也可以通过一步步单击方式进入
        self.driver.find_element(*MainPageLocators.menu_admin_viewAdminModule_btn).click()
        self.driver.find_element(*MainPageLocators.menu_admin_Job).click()
        self.driver.find_element(*MainPageLocators.menu_admin_viewJobTitleList).click()

    # 方法
    # 1. 单击 add_jobtitle 的按钮
    def click_add_jobtitle(self):
        element_click = self.driver.find_element(*JobTitlePageLocators.btnAdd_btn)
        element_click.click()

    # 2. 输入 jobtitle
    def enter_jobTitle(self, jobtitle):
        # 智能等待加载这个元素
        # 等待加载用户名这个文本框
        WebDriverWait(self.driver, 10) \
            .until(lambda driver: driver.find_element(*JobTitlePageLocators.jobTitle_jobTitle_text))
        # 用户名元素定位
        element = self.driver.find_element(*JobTitlePageLocators.jobTitle_jobTitle_text)
        # 单击
        element.click()
        # 清空文本框
        element.clear()
```

```python
        #输入传入的内容
        element.send_keys(jobtitle)

    #2. 输入 jobDescription
    def enter_jobDescription(self, description):
        WebDriverWait(self.driver, 10) \
            .until(lambda driver:
driver.find_element(*JobTitlePageLocators.jobTitle_jobDescription_text))
        element = self.driver.find_element(*JobTitlePageLocators.jobTitle_jobDescription_text)
        element.clear()
        element.send_keys(description)

    #3. 输入上传文件名 jobTitle_jobSpec_upload
    def enter_jobSpec_upload(self, upload_filename):
        WebDriverWait(self.driver, 10) \
            .until(lambda driver: driver.find_element(*JobTitlePageLocators.jobTitle_jobSpec_upload))
        element = self.driver.find_element(*JobTitlePageLocators.jobTitle_jobSpec_upload)
        element.clear()
        element.send_keys(upload_filename)

    #4. 输入注释文本 note_text
    def enter_note_text(self, note_text):
        WebDriverWait(self.driver, 10) \
            .until(lambda driver:
driver.find_element(*JobTitlePageLocators.jobTitle_note_text))
        element = self.driver.find_element(*JobTitlePageLocators.jobTitle_note_text)
        element.clear()
        element.send_keys(note_text)

    #5. 单击 addjobtitle

    def click_addjobtitle_submit(self):
        element_click = self.driver.find_element(*JobTitlePageLocators.btnSave_submit)
        element_click.click()

    #6. 返回要验证的文本
    def result_addjobtitle_success(self):
    #presene_of_element_located(locator),locator 需要元组格式
        WebDriverWait(self.driver,10)\
            .until(EC.presence_of_element_located(MainPageLocators.welcome_link))
        return self.driver.find_element(*MainPageLocators.welcome_link).text
```

每个功能都封装了智能等待和连贯操作的方法,其他功能可以根据上面代码编写。

9.4.5 提高代码的复用性和灵活性——封装

基本业务代码编写完成后,为提高代码的复用性和灵活性,可以封装几个常用的函数。首先需要获得当前路径,在读入浏览器驱动文件和数据文件时不能把路径写死,而且所编写的代码最终需要在测试环境下运行,路径也需要获得相对路径。

1. 获得路径 get_path.py

代码如下:

```python
#File: get_path.py

import os


def get_par_path():
    root_path = os.path.abspath(os.path.dirname(__file__)).split('Utils')[0]
    #返回路径为根路径
    return root_path
```

2. 读取 yaml 数据

read_yaml.py 读取 yaml 文件,yaml 文件是配置文件,其类型为数据文件类型。

代码如下:

```python
#File: read_yaml.py
import yaml
import os


def get_yaml_data(yaml_file):
    with open(yaml_file, 'r', encoding="UTF-8")as file:
        file_data = file.read()

    data = yaml.load(file_data)
    return data
```

3. 配置日志文件 log.py

log 日志文件可以使用 Python 自带的配置文件,也可以用第三方的 logbook,可以通过配置写在文件中,也可暂时输出到控制台。pytest 也有对应的 log 配置。

代码如下:

```python
#File: log.py
from logbook import Logger, StreamHandler,FileHandler
import sys
```

```python
class conf():
    @staticmethod
    def logcon():
        log = Logger('HRM系统测试自动化日志')
        StreamHandler(sys.stdout).push_application()
        #FileHandler("logs/pytest_log.log").push_application()
        return log
```

4. config.yaml 登录数据配置

配置文件是相对变化的数据，与测试数据不同，所以我们可以写在单独文件中。下面只是以管理员登录数据配置为例。

代码如下：

```
baseURL: http://81.70.24.116/orangehrm
username: admin
password: root1234
```

5. 文件配置 pytest.ini

这是 pytest 框架的各种配置文件，下面代码是 log 日志格式级别的配置，陆续编写会增加这个文件的配置。

代码如下：

```ini
[pytest]
log_cli = 1
log_cli_level = INFO
log_cli_format = %(asctime)s [%(levelname)8s] %(message)s (%(filename)s:%(lineno)s)
log_cli_date_format = %Y-%m-%d %H:%M:%S
```

9.4.6 编写测试用例

首先编写共享数据的 conftest.py 文件，把打开浏览器方法、添加日志、关闭浏览器方法写在 addfinalizer 方法中，无论执行正确还是错误最终都执行关闭浏览器的方法。

1. 编写 conftest.py

代码如下：

```python
#File: conftest.py
import pytest,os

from ..Utils.get_path import get_par_path
from ..Utils.log import conf
from selenium import webdriver
import allure
```

```python
import sys

sys.path.append('..')

@allure.feature('打开浏览器')
@pytest.fixture(scope="session", autouse=True)
def init_driver(request):
    log = conf.logcon()
    log.info('setup_class')
    driver_path = os.path.join(get_par_path(), "driver/Chromedriver")
    driver = webdriver.Chrome(executable_path=driver_path)


    def close_browser():
        driver.quit()

    # 无论执行正确还是错误最终都执行关闭浏览器的方法
    request.addfinalizer(close_browser)
    return driver
```

2．编写添加职称模块的测试方法

数据提前导入时配置数据和登录功能使用 fixture 功能依赖注入。按业务逻辑调用测试步骤，最后添加断言和截图。添加多条数据时使用 fixture 和 parametrize 组合实现。数据是从 yaml 文件读取的，读取的每条数据都是字典类型，可参考 9.4.8 节，如图 9-13 所示。右上角取回的一条数据是{'case':{'jobTitle':'title1','jobDescription':'desc9','note_text':'note9'}}。

使用 Allure 框架中的 step 功能、feature 功能、with allure.step 上下文功能及 allure.attach.file 附加文件功能定制报告，代码如下：

```python
# File: test_add_jobTitle.py
from datetime import datetime

import pytest,os


from pages.add_jobtitle_method import AddJobtitlePage
from pages.login_method import LoginPage
from ..Utils.get_path import get_par_path
from ..Utils.log import conf
from ..Utils.read_yml import get_yaml_data
from selenium import webdriver
from pages.base_public_method import BasePage
import allure
```

```python
with allure.step("0. 从数据文件中读取添加职称数据信息"):
    yaml_path = os.path.join(get_par_path(), "datafile/addjob_data.yaml")
    test_data = get_yaml_data(yaml_path)


class TestClass(object):

    @allure.step("从配置文件中读取登录数据")
    @pytest.fixture()
    def login_data(self):
        self.log = conf.logcon()
        self.log.info("read config.yaml")
        yaml_path = os.path.join(get_par_path(), "config/config.yaml")
        test_data = get_yaml_data(yaml_path)
        return test_data

    @allure.feature("登录功能")
    @allure.step("使用管理员身份登录")
    def test_login(self, init_driver, login_data):
        self.log.info("login")
        init_driver.get(login_data['baseURL'] + "/symfony/web/index.php/auth/login")
        init_driver.maximize_window()
        init_driver.implicitly_wait(30)
        base_page = BasePage(init_driver)
        with allure.step("初始化登录页"):
            login_page = LoginPage(init_driver)
        with allure.step("输入用户名及密码后单击登录"):
            login_page.enter_username(login_data['username'])
            login_page.enter_password(login_data['password'])
            login_page.click_login()
        with allure.step("断言 admin 是否登录成功并截图"):
            assert 'Admin' in login_page.result_login_success()
            pic_path = os.path.join(get_par_path(), "shootpicture/")
            base_page.save_picture(pic_path + str(datetime.now()) + 'login.png')

    @allure.step("0. 这是初始化数据")
    @pytest.fixture()
    def get_data(self, request):
        value = request.param
        return value

    @allure.feature("添加职称功能")
    @pytest.mark.parametrize("get_data", test_data, indirect=True)
    def test_2add_jobtitle(self, init_driver, get_data):

        job = get_data['case']
```

```python
        jobTitle = job["jobTitle"]
        jobDescription = job["jobDescription"]
        note_text = job["note_text"]
        base_page = BasePage(init_driver)
        addjobtitle_page = AddJobtitlePage(init_driver)
        with allure.step("1. 进入职称页"):
            addjobtitle_page.click_movetoelement_job()
            addjobtitle_page.click_add_jobtitle()
        with allure.step("2. 输入信息"):
            addjobtitle_page.enter_jobTitle(jobTitle)
            addjobtitle_page.enter_jobDescription(jobDescription)
        with allure.step("3. 上传文件"):
            data_path = os.path.join(get_par_path(), "datafile")
            addjobtitle_page.enter_jobSpec_upload(data_path + "/upload.log")
        with allure.step("4. 提交保存"):
            addjobtitle_page.enter_note_text(note_text)
            addjobtitle_page.click_addjobtitle_submit()
        with allure.step("5. 断言添加成功并截图"):
            assert jobTitle in init_driver.page_source
            pic_path = os.path.join(get_par_path(), "shootpicture/")
            pic_name = pic_path + str(datetime.now()) + '_addjob.png'
            base_page.save_picture(pic_name)
            allure.attach.file(pic_name, attachment_type=allure.attachment_type.PNG)
```

9.4.7 测试执行

使用 pytest 和 allure 命令执行测试用例,执行的命令如下:

```
pytest -s -v test_add_jobTitle.py --alluredir=./result
```

执行结果显示如下:

```
pytest -s -v test_add_jobTitle.py --alluredir=./result

test_add_jobTitle.py::TestClass::test_login [2020-10-07 08:13:57.217683] INFO: HRM 系统
测试自动化日志 class
[2020-10-07 08:13:59.217387] INFO: HRM 系统测试自动化日志: read config.yaml
[2020-10-07 08:13:59.220480] INFO: HRM 系统测试自动化日志: login
PASSED
test_add_jobTitle.py::TestClass::test_2add_jobtitle[get_data0] PASSED
test_add_jobTitle.py::TestClass::test_2add_jobtitle[get_data1] PASSED
test_add_jobTitle.py::TestClass::test_2add_jobtitle[get_data2] PASSED

============================ 4 passed in 19.31s ==================
```

9.4.8 生成 Allure 报告

在 9.4.7 节执行完 pytest -s -v test_add_jobTitle.py --alluredir=./result 命令后,执行的数据会被保存到 result 文件夹中,再执行下面命令会生成 Allure 测试报告。

```
allure serve ./result
```

执行结果如图 9-12 所示,截图的效果如图 9-13 所示,报告的效果与代码的匹配需要读者自行体会。

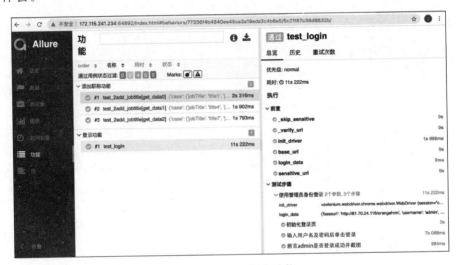

图 9-12 Allure 报告功能

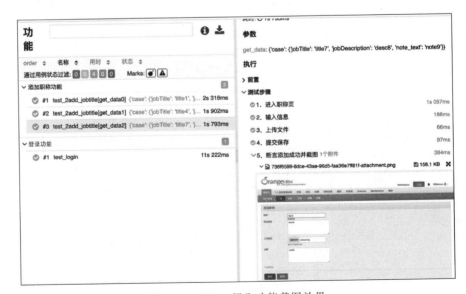

图 9-13 Allure 报告功能截图效果

9.5 Web 自动化测试本地环境持续集成

我们使用 Jenkins 软件将我们的测试代码自动化执行。Jenkins 是一款开源 CI&CD 软件,用于自动化完成各种任务,包括构建、测试和部署软件。Jenkins 支持各种运行方式,可通过系统包、Docker 或者通过一个独立的 Java 程序运行。

Jenkins 的使用方法需要读者自行学习,本书不详细说明。

9.5.1 Jenkins 2 实现自动化执行测试及持续集成流程

不同公司根据自身的实际情况,采用不同的工具实现持续集成的过程,如图 9-14 所示。

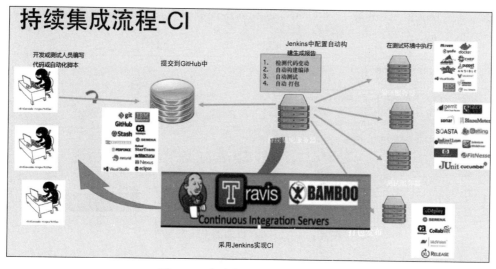

图 9-14 自动化测试持续集成流程

可以将测试自动化脚本上传到 GitHub,在 Jenkins 中配置在本机中进行测试并使用 allure 生成报告。

常见的配置过程有 3 种:第 1 种简单上传本地测试脚本,配置自由风格项目,定时或运行出结果;第 2 种使用 pipeline 流水线的方式配置执行;第 3 种使用 BlueOcean 方式配置执行,这种方式是可以显示流程步骤的 pipeline。

9.5.2 使用自由风格配置 Python 自动化测试

Jenkins 的下载、安装、启动及基本插件加载,可自行学习,本机采用的版本是 2.258。Jenkins 安装在本机,代码也在本机工程中,而执行测试部署在云服务器上(以下截图是一个示例项目,配置相同,但执行结果不同)。

(1) 启动 Jenkins。

在 war 包所在目录下执行命令 java -jar jenkins.war --httpPort=8088,登录主界面。

（2）安装 Allure 插件。

Manage Jenkins（Manage plugins）在可选插件中搜索 Allure，然后进行安装，安装完成后可在已安装处查看，如图 9-15 所示。

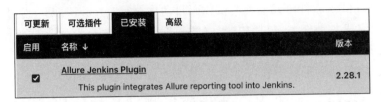

图 9-15　Allure 插件安装

（3）在全局工具配置 Allure 客户端。

Manage Jenkins-Global Tool Configuration 中配置或自动安装 Allure Commandline 程序，如图 9-16 所示。

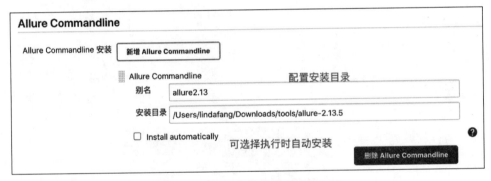

图 9-16　Allure 客户端的安装及配置

（4）新建一个 FreeStyle（自由风格）的 item（工作）。

（5）配置源码管理 Git。

在"源码管理"下，选择 Git，如图 9-17 所示。如果选择的时候没有显示 Git 账户，则需要添加一个，在添加界面输入自己 Git 仓库登录的账号和密码。将自动化测试代码上传到 GitHub 上，此步骤省略。如果 Jenkins 中 Git 插件未添加，需要在全局工具配置中自行添加。

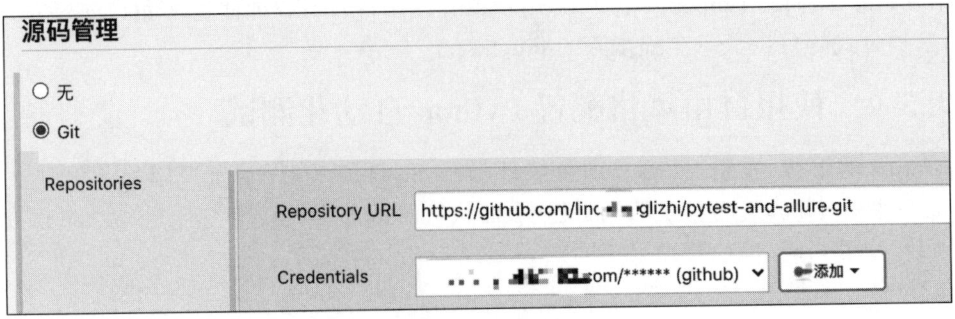

图 9-17　Git 的配置

（6）配置构建步骤。

在"构建"中选择增加构建步骤，不同操作系统选择不同执行方式，如图9-18所示。

图 9-18　Windows系统构建执行步骤选择

- Windows 系统下配置

在 Windows 系统中选择 batch command 执行，输入命令 pytest -s -v --alluredir="allure-results"，如图 9-19 所示。

图 9-19　Windows系统构建的命令

- Linux 或 Mac 系统下的配置

在 Linux 或 Mac 系统下选择 shell，如图 9-20 所示，并输入命令 pytest -s -v --alluredir="allure-results"，如图 9-21 所示。

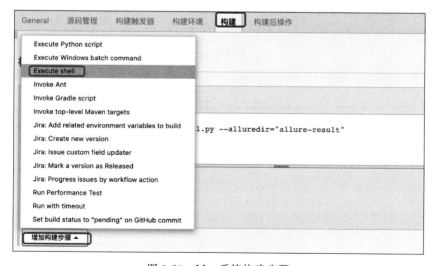

图 9-20　Mac系统构建步骤

```
构建
Execute shell
命令  cd  excise/allureDemo
      pytest -s -v test_allure_all.py --alluredir="allure-result"

查看 可用的环境变量列表
```

图 9-21　Mac 系统下的执行命令

（7）配置构建后操作。

在"构建后操作"中配置 path 路径，如果生成的报告没有数据，则通常是路径配置有问题，如图 9-22 所示。

```
构建后操作
Allure Report
Disabled  ☐
Results:
       Path  allure-result
       新增
```

图 9-22　构建后的操作

（8）运行构建。

保存配置并运行构建，进入某次构建如图 9-23 所示。

图 9-23　执行构建结果

(9)查看控制台输出。

控制台输出通常用于检查执行过程中究竟哪里出错,可以看到执行的每个步骤,包括 Git 下载代码的过程、pytest 执行测试过程,以及生成报告的结果。

结果如下:

```
Started by user admin
Running as SYSTEM
Building in workspace /Users//.jenkins/workspace/python2001
using credential 979e94f1-07e2-4e46-9a62-718671dc66fd
 > git rev-parse --is-inside-work-tree # timeout=10
Fetching changes from the remote Git repository
 > git config remote.origin.URL https://github.com/lizhi/pytest-and-allure.git # timeout=10
Fetching upstream changes from https://github.com/lizhi/pytest-and-allure.git
 > git --version # timeout=10
 > git --version # 'git version 2.24.3 (Apple Git-128)'
using GIT_ASKPASS to set credentials GitHub
 > git fetch --tags --force --progress -- https://github.com/lizhi/pytest-and-allure.git +refs/heads/*:refs/remotes/origin/* # timeout=10
 > git rev-parse refs/remotes/origin/master^{commit} # timeout=10
 > git rev-parse refs/remotes/origin/origin/master^{commit} # timeout=10
Checking out Revision 808209f3e6f8a52e3388db95e1716edaf1f65653 (refs/remotes/origin/master)
 > git config core.sparsecheckout # timeout=10
 > git checkout -f 808209f3e6f8a52e3388db95e1716edaf1f65653 # timeout=10
Commit message: "unittest and pytest demo"
 > git rev-list --no-walk 808209f3e6f8a52e3388db95e1716edaf1f65653 # timeout=10
[python2001] $ /bin/sh -xe /var/folders/kc/8vxs08p92wgf9ttn0b11lpvr0000gn/T/jenkins3978875084151115900.sh
+ cd excise/allureDemo
+ pytest -s -v test_allure_all.py --alluredir=allure-result
========================== test session starts ==========================
platform darwin -- Python 3.6.8, pytest-6.0.1, py-1.9.0, pluggy-0.13.1 -- /Library/Frameworks/Python.framework/Versions/3.6/bin/python3.6
cachedir: .pytest_cache
sensitiveURL: .*
'http://localhost:8080/job/python2001/12/', 'NODE_NAME': 'master', 'JOB_NAME': 'python2001',
'BUILD_TAG': 'jenkins-python2001-12', 'EXECUTOR_NUMBER': '1', 'JENKINS_URL':
'http://localhost:8080/', 'JAVA_HOME': '/Library/Java/JavaVirtualMachines/JDK1.8.0_201.JDK/Contents/Home', 'WORKSPACE': '/Users//.jenkins/workspace/python2001', 'GIT_COMMIT':
'808209f3e6f8a52e3388db95e1716edaf1f65653', 'GIT_URL': 'https://github.com/lizhi/pytest-and-allure.git', 'GIT_BRANCH': 'origin/master', 'Base URL': '', 'Driver': None, 'Capabilities': {}}
rootdir: /Users//.jenkins/workspace/python2001/excise/allureDemo
plugins: ...
collecting ... collected 4 items

test_allure_all.py::test_case_19688[letter] hello hello world hello world
PASSED
```

```
test_allure_all.py::test_case_19688[decimal123] hello 4 54
PASSED
test_allure_all.py::test_case_19688[unicode] hello 笔者不是超人 笔者是超人
PASSED
test_allure_all.py::test_case_19688[mix] hello 888 笔者是超人
PASSED

========================= 4 passed in 0.07s =========================
[python2001] $ /Users//Downloads/tools/allure - 2.13.5/bin/allure generate /Users//.
jenkins/workspace/python2001/allure-result -c -o /Users//.jenkins/workspace/python2001/
allure-report
Report successfully generated to /Users//.jenkins/workspace/python2001/allure-report
Allure report was successfully generated.
Creating artifact for the build.
Artifact was added to the build.
Finished: SUCCESS
```

(10) 查看 Allure 报告生成情况。

可在如图 9-24 所示的长方形框处进入。进入后如图 9-25 所示,可查看报告的具体内容。此报告内容不是本书中 Web 自动化脚本执行所生成的报告。

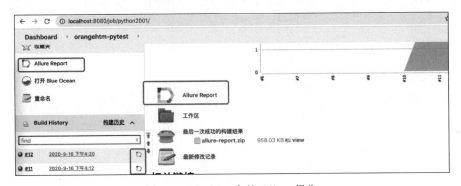

图 9-24　Jenkins 中的 Allure 报告

9.5.3　使用 pipeline 配置 Python 自动化测试

(1) pipeline 介绍。

pipeline 是 Jenkins 2.X 最核心的特性,帮助 Jenkins 实现从 CI 到 CD 与 DevOps 的转变。pipeline 是一套运行于 Jenkins 上的工作流框架,将原本独立运行于单个或者多个节点的任务连接起来,实现单个任务难以完成的复杂流程编排与可视化。

(2) 新建一个 pipeline(流水线)的工作。

(3) 使用 pipeline script 建立流水线。

在 script 中粘贴下面的脚本,根据自己的实际情况进行修改。脚本的主要含义:选择执行环境,将 agent 设置为 any,目前是本机,测试环境标志为 YES,执行测试阶段。初始化定义构建环境为 true,用户为 linda,在构建环节确认了 node 和 npm 是否安装成功,在测试阶段确认 Python 的版本并通过 pytest 执行测试脚本，在部署阶段只输出提示。这样可以定制公司所需执行的各个环节，细化具体过程的命令，在这里只是抛砖引玉。编写 script 脚本的语法可参考流水线语法。如图 9-26 所示，左下角为流水线语法。

图 9-25　Allure 报告具体内容

代码如下：

```
pipeline {
    agent any

    environment {
        ENVIRONMENT_TEST_FLAG = 'YES'
    }
    stages {
        stage('Init') {
            steps {
                script {
                    BUILD_EXPRESSION = true
                    DEPLOY_USER = 'linda'

                }
            }
        }
        stage('Build') {
            steps {
                script {
                    if ( BUILD_EXPRESSION ) {
```

```
                    sh 'echo Build stage ...'
                    sh 'node -v'
                    sh 'npm -v'
                }
            }
        }
    }
    stage('Test'){
        steps {
            script {
                if ( ENVIRONMENT_TEST_FLAG == 'YES' ) {
                    sh 'python --version'
                    sh 'pytest -s -v /Users/PyCharmProjects/py_techDemo/test/test_cacu1.py'
                }
            }
        }
    }

    stage('Deploy') {
        steps {
            script {
                if ( DEPLOY_USER == 'linda' ) {
                    sh 'echo Deploy stage ...'
                }
            }
        }
    }
}
```

（4）构建后的效果。

将光标悬停在阶段上查看提示信息，如图 9-26 所示。

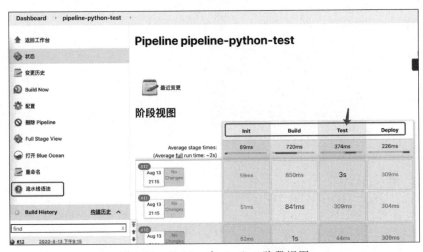

图 9-26　Jenkins 中 pipeline 阶段视图

(5)由控制台查看的结果。

```
Started by user admin
Running in Durability level: MAX_SURVIVABILITY
[Pipeline] Start of Pipeline
[Pipeline] node
Running on Jenkins in /Users//.jenkins/workspace/pipeline-python-test
[Pipeline] {
[Pipeline] withEnv
[Pipeline] {
[Pipeline] stage
[Pipeline] { (Init)
[Pipeline] script
[Pipeline] {
[Pipeline] }
[Pipeline] //script
[Pipeline] }
[Pipeline] //stage
[Pipeline] stage
[Pipeline] { (Build)
[Pipeline] script
[Pipeline] {
[Pipeline] sh
+ echo Build stage ...
Build stage ...
[Pipeline] sh
+ node -v
v13.12.0
[Pipeline] sh
+ npm -v
6.14.4
[Pipeline] }
[Pipeline] //script
[Pipeline] }
[Pipeline] //stage
[Pipeline] stage
[Pipeline] { (Test)
[Pipeline] script
[Pipeline] {
[Pipeline] sh
+ python --version
Python 2.7.16
[Pipeline] sh
+ pytest -s -v /Users/PyCharmProjects/py_techDemo/test/test_cacu1.py
========================= test session starts =============================
```

```
platform darwin -- Python 3.6.8, pytest-6.0.1, py-1.9.0, pluggy-0.13.1 -- /Library/
Frameworks/Python.framework/Versions/3.6/bin/python3.6
cachedir: .pytest_cache
sensitiveURL: .*
metadata: {'Python': '3.6.8', 'Platform': 'Darwin-19.6.0-x86_64-i386-64bit', 'Packages':
{'pytest': '6.0.1', 'py': '1.9.0', 'pluggy': '0.13.1'}, 'Plugins': {'rerunfailures': '9.0',
'ordering': '0.6', 'allure-pytest': '2.8.17', 'selenium': '1.17.0', 'freezegun': '0.4.1',
'metadata': '1.8.0', 'bdd': '3.2.1'}, 'BUILD_NUMBER': '12', 'BUILD_ID': '12', 'BUILD_URL': 'http://
localhost:8087/job/pipeline-python-test/12/', 'NODE_NAME': 'master', 'JOB_NAME': 'pipeline-
python-test', 'BUILD_TAG': 'jenkins-pipeline-python-test-12', 'EXECUTOR_NUMBER': '0',
'JENKINS_URL': 'http://localhost:8087/', 'JAVA_HOME': '/Library/Java/JavaVirtualMachines/
JDK1.8.0_201.JDK/Contents/Home', 'WORKSPACE': '/Users/linng/.jenkins/workspace/pipeline-
python-test', 'Base URL': '', 'Driver': None, 'Capabilities': {}}
rootdir: /Users/
plugins: rerunfailures-9.0, forked-1.0.2, pep8-1.0.6, apiritif-0.9.4, variables-1.9.
0, emoji-0.2.0, flakes-4.0.0, base-URL-1.4.2, assume-1.2.2,
collecting ... collected 1 item

../../../PyCharmProjects/py_techDemo/test/test_cacu1.py::TestCalss::test_add 在每个测试方法
前执行
PASSED

========================== 1 passed in 0.05s ==========================
[Pipeline] }
[Pipeline] //script
[Pipeline] }
[Pipeline] //stage
[Pipeline] stage
[Pipeline] { (Deploy)
[Pipeline] script
[Pipeline] {
[Pipeline] sh
+ echo Deploy stage ...
Deploy stage ...
[Pipeline] }
[Pipeline] //script
[Pipeline] }
[Pipeline] //stage
[Pipeline] }
[Pipeline] //withEnv
[Pipeline] }
[Pipeline] //node
[Pipeline] End of Pipeline
Finished: SUCCESS
```

9.5.4 使用 BlueOcean 配置 Python 自动化测试

(1) BlueOcean 介绍。

BlueOcean 是 Jenkins 团队从用户体验的角度出发，专为 Jenkins Pipeline 重新设计的一套 UI。

（2）启动 Jenkins。

（3）在插件管理中安装 BlueOcean。

安装完成后的效果如图 9-27 所示。左下角会出现"打开 Blue Ocean"选项。

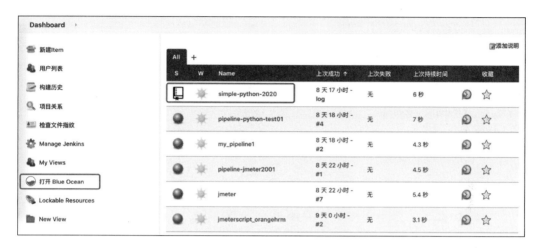

图 9-27　BlueOcean 的控制台

（4）在面板中打开 BlueOcean，进入界面。

进入界面后，会显示所有流水线，同时可以创建流水线，如图 9-28 所示。

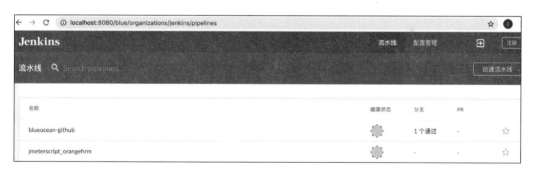

图 9-28　BlueOcean 创建的流水线

（5）将测试代码上传到 GitHub 上，如图 9-29 所示。

可以在 PyCharm 中将代码提交到 GitHub，具体的配置及命令可自行学习。

（6）创建流水线，选择源码管理为 GitHub，如图 9-30 所示。

（7）设置 Personal access tokens。

如果未连接 GitHub，则需要进入 GitHub，选择 Settings→Developer settings，设置 Personal access tokens，允许 Jenkins 访问 GitHub 需要的证书 Token，如图 9-31 所示。

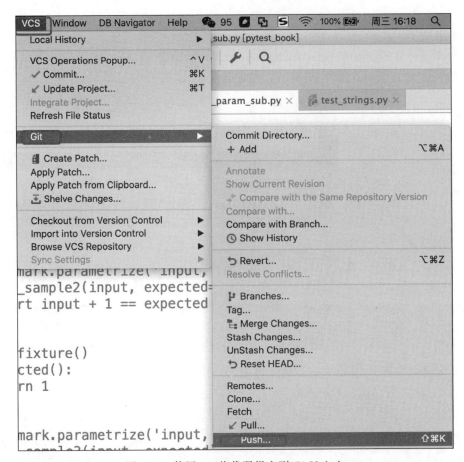

图 9-29 使用 Git 将代码提交到 GitHub 上

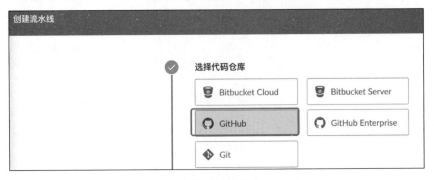

图 9-30 选择源码管理为 GitHub

（8）选择要执行的测试脚本项目。

配置完成后可以看到自己 GitHub 上的用户名显示在流水线配置中，选择此用户后会出现该用户上传的项目，选择要执行的测试脚本项目，如图 9-32 所示。

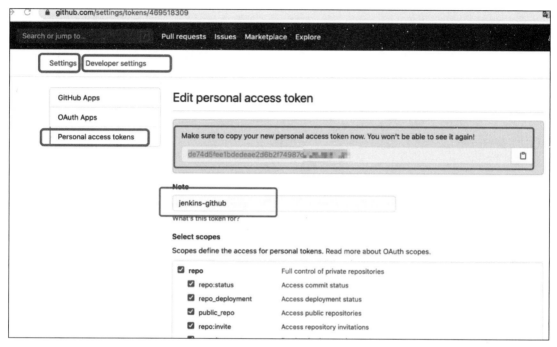

图 9-31　BlueOcean 建立 Jenkins 与 GitHub 的联系

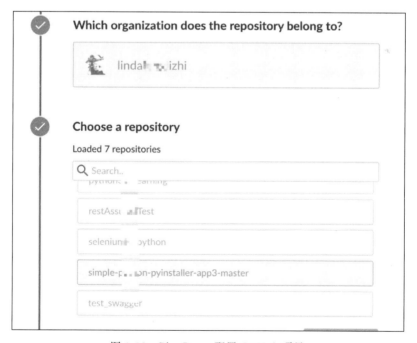

图 9-32　BlueOcean 配置 GitHub 项目

（9）简单配置流水线可以在各阶段中设置阶段名（这步可选，不是必须），添加步骤如图 9-33 所示。

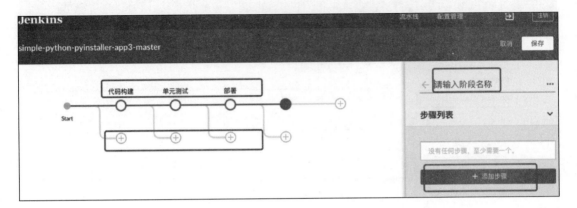

图 9-33　BlueOcean 配置流水线阶段名

（10）使用 Jenkinsfile 配置流水阶段。

可以在上传的项目中加入 Jenkinsfile 流水线脚本文件，这个文件是流水线的配置脚本，这是另一种配置流水线的方式。样例如图 9-34 所示。

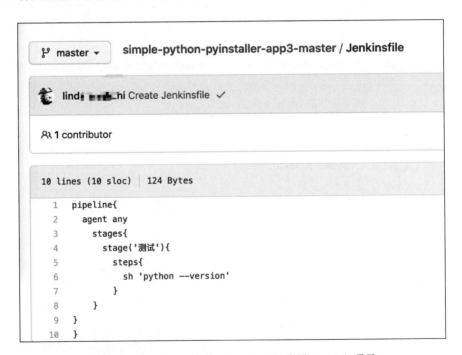

图 9-34　BlueOcean 中使用 Jenkinsfile 配置 GitHub 项目

（11）完成脚本并执行，可以查看具体的执行情况，如图 9-35 所示。

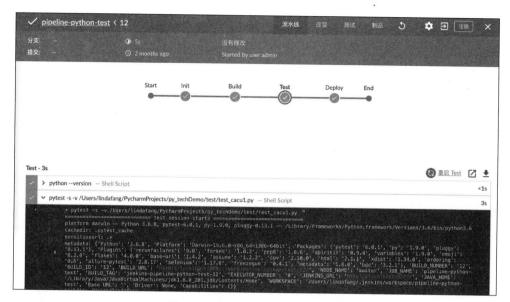

图 9-35　BlueOcean 整体执行情况

9.6　本章小结

本章主要针对一个 Web 项目进行自动化测试及持续集成。
（1）使用 Selenium 框架编写基础代码。
（2）使用 POM 将页面代码与测试代码分离。
（3）使用读取 yaml 的测试用例文件和 pytest 中的参数化实现代码与测试数据分离。
（4）使用 pytest 和 Allure 的特点定制精美报告。
（5）将测试代码上传到 GitHub 上。
（6）在 Jenkins 中使用自由风格，采用 pipeline 和 blueOcean 执行测试脚本。

第 10 章 App 自动化测试项目实践

10.1 App 自动化测试框架选择

操作 App 和 App 中的元素的框架可以使用 Appium，Appium 是一个开源的自动化测试框架，支持跨平台，支持多种编程语言，可用于原生、混合和移动 Web 应用程序，使用 webdriver 驱动 iOS、Android 应用程序。

基于框架选择的原则，运行测试用例，以及断言管理的框架我们选择 pytest。

数据驱动我们选择 pytest.mark.paramlize 和 yaml/csv 文件的读取实现。

生成测试报告我们选择 pytest 的插件 Allure 实现。

10.2 App 自动化测试环境的搭建

10.2.1 安装和验证 Java JDK——Windows 系统

1. 安装 JDK（Java 8 以上版本）

JDK 下载网址：https://www.oracle.com/java/technologies/javase/javase-JDK8-downloads.html。

打开 JDK 下载网址，以图 10-1 所示的方式下载想要的版本。

2. 安装步骤

双击下载好的 JDK 并安装，单击"下一步"按钮，可以修改安装路径，然后单击"下一步"按钮，并等待安装完成。

3. 配置环境变量 JAVA_HOME

右击"计算机"→"属性"→"高级系统设置"，单击"环境变量"按钮，在系统变量中单击"新建"按钮，设置变量名：JAVA_HOME，设置变量值：JDK 的安装路径，最后单击"确定"按钮，如图 10-2 所示。

4. 配置环境变量 CLASSPATH

新建一个系统变量，将变量名设置为 CLASSPATH，然后需要设置变量值，例如将变量值设置为 C:\Program Files\Java\JDK1.8\lib。

图 10-1　Java JDK 下载

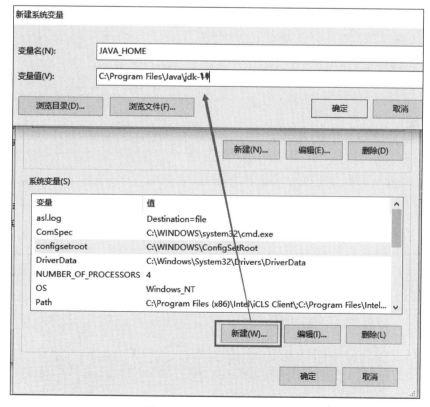

图 10-2　Java 环境变量配置

5．配置环境变量 Path

在 Path 上添加 Java 可执行文件路径。在系统变量中找到 Path 并选择，单击"编辑"→"新建"按钮并添加 JDK 的 bin 文件夹所在目录，例如：C:\Program Files\Java\JDK1.8\bin。如图 10-3 所示。

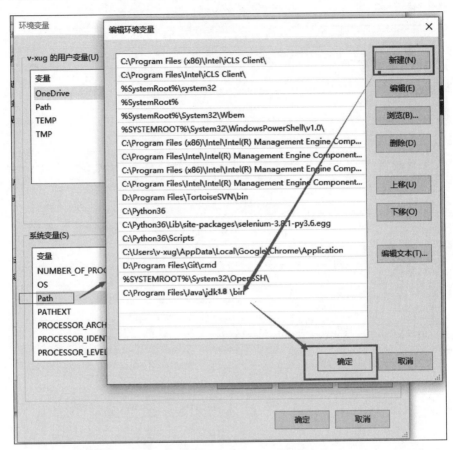

图 10-3　Java 环境变量 Path 配置

6．在 Windows 系统验证安装成功

完成以上步骤后 JDK 便安装完成了，接下来验证环境变量是否配置成功，打开 cmd（Window＋R）并输入 java -version，然后按回车键，看到以下信息就表明我们的环境变量配置成功了，如图 10-4 所示。

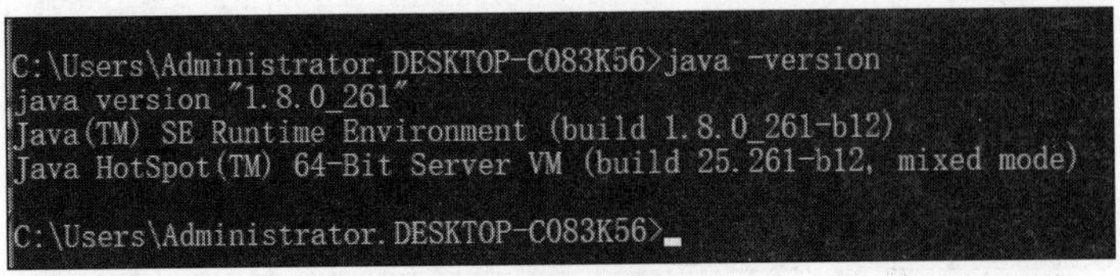

图 10-4　验证安装成功信息

7. 安装和验证 Java JDK——Mac

在 Mac 系统上安装略，验证方式与 Windows 系统一样，如图 10-5 所示。

```
lindafang@cpe-172-115-248-82 ~ % java -version
java version "1.8.0_201"
Java(TM) SE Runtime Environment (build 1.8.0_201-b09)
Java HotSpot(TM) 64-Bit Server VM (build 25.201-b09, mixed mode)
```

图 10-5　Mac 中验证安装成功信息

10.2.2　安装和验证 Node.js

安装 Node.js 之前，我们需要知道，为什么需要安装 Node.js？因为 Appium 是由 Node.js 实现的，Node.js 相当于 Appium 的解释器，Node.js 下载网址：https://node.js.org/en/download/current/。

如果读者的计算机是 Windows 64 位的系统，则可以直接下载 Windows 64 位的 msi 程序文件，下载完成后双击安装即可，所以笔者在此就不具体说明安装步骤了，安装后，检查是否安装成功。

在 cmd 终端输入命令，执行结果如图 10-6 所示。这说明 Node.js 已经安装成功。

```
node -v
npm -version
```

```
:\Users\1AS>node -v
14.7.0
:\Users\1AS>npm -v
6.14.7
```

图 10-6　验证 node 和 npm 安装成功信息

10.2.3　安装 Android SDK

因为有时需要使用安卓模拟器进行代码操作，所以选择安卓测试环境。

1. 下载 Android SDK

安卓官网已经没有单独的 Android SDK 的下载超链接，官方推荐下载包含 Android SDK 的 Android Studio。下载网址：https://developer.android.com/studio/index.html?hl=zh-cn。注意要下载和安卓系统相对应的版本。将下载好的 Android SDK 解压，便可得到目录，建议解压路径不包含中文。

2. 设置 Android 环境变量

以笔者的计算机为例，选择计算机→右击菜单→属性→高级→环境变量→系统变量→新建 ANDROID_HOME，如图 10-7 所示。

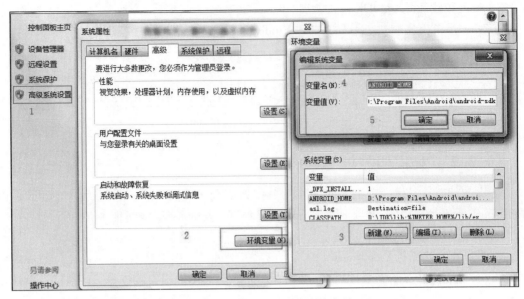

图 10-7　设置 Android 的环境变量

3. 配置 Path 环境变量

找到 Path 变量，编辑添加，在后面把 platform-tools 及 tools 添入，变量值（路径）之间使用分号隔开，两个变量值分别为％ANDROID_HOME％\platform-tools；％ANDROID_HOME％\tools，或者直接添加绝对路径，如图 10-8 所示。

图 10-8　android 环境变量 path 的设置

4. 验证环境变量

设置好后，在 cmd 下运行命令 adb version，验证环境搭建是否成功。

10.2.4 安装模拟器或连接真机

1. 连接模拟器

夜神模拟器官网：https://www.yeshen.com/，安装好模拟器后，打开模拟器，然后打开 cmd 终端，切换 cmd 的工作目录到夜神模拟器安装目录的 bin 目录下（或者把这个目录添加到环境变量，但是最好不要添加到环境变量，因为 bin 目录下也存在一个 adb.exe 可执行文件，这会和 Android SDK 中的 adb.exe 冲突），执行命令 nox_adb.exe connect 127.0.0.1:62001。

执行命令 adb devices，如果出现下面的结果，表明连接成功，如图 10-9 所示。

```
D:\Program Files\Nox\bin>nox_adb.exe connect 127.0.0.1:62001
already connected to 127.0.0.1:62001

D:\Program Files\Nox\bin>adb devices
List of devices attached
127.0.0.1:62001 device
```

图 10-9　连接模拟器

如果连接失败，则需要尝试重启模拟器，如果还不行，则需要检查 SDK 版本和模拟器版本是否一致。

2. 连接真机

将 Android 设备与要运行 adb 的计算机连接到同一个局域网，例如连到同一个 WiFi。将设备与计算机通过 USB 线连接，Android 设备的开发者选项和 USB 调试模式已开启。可以到"设置"→"开发者选项"→"Android 调试"查看。如果在设置里找不到开发者选项，可以在"设置"→"关于手机"处连续单击"版本号"7 次。执行命令 adb devices，如果看到 xxxxxx device，则表示连接成功。

10.2.5 安装 appium-desktop

下载网址：https://github.com/appium/appium-desktop/releases/，如图 10-10 所示。

安装比较简单，双击可执行文件，然后等待安装完成，中间不需要任何设置。安装完成后桌面会生成一个紫色的 appium 图标，双击打开。默认监听 0.0.0.0:4723，单击"启动服务器 v1.15.1"按钮，启动服务，如图 10-11 所示。

执行脚本时 appium 会有日志，在 App 自动化测试中环境和执行的问题有时会很多，通过日志的提示进行调试，如图 10-12 所示。之后启动模拟器，这样就可以准备测试脚本了。

10.2.6 安装 appium-client

如果使用 Python 开发 Appium 的自动化测试脚本，肯定离不开 Appium 的 API 库，所以说这里的 Appium 客户端就是 Python 代码使用的库，用来连接 Appium Server 及操作 App。

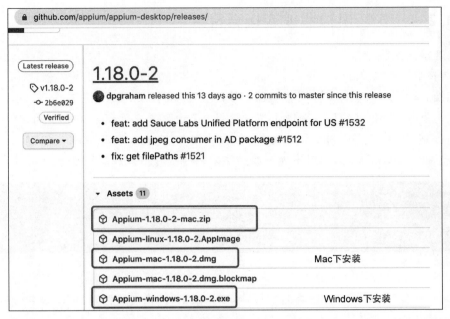

图 10-10 appium-desktop 的下载网址

图 10-11 appium-desktop 的服务启动

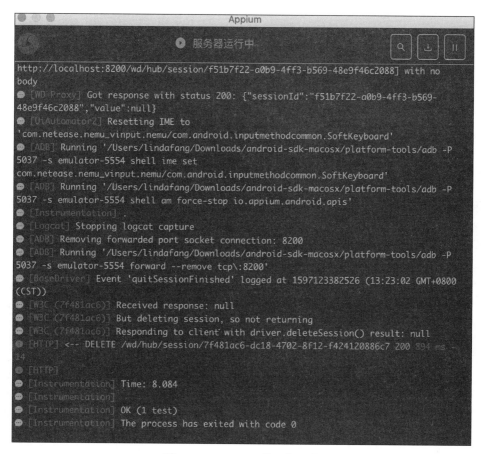

图 10-12　Appium 的运行日志

安装 Appium Client 非常简单，只要已经配置好了 Python 环境，就可以直接使用 pip 命令安装了，在 cmd 客户端执行 pip install Appium-Python-Client 即可完成安装，如图 10-13 所示。Appium-Python 第三方库参考资料网址：https://pypi.org/project/Appium-Python-Client/。

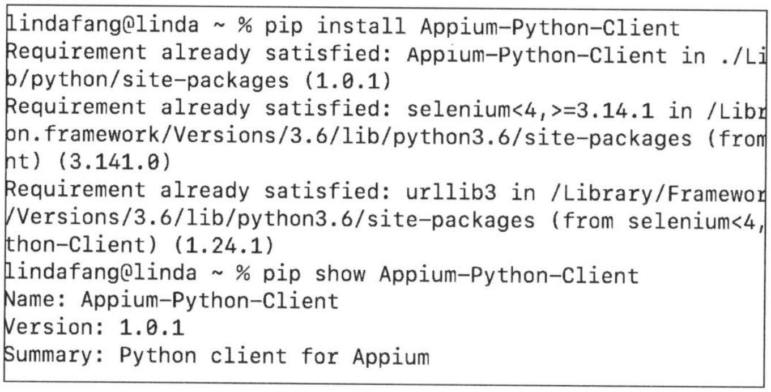

图 10-13　appium-client 的安装过程

10.2.7 appium-doctor 环境检查

打开 cmd 命令窗口执行命令 npm install -g appium-doctor，安装 appium-doctor，如果安装不成功，可以换个淘宝的镜像进行安装。打开 cmd 命令窗口执行下面命令：

```
npm install  -g appium-doctor  --registry=https://registry.npm.taobao.org
cnpm install appium-doctor -g
```

Windows 系统下执行 appium-doctor 后的结果如图 10-14 所示。

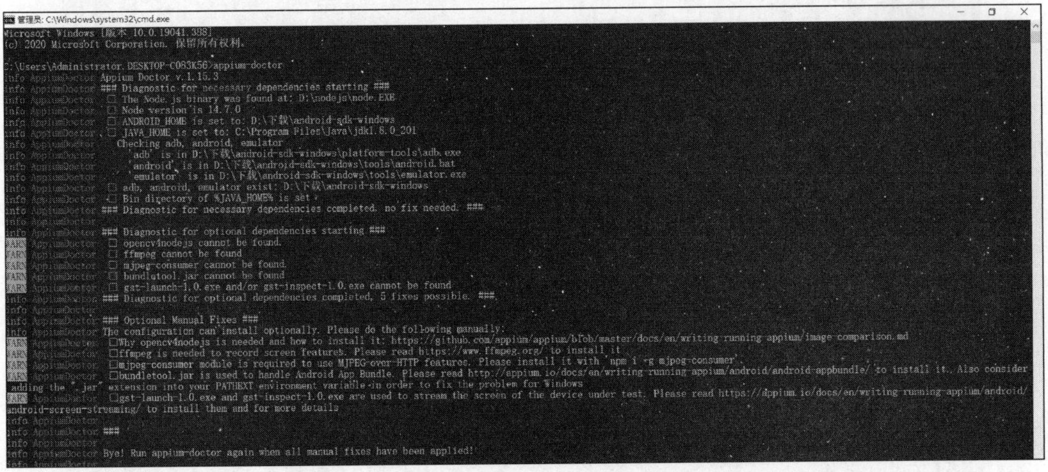

图 10-14 appium-doctor 在 Windows 系统执行

Mac 系统下执行 appium-doctor 后的结果如图 10-15 所示。

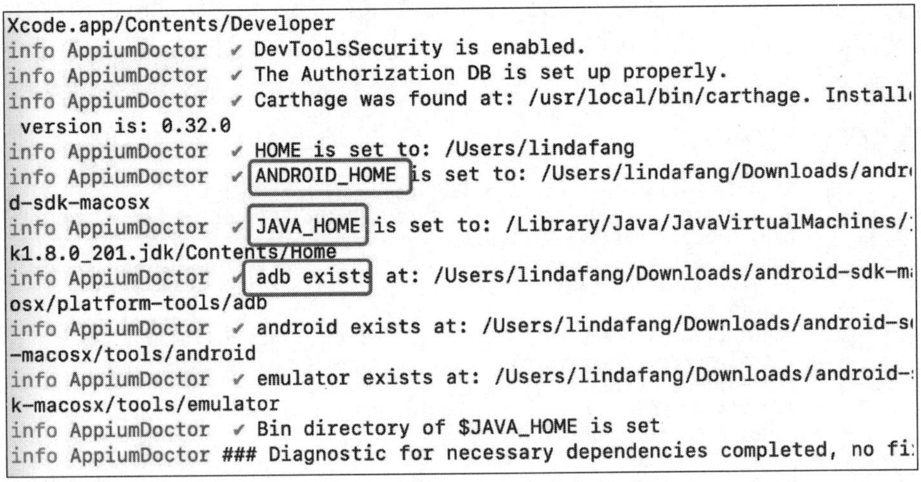

图 10-15 appium-doctor 在 Mac 系统执行

10.3 使用 pytest 和 Allure 建立 App 自动化混合框架

当需要进行大量的 Python 脚本编写时我们需要 Python 的 IDE 工具。下面的操作在 PyCharm 中操作。新建立项目 Pappium_yuodao_demo，最后复制到 pytest_book。

10.3.1 安装所需要的包和插件

Appium：操作手机和浏览器，以及定位页面元素和操作元素。

```
pip install Appium-python-client
```

pytest：Python 的测试框架，执行测试用例及断言。

```
pip install pytest
```

pytest-xdist：并发执行测试用例。

```
pip install pytest-xdist
```

allure-pytest：生成精美报告。

```
pip install allure-pytest
```

10.3.2 建立目录结构

项目名：Pappium_yuodao_demo，建立以下文件夹或包。代码需要建立包，其他可以是文件夹。页面代码与测试代码分离，数据与代码分离，其他资源及报告单独建立。可以根据公司实际情况建立，本书所编写的练习 demo，建立的目录结构如图 10-16 所示。

10.3.3 连接 App 的配置及启动 App

将要测试的 App 放在 Papp 包中，下面是连接 App 的调试。连接手机的描述 platformName 是手机平台，deviceName 的值执行 adb devices 的结果，如果是夜神模拟器，则其值一般是 127.0.0.1:62001，automationName 是 Android 自带的自动化库，platformVersion 的值是手机的版本，App 是要测试的 App 的文件位置及文件名，unicodeKeyboard 的值为 True，使用 unicodeKey 键盘，resetKeyboard 的值为 True，使用完恢复键盘设置。

图 10-16 目录结构

Appium 封装了 webdriver，通过 Appium 远程调用 Remote。

代码如下：

```python
import time

from appium import webdriver
desired_caps = {
"platformName": "android",
#使用 adb devices 查看的结果
    "deviceName": "127.0.0.1:62001",
    "automationName":"UiAutomator2",
    "platformVersion": "6.0",
    "app":"apps/youdao_note.apk",
    "noReset":True,
    "unicodeKeyboard": True,
    "resetKeyboard": True,
}

driver = webdriver.Remote("http://127.0.0.1:4723/wd/hub", desired_caps)
time.sleep(2)
driver.quit()
```

连接成功后，可以针对手机中各元素进行定位，要熟练使用工具进行定位。

10.3.4　使用各种工具进行元素定位

至少有两种定位工具，但经常使用 Appium 自带的工具。

1. 使用 uiautomatorview 定位

在所安装的 Android SDK 路径下的 tools 中 uiautomatorviewer 程序启动后可定位 App 各元素的位置。启动后的效果如图 10-17 所示，这种方法目前较少使用。

2. 使用 appium-desktop 定位

在运行的服务中单击下面红色框，以此启动检查器会话，如图 10-18 所示。

首先在 Appium 的查看中添加 App 的连接信息，然后启动会话。添加信息可以一个一个地编辑，也可在右侧 json 串中编辑或粘贴也可，最好保存下来，最后启动会话。按图 10-19 所示箭头的顺序执行。

Appium 的定位如图 10-20 所示，将鼠标移动到最上面的功能键后会有具体的功能说明，根据说明即可按功能进行定位、查看及操作。图 10-20 中菜单左边一组 第 1 个图标是选择，第 2 个图标是滑动，先单击一个位置，再单击一个位置就从第 1 个位置滑动到第 2 个位置了。第 3 个是单击坐标点，可以确定单击位置的坐标。

图 10-20 右侧包含查找、选择器，以及可以进行的操作，通过这些功能可以进行操作，并且可以定位需要的元素，例如先在左侧选择 Accessibility，这个元素的定位方式会在右侧

第 10 章　App 自动化测试项目实践

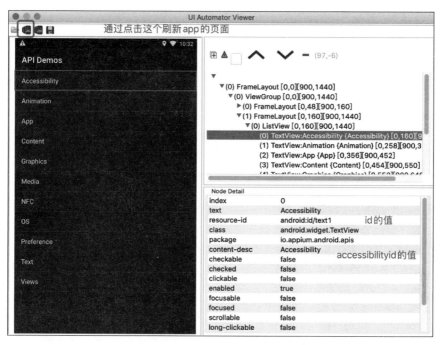

图 10-17　uiautomatorviewer 元素定位

图 10-18　Appium 定位的启动

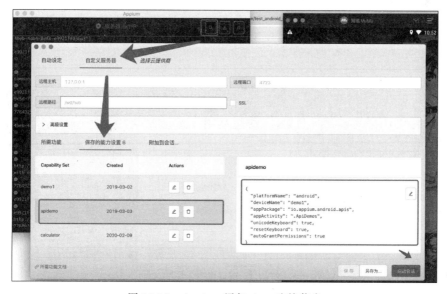

图 10-19　Appium 添加 App 连接信息

"查找"和"选择器"中获得,也就是定位方式是 Appium 的独有方式 accessibilityid,具体的值是 Accessibility。如果需要进入 Accessibility,则可以通过右侧"选定的元素"下面的"单击"进入。

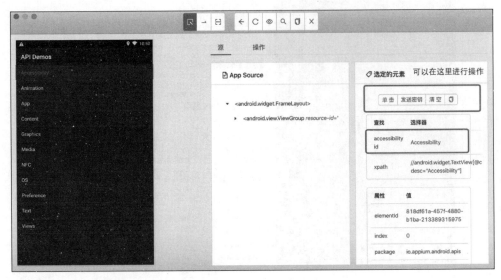

图 10-20　Appium 的定位

如果选择器只有 XPath 定位,则可能需要自己优化 XPath。

图 10-20 中菜单右边一组 图标,第 1 个表示返回,第 2 个表示刷新,第 3 个表示录制脚本,第 4 个表示查找,第 5 个表示将 xml 源码复制到剪切板(可以通过录制脚本并导出简单的代码来借鉴定位方式)。

选择搜索元素,通过右侧的定位,试验是否能成功定位,如果定位的结果为 1,则表示可以直接定位,如果搜索结果>1,则表示可通过 list[index]的方式获取第 index+1 个元素的定位,如图 10-21 所示。

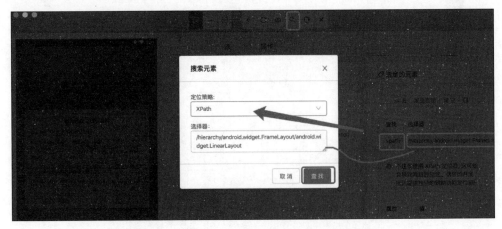

图 10-21　Appium 的定位,通过 XPath 搜索元素

选中元素后，单击"查找"按钮，如图 10-22 所示，如果查找的元素的个数为 1，则选中这个元素，然后单击这个元素，之后单击"完成"按钮。可以直接测试是否能进行各种操作。这种定位测试最适合定位查找元素个数超过 1 个时，具体选择哪个可以通过具体情况确定。

图 10-22　Appium 的定位，搜索出结果

注意：如果有实在定位不了的元素，开发人员增加 id 属性就可以了。

10.3.5　使用 PO 方式建立元素定位 locators 类

编写代码的第 1 步为建立元素定位。

在 pageobjects 包下创建 locators.py 文件，编写打开和添加笔记的元素位置。

代码如下：

```
from appium import webdriver
from appium.webdriver.common.mobileby import MobileBy


class FirstOpenPageLocators(object):
    btn_ok = (MobileBy.ID, "com.youdao.note:id/btn_ok")
    permission_allow = (MobileBy.ID, "com.android.packageinstaller:id/permission_allow_button")

class AddNotePageLoacators(object):
    add_note = (MobileBy.ID, "com.youdao.note:id/add_note")
    add_icon = (MobileBy.ID, "com.youdao.note:id/add_icon")
    note_title = (MobileBy.ID, "com.youdao.note:id/note_title")
    note_editText = (MobileBy.CLASS_NAME, "android.widget.EditText")
    submit_complete = (MobileBy.ID, "com.youdao.note:id/actionbar_complete_text")

class MePageLocators(object):
    pass
```

10.3.6　使用 PO 方式建立元素操作方法基类

编写代码的第 2 步,建立元素操作公共方法。

在 pageobjects 下创建 basepage.py 文件,建立一些公共的方法、初始化方法、输入文本方法、单击方法,以及滑动方法的封装。这次我们将输入文本和单击方法都进行了再次封装。

初始化方法中的 self.driver：WebDriver＝driver,冒号后面是 driver 的返回类型,确定了类型后 driver 中的方法才能调用,也就是可以通过圆点点出来。

self.driver.find_element(＊locator)使用 find_element 封装方法,这种方法的参数采用元组的形式,前面是定位,后面是定位的值。＊是将元组类型拆分为两个参数。

滑动方法的封装通过获得手机屏幕的长和宽,相对定位,这样可以适合所有的机型。这种兼容性测试必须提高代码的灵活性。

代码如下：

```python
from appium import webdriver
from selenium.webdriver.support.ui import WebDriverWait
from selenium.webdriver.support import expected_conditions as EC
from appium.webdriver.webdriver import WebDriver
from appium.webdriver import WebElement
import logging

class BasePage(object):
    def __init__(self, driver):
        # init
        # self
        # 设置 driver 的类型,这样所有方法在没有实例时也能点出来
        self.driver: WebDriver = driver

    def set_value(self, locator, value):

        if(WebDriverWait(self.driver, 10).until(
            EC.presence_of_element_located(locator)
        )):
            # 参数传递,元组格式与非元组格式
            element = self.driver.find_element(*locator)
            element.click()
            element.clear()
            element.send_keys(value)
        else:
            logging.error("无法输入信息,找不到元素")

    def click_thing(self, locator):
```

```python
        if(WebDriverWait(self.driver, 10).until(
            EC.presence_of_element_located(locator)
        )):
            element = self.driver.find_element(*locator)
            element.click()
        else:
            logging.error("无法单击,找不到元素")

    def save_pic(self, filepath):
        self.driver.get_screenshot_as_file(filepath)

    def swipe_up(self, driver):
        #兼容性,适合所有手机,获得屏幕大小,即宽和高
        width = driver.get_window_size()['width']
        height = driver.get_window_size()['height']
        #swipe(x=起始点x的坐标,y=起始点y的坐标,x1=滑动到的点的x坐标,
        # y1=滑动到的点的y坐标)
        #0,0的点是在屏幕左上角
        self.driver.swipe(width/2, height/3*2, width/2, height/3*1)

    def swipe_down(self, driver):
        width = driver.get_window_size()['width']
        height = driver.get_window_size()['height']
        self.driver.swipe(width / 2, height / 3, width / 2, height / 3 * 2)

    def swipe_right(self, driver):
        width = driver.get_window_size()['width']
        height = driver.get_window_size()['height']
        self.driver.swipe(width / 10, height / 2, width / 10 * 9, height / 2)

    def swipe_left(self, driver):
        width = driver.get_window_size()['width']
        height = driver.get_window_size()['height']
        self.driver.swipe(width / 10 * 9, height / 2, width / 10, height / 2)
```

10.3.7 使用 PO 方式建立每个页面或功能的元素操作方法类

编写代码的第 3 步,建立每个元素的操作方法,通过调用便可获得元素位置信息。

在 pageobjects 下创建 addnotepage.py 文件。继承基类的公共方法,将每个页面创建为一个类,为每个页面的操作建立方法,方法的步骤通过封装调用(大家一定要根据实际情况进行封装,避免过度封装)。

代码如下:

```python
from .BasePageM import BasePage
from .locators import FirstOpenPageLocators
```

```python
from .locators import AddNotePageLoacators

class FirstOpenPage(BasePage):
    # 继承的类 BasePage 是笔者创建的父类,父类中的所有方法笔者都能使用
    def permission_allowandok(self):
        BasePage.click_thing(self,FirstOpenPageLocators.btn_ok)
        BasePage.click_thing(self,FirstOpenPageLocators.btn_ok)
        BasePage.click_thing(self,FirstOpenPageLocators.permission_allow)

class AddNotePage(BasePage):
    def add_note_ok(self,title,editText):
        BasePage.click_thing(self,AddNotePageLoacators.add_note)
        BasePage.click_thing(self, AddNotePageLoacators.add_icon)
        BasePage.set_value(self, AddNotePageLoacators.note_title,title)
        BasePage.set_value(self,AddNotePageLoacators.note_editText,editText)
        BasePage.click_thing(self, AddNotePageLoacators.submit_complete)
        # BasePage.save_pic(self, "1.png")
```

10.3.8 结合 pytest 的特性建立公共数据共享文件 conftest.py

编写代码的第 4 步,建立测试文件中的共享文件。

在项目根目录下创建 conftest.py 文件,把初始化浏览器驱动方法作为 fixture 进行依赖注入。

代码如下:

```python
# File: conftest.py
from utils import desired_caps
from appium import webdriver
import pytest
# 在此包下所有文件共享的数据都写在这里
from utils.get_path import PATH
from utils.get_yaml import get_yaml_data
import logging


@pytest.fixture(scope = 'session')
def run_app(request):
    desired_cap_value = desired_caps.get_desired_caps()
    driver = webdriver.Remote(get_yaml_data(PATH("../config/conf.yaml"))['base_URL'], desired_cap_value)

    def close_app():
        driver.quit()

    request.addfinalizer(close_app)
    return driver
```

10.3.9 使用 PO 的方式建立测试类

编写代码的第 5 步,建立测试用例。

在 testcases 建立 addnotetest 用例,按测试逻辑调用,打开手机,使用 PO 的方式初始化并添加笔记页、标题和内容。

先打开手机的初始化页面,同样使用 fixture 依赖注入进行自动加载,范围是 module,也就是初始化,打开手机,同意权限,一个文件只运行一次,建立添加笔记的测试用例,使用参数化读取 yaml 文件的方式实现。

测试用例中使用 parametrize,简单文件使用 yaml 的安全加载文件。参数名是 add,在测试方法参数中加上 add,把数据赋值给 note_title,如果数据需要解析,则可通过这个进行转换。例如所写的 yaml 文件返回的类型是字典类型,则可通过 data[key] 的方式获得正确值。

Allure 框架的使用可以定制添加 feature 功能、step 步骤、级别、附件等。

代码如下:

```python
#File: test_add_note_pytest.py
import pytest,allure
from pageobject import addnotepage
import yaml
from datetime import datetime
from utils.get_path import PATH


@allure.feature("笔记功能")
class TestNotePytest(object):
    @allure.step("初始化,打开手机,同意权限")
    @pytest.fixture(scope = 'module', autouse = True)
    def agree_init(self, run_app):
        #初始化 driver,文件级通用
        #先初始化页面
        ok = addnotepage.FirstOpenPage(run_app)
        #操作的方法
        ok.permission_allowandok()


    @allure.story("添加笔记功能")
    @pytest.mark.regression
    @allure.severity("critical")
    @pytest.mark.parametrize("add", yaml.safe_load(open(PATH("../Testdatas/" + "addnotedata.yaml"),'r',encoding = 'utf8')))
    def test_addnote(self,add,run_app):
        note_title = add
```

```python
        #初始化,添加笔记页
        with allure.step("初始化,添加页面"):
            addnote = addnotepage.AddNotePage(run_app)
        #添加标题和内容
        addnote.add_note_ok(note_title)
        #截图
        addnote.save_pic(PATH("../shotpicture/" + str(datetime.now())) + "addnote.png")

    @pytest.mark.sanity
    #@pytest.mark.skip
    def test_deletenote(self):
        pass


if __name__ == '__main__':
    pytest.main()
```

10.3.10 使用 yaml 文件及 pytest 中的 parametrize 作为数据驱动程序

编写代码的第 6 步,编写数据文件。

这步需要在第 5 步之前建立,在 testdatas 文件夹中创建 addnotedata.yaml 文件,此文件只是示例,这是笔记标题数据。

```
- titlename
- mydaily
- 我的日记
```

10.3.11 使用 Allure 标签定制报告

编写代码的第 7 步,使用 allure 标签定制报告。

使用@allure.feature("笔记功能")添加主功能,使用 @allure.story("添加笔记功能")添加分支功能,使用@allure.step("初始化")或 with allure.step 添加步骤,使用@allure.severity("critical")添加等级。具体见 10.3.9 节所讲内容。执行下面命令即可执行用例,也可以把执行写在另一个单独文件中。

```
pytest -s -v test_add_note_pytest.py --alluredir=../reports
allure serve ../reports
```

运行后的结果如图 10-23 所示。

第10章 App自动化测试项目实践 325

图 10-23　Appium＋Allure 的报告

10.3.12　封装的一些公共的方法

编写代码的第 8 步，这步不是必备的。为增加复用性和灵活性，封装常用的方法包括获得路径、读取 yaml 和 csv 文件、手机连接信息等。

(1) 封装获得当前文件所有地址和文件名。

代码如下：

```
#File: get_path.py
import os
def PATH(file):
    #dirname: 获得当前文件的相对目录
    #join: 把两个字符串组合在一起
    #abspath: 获得绝对路径
    return os.path.abspath(os.path.join(os.path.dirname(__file__),file))
```

(2) 封装 desired_capabilities。

代码如下：

```
#File: desired_caps.py
from utils.get_path import PATH


def get_desired_caps():
    desired_caps = {
        "platformName": "android",
```

```
            "deviceName": "emulator-5554",
            #"platformVersion": "6.0",
            "app": PATH("../Papp/youdaonote_android_6.7.18_youdaoweb.apk"),
            #"unicodeKeyboard": True,
            #"resetKeyboard":False,
        }
        return desired_caps
```

（3）封装读取 yaml 文件的方法，如果数据文件简单，则可直接读取。
代码如下：

```
#File: get_yaml.py
import yaml


def get_yaml_data(yaml_file):
    #打开 yaml 文件
    with open(yaml_file, 'r', encoding="UTF-8") as file:
        file_data = file.read()

    data = yaml.load(file_data)
    return data
```

（4）封装读取 csv 文件的方法。
代码如下：

```
#File: get_csvdata.py
import csv


def get_csv_data(file_name):
    rows = []
    with open(file_name, 'r') as f:
        reader = csv.reader(f)
        next(reader, None)
        for row in reader:
            rows.append(row)
        return rows
```

10.3.13　在文件中读取配置文件数据

编写代码的第 9 步，编写配置文件及读取。
在 config 添加 config.yaml 文件及要配置的内容。

```
base_URL: http://127.0.0.1:4723/wd/hub
```

在测试用例中使用读取 yaml 的方式读取每个值,例如设置路径等,在 conftest.py 文件中读取 Appium 的连接地址,参考 10.3.8 节中代码实现。

```
get_yaml_data(PATH("../config/conf.yaml"))['base_URL']
```

10.3.14　在测试用例中添加 log 日志

编写代码的第 10 步,添加日志。

添加日志是为了在出现问题时能快速定位问题。可以使用 pytest 的 log,在 pytest.ini 中进行配置后,在关键步骤中编写日志,这样就可以使用了。

代码如下:

```
[pytest]
log_cli = 1
log_cli_level = INFO
log_cli_format = %(asctime)s [%(levelname)8s] %(message)s (%(filename)s:%(lineno)s)
log_cli_date_format = %Y-%m-%d %H:%M:%S
```

在测试文件中添加日志,这是封装的 BasePage 页,参考 10.3.6 节实现。

代码如下:

```
import logging

log = logging.getLogger("linda's test")

class BasePage(object):

    def __init__(self, driver):

        # 设置 driver 的类型,所有方法在没有实例时也能点出来
        self.driver: WebDriver = driver
        log.info("这是个初始化驱动!")
```

执行的结果大致如下,可以修改设置,以便让日志保存到文件。

```
------------------------ live log setup ------------------------
2020-11-13 10:10:44 [    INFO] setup_class() (test_log.py:23)
2020-11-13 10:10:44 [    INFO] xxxxxxxxxxxxxxx (test_log.py:25)
------------------------ live log call ------------------------
2020-11-13 10:10:44 [    INFO]
```

```
setup_method() (test_log.py:32)
2020 - 11 - 13 10:10:54 [    INFO] - test_7() (test_log.py:51)
2020 - 11 - 13 10:10:54 [    INFO] ffjiafuiodafdfj__teardown (test_log.py:45)
PASSED                                                                 [ 33 % ]
----------------------- live log call ------------------------------
2020 - 11 - 13 10:10:54 [    INFO]
setup_method() (test_log.py:32)

>>>>>>>> PDB set_trace (IO - capturing turned off) >>>>>>>>>>>>>>>>>>>>
> /Users/lindafang/PyCharmProjects/AppiumAutoDemo/excise/test_log.py(58)test_4()
-> log.info(' - test_4()')
(Pdb) < built - in function dir >
(Pdb) < Logger excise.test_log (INFO)>
(Pdb) 2020 - 11 - 13 10:11:26 [    INFO] ffjiafuiodafdfj__teardown (test_log.py:45)
--------------------- live log sessionfinish ------------------------
2020 - 11 - 13 10:11:26 [    INFO] teardown_class() (test_log.py:29)

!!!!!!!!!!!!!!!!!! _pytest.outcomes.Exit: Quitting debugger !!!!!!!!!!!!!!!!!!
========================= 1 passed in 42.08s =================================
Process finished with exit code 0
```

10.3.15 组织测试用例(添加运行标记)

编写代码的第 11 步,组织测试用例。

不同类型的测试用例可以使用 mark 进行标记。这个 mark 需要在 pytest.ini 中进行配置,否则会警告。同时在运行时可以在执行命令中通过-m 参数选择不同的测试用例来组织测试用例的执行。

(1) 在每个测试方法上添加 mark。

```
@pytest.mark regression
@pytest.mark.sanity
```

详细见 10.3.6 节和 10.3.7 节。

(2) 在 pytest.ini 中添加 markers 输入。

代码如下:

```
pytest.ini
[pytest]
markers =
    regression
    sanity
```

（3）在运行时选择不同组。

```
Pytest -s -m "sanity"
Pytest -s -m "regression"
Pytest -s -m "sanity and regression"
Pytest -s -m "sanity or regression"
```

10.3.16　实现持续集成——在 Jenkins 运行测试代码

与 Web 自动化持续集成类似，使用 pipeline 的方式可以直接在本机执行脚本。Jenkins 技术详情见 9.5.3 节。pipeline 的 script 如图 10-24 所示。

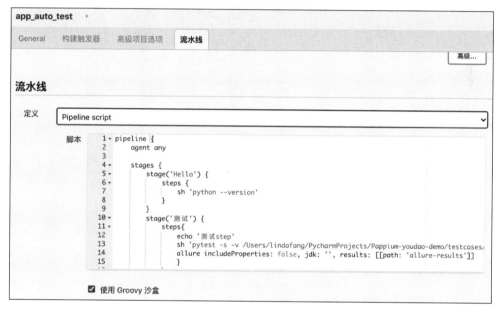

图 10-24　pipeline script

在 Linux 和 Mac 系统下使用 sh 执行命令，在 Windows 系统下使用 bat 执行命令。script 代码如下：

```
pipeline {
    agent any
    stages {
        stage('Hello') {
            steps {
                sh 'python --version'
            }
        }
        stage('测试') {
```

```
            steps{
                echo '测试 step'
                sh 'pytest -s -v /Users/Pappium-youdao-demo/testcases/test_add_note_pytest.py --alluredir=./allure-results'
                allure includeProperties: false, JDK: '', results: [[path: 'allure-results']]
            }
        }
    }}
```

在 Jenkins 中控制台的信息显示如下：

```
Started by user admin
Running in Durability level: MAX_SURVIVABILITY
[Pipeline] Start of Pipeline
[Pipeline] node
Running on Jenkins in /Users/lindafang/.jenkins/workspace/app_auto_test
[Pipeline] {
[Pipeline] stage
[Pipeline] { (Hello)
[Pipeline] sh
+ python --version
Python 2.7.16
[Pipeline] }
[Pipeline] //stage
[Pipeline] stage
[Pipeline] { (测试)
[Pipeline] echo
测试 step
[Pipeline] sh
+ pytest -s -v /Users/lindafang/PyCharmProjects/Pappium-youdao-demo/testcases/test_add_note_pytest.py --alluredir=./allure-results
============================= test session starts =====================
platform darwin -- Python 3.6.8, pytest-6.0.1, py-1.9.0, pluggy-0.13.1 -- /Library/Frameworks/Python.framework/Versions/3.6/bin/python3.6
cachedir: .pytest_cache
sensitiveURL: .*
metadata: {'Python': '3.6.8', 'Platform': 'Darwin-19.6.0-x86_64-i386-64bit', 'Packages': {'pytest': '6.0.1', 'py': '1.9.0', 'pluggy': '0.13.1'}, 'Plugins': {'rerunfailures': '9.0', 'bdd': '3.2.1'}, 'BUILD_NUMBER': '7', 'BUILD_ID': '7', 'BUILD_URL': 'http://localhost:8080/job/app_auto_test/7/', 'NODE_NAME': 'master', 'JOB_NAME': 'app_auto_test', 'BUILD_TAG': 'jenkins-app_auto_test-7', 'EXECUTOR_NUMBER': '1', 'JENKINS_URL': 'http://localhost:8080/', 'JAVA_HOME': '/Library/Java/JavaVirtualMachines/JDK1.8.0_201.JDK/Contents/Home', 'WORKSPACE': '/Users/lindafang/.jenkins/workspace/app_auto_test', 'Base URL': '', 'Driver': None, 'Capabilities': {}}
rootdir: /Users/lindafang/PyCharmProjects/Pappium-youdao-demo, configfile: pytest.ini
```

```
plugins: rerunfailures-9.0, forked-1.0.2, pep8-1.0.6, apiritif-0.9.4, variables-1.9.0,
emoji-0.2.0, flakes-4.0.0, base-URL-1.4.2, allure-pytest-2.8.18, cov-2.10.0, html-
2.1.1, xdist-1.34.0, ordering-0.6, assume-2.3.3, selenium-1.17.0, freezegun-0.4.1,
metadata-1.8.0, bdd-3.2.1
collecting ... collected 4 items

../../../PyCharmProjects/Pappium-youdao-demo/testcases/test_add_note_pytest.py::
TestNotePytest::test_addnote[titlename]
-------------------------- live log setup --------------------------
2020-11-13 13:57:35 [         INFO] 这是初始化驱动! (basepage.py:22)
-------------------------- live log call --------------------------
2020-11-13 13:57:36 [         INFO] 这是初始化驱动! (basepage.py:22)
PASSED
../../../PyCharmProjects/Pappium-youdao-demo/testcases/test_add_note_pytest.py::
TestNotePytest::test_addnote[mydaily]
-------------------------- live log call --------------------------
2020-11-13 13:57:45 [         INFO] 这是初始化驱动! (basepage.py:22)
PASSED
../../../PyCharmProjects/Pappium-youdao-demo/testcases/test_add_note_pytest.py::
TestNotePytest::test_addnote[\u6211\u7684\u65e5\u8bb0]
-------------------------- live log call --------------------------
2020-11-13 13:57:53 [         INFO] 这是初始化驱动! (basepage.py:22)
PASSED
../../../PyCharmProjects/Pappium-youdao-demo/testcases/test_add_note_pytest.py::
TestNotePytest::test_deletenote PASSED

==================== 4 passed in 34.21s ============================
[Pipeline] allure
[app_auto_test] $ /Users/lindafang/Downloads/tools/allure-2.13.5/bin/allure generate /
Users/lindafang/.jenkins/workspace/app_auto_test/allure-results -c -o /Users/lindafang/.
jenkins/workspace/app_auto_test/allure-report
Report successfully generated to /Users/lindafang/.jenkins/workspace/app_auto_test/allure-report
Allure report was successfully generated.
Creating artifact for the build.
Artifact was added to the build.
[Pipeline] }
[Pipeline] //stage
[Pipeline] }
[Pipeline] //node
[Pipeline] End of Pipeline
Finished: SUCCESS
```

Jenkins运行主页如图10-25所示。可以显示日志和各阶段情况,可用于查看运行是否成功。

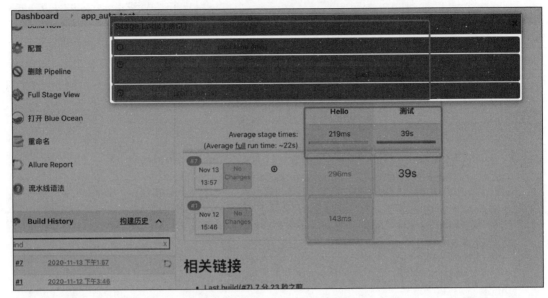

图 10-25　显示主页日志及阶段情况

　　Allure 报告如图 10-26 所示。可以看到运行测试用例数量、趋势、场景、运行器、环境、类别等。可以进入不同的目录查看具体数据。

图 10-26　Allure 报告

附：本次运行的部分 appium 日志翻译

```
[info] [35m[Appium][39m Welcome to Appium v1.15.1     版本
...
[info] [35m[Appium][39m Appium REST http interface listener started on 0.0.0.0:4723    侦听
端口
[info]下面是 App 连接的信息
[90m{"capabilities":{"firstMatch":[{"platformName":"android","appium:deviceName":
"emulator-5554","appium:app":"/Users/lindafang/PyCharmProjects/Pappium-youdao-demo/
Papp/youdaonote_android_6.7.18_youdaoweb.apk"}]},"desiredCapabilities":{"platformName":
"android","deviceName":"emulator-5554","app":"/Users/lindafang/PyCharmProjects/Pappium-
youdao-demo/Papp/youdaonote_android_6.7.18_youdaoweb.apk"}}
....
之后是将信息在手机中设置的过程
[warn] [35m[Appium][39m   Setting 'automationName=UiAutomator2' by default and using the
....
建立本次 session 会话
 Session created with session id: d51bb288-2f22-47b5-abad-7dc5b35d15c2
......[ADB]下面是与手机沟通的 adb 命令，连接手机，安装 App,启动主 activity,添加 io.appium.
settings,验签
[info] [35m[ADB][39m Using 'adb' from '/Users/lindafang/Downloads/android-sdk-Mac OS X/
platform-tools/adb'
[debug] [35m[ADB][39m Connected devices: [{"udid":"emulator-5554","state":"device"}]
[info] [35m[ADB][39m Package name: 'com.youdao.note'
[info] [35m[ADB][39m Main activity name: 'com.youdao.note.activity2.SplashActivity'
[debug] [35m[AndroidDriver][39m Parsed package and activity are: com.youdao.note/com.youdao.
note.activity2.SplashActivity
[debug] [35m[ADB][39m Getting package info for 'io.appium.settings'
[debug] [35m[ADB][39m Signer #1 certificate SHA-256 digest:
a40da80a59d170caa950cf15c18c454d47a39b26989d8b640ecd745ba71bf5dc
[debug]  [35m [ADB] [39m '/Users/lindafang/PyCharmProjects/Pappium-youdao-demo/Papp/
youdaonote_android_6.7.18_youdaoweb.apk' is already signed.
[debug] [35m[ADB][39m 'com.youdao.note' is installed

......通过代理协议发请求
[debug] [35m[WD Proxy][39m Proxying [GET /status] to [GET http://localhost:8200/wd/hub/
status] with no body
[info] [35m[WD Proxy][39m Got an unexpected response with status undefined: {"errno":
"ECONNRESET","code":"ECONNRESET","syscall":"read"}[debug] [35m[WD Proxy][39m Matched '/
status' to command name 'getStatus'
[debug] [35m[WD Proxy][39m Proxying [POST /session] to [POST http://localhost:8200/wd/hub/
session] with body: {"capabilities":{"firstMatch":[{"platform":"LINUX","webStorageEnabled":
false," takesScreenshot ": true," JavaScriptEnabled ": true," DatabaseEnabled ": false,
"networkConnectionEnabled":true,"locationContextEnabled":false,"warnings":{},"desired":
```

```
{"platformName":"android","deviceName":"emulator-5554","app":"/Users/lindafang/
PyCharmProjects/Pappium-youdao-demo/Papp/youdaonote_android_6.7.18_youdaoweb.apk"},
"platformName":"android","deviceName":"emulator-5554","app":"/Users/lindafang/
PyCharmProjects/Pappium-youdao-demo/Papp/youdaonote_android_6.7.18_youdaoweb.apk",
"deviceUDID":"emulator-5554","appPackage":"com.youdao.note"}],"alwaysMatch":{}}}
```

……findElement命令通过协议发请求,定位具体信息并进入操作

```
[debug] [35m[WD Proxy][39m Matched '/element' to command name 'findElement'
[debug] [35m[WD Proxy][39m Proxying [POST /element] to [POST http://localhost:8200/wd/hub/
session/247afc67-30a1-4c52-bb55-51f0e09bb737/element] with body: {"strategy":"id",
"selector":"com.youdao.note:id/btn_ok","context":"","multiple":false}[debug] [35m[WD
Proxy][39m Got response with status 200: {"sessionId":"247afc67-30a1-4c52-bb55-
51f0e09bb737","value":{"ELEMENT":"311f9678-3919-4fed-931d-1dd62ae5a446","element-
6066-11e4-a52e-4f735466cecf":"311f9678-3919-4fed-931d-1dd62ae5a446"}}
[info] [35m[HTTP][39m [90m{"text":"笔者的日记","value":["我","的","日","记"],"id":
"02da26c1-bcb2-47ec-88e1-b8620c43ce81"}[39m
[debug] [35m[W3C (d51bb288)][39m Calling AppiumDriver.setValue() with args: [["我","的",
"日","记"],"02da26c1-bcb2-47ec-88e1-b8620c43ce81","d51bb288-2f22-47b5-abad-
7dc5b35d15c2"]
```

……销毁和删除会话

```
[debug] [35m[BaseDriver][39m Event 'quitSessionRequested' logged at 1605245955170 (13:39:15
GMT+0800 (CST))
[info] [35m[Appium][39m Removing session d51bb288-2f22-47b5-abad-7dc5b35d15c2 from
our master session list
[debug] [35m[UiAutomator2][39m Deleting UiAutomator2 session
[info] [35m[HTTP][39m [37m<-- DELETE /wd/hub/session/d51bb288-2f22-47b5-abad-
7dc5b35d15c2 [39m[32m200[39m [90m623 ms - 14[39m
[debug] [35m[Instrumentation][39m OK (1 test)[debug] [35m[Instrumentation][39m The process
has exited with code 0
```

第 11 章 行为驱动开发(BDD)实现自动化测试

11.1 什么是 BDD

在软件工程中,行为驱动开发(Behavior-Driven Development,BDD)是一种敏捷开发流程。减少传统测试过程中由于技术背景能力,以及非技术与商业参与者之间业务理解不同而导致的问题。BDD 关注的核心是设计,其要求在设计测试用例的时候对系统进行定义,倡导使用通用的语言将系统的行为描述出来,将系统设计和测试用例结合起来,从而驱动开发工作。

简单来说就是技术人员编写的测试用例让老板和客户及用户能够看明白,同时也展示自己的工作量、思路及全面统筹的能力。

11.2 BDD 开发过程

产品经理(业务人员)通过具体的用户使用场景来告诉软件需求人员想要什么样的软件产品。用软件产品的使用场景来描述软件需求可以尽可能地避免相关人员错误理解软件需求或增加自己的主观想象。

软件需求分析人员(BA)和研发团队(研发人员、测试人员)一起对产品经理(业务人员)的用户使用场景进行分析,并梳理出具体的软件产品使用场景,这些场景使用结构化的自然语言进行描述,例如中文、英文等。

研发团队使用 BDD 工具把用户使用场景文件转化为可执行的自动化测试代码,研发人员运行自动化测试用例来验证开发出来的软件产品是否符合用户使用场景的验收要求。

测试人员可以根据自动化测试结果开展手工测试和探索性测试。

产品经理(业务人员)可以实时查看软件研发团队的自动化测试结果和 BDD 工具生成的测试报告,确保软件实现符合产品经理(业务人员)的软件期望,如图 11-1 所示。

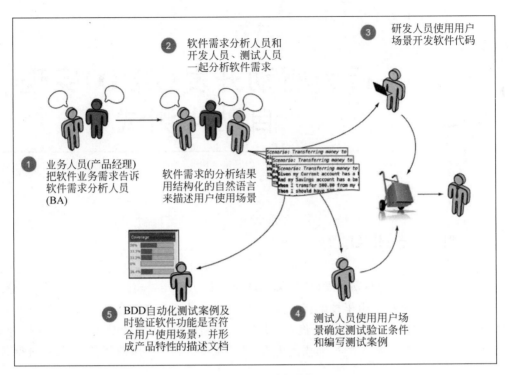

图 11-1 BDD 开发过程

11.3 BDD 的功能和作用

（1）从"测试"思维转变为"行为"思维。
（2）业务利益相关者、业务分析师、质量保证团队和开发人员之间的协作。
（3）语言无处不在，很容易描述。
（4）商业价值驱动。
（5）通过非技术利益相关者可以理解的自然语言来扩展测试驱动开发（TDD）。

诸如 Cucumber 或 JBehave 之类的 BDD 框架是一个工具，充当了业务和技术语言之间的"桥梁"。

BDD 很流行，可以用于单元级别的测试用例和 UI 级别的测试用例。诸如 RSpec（针对 Ruby）或 .NET 之类的工具（如 MSpec 或 SpecUnit）在采用 BDD 方法的单元测试中很流行。或者，可以编写有关 UI 交互的 BDD 风格的规范。假设你正在构建 Web 应用程序，则可能会使用诸如 WatiR / WatiN 或 Selenium 之类的浏览器自动化库，并使用笔者刚提到的框架之一，诸如 Cucumber（对于 Ruby）或 SpecFlow（对于 .NET）。我们本章讲解的是与 pytest 集成的 BDD。

这些框架工具为了避免瞎子摸象各自为政，从而出现沟通上的问题，如图 11-2 所示，所以采用通用语言让大家都可以理解。

图 11-2　瞎子摸象各自为政地沟通问题

11.4　中国 BDD 现状

Allure 报告的出现,使以前很多看不明白的结果可以通过各种描述和说明尽量使人看明白,所以感觉 BDD 有点可有可无。刚开始时,BDD 只有英文的写法,不太适合中国人的习惯,但是现在有些模块有了汉语的写法,但很多中国的技术公司还未将技术提高到如此精细化的程度,从而导致这个行为驱动开发未流行起来。

11.5　pytest-bdd 实现 BDD 开发

pytest-bdd 是 Gherkin 语言的一个 BDD 测试框架,类似于 behave 和 cucumber,BDD 框架 pytest-bdd 不需要单独的运行程序,它可以统一单元测试和功能测试,减轻连续集成服务器配置的负担,并允许重用测试。pytest-bdd 作为 pytest 的一个插件,所有 pytest 的功能和插件都可以用于 pytest-bdd。

为单元测试编写的 pytest 固定模块可以通过依赖项注入重新用于功能步骤中的设置和操作。允许对需求进行真正的 BDD 说明,而无须维护任何包含 Gherkin 命令性声明的上下文对象。

11.5.1　pytest-bdd 安装

pytest-bdd 安装命令如下:

```
pip install pytest-bdd
```

11.5.2 pytest-bdd 的项目结构

pytest-bdd 的项目结构实际上是非常灵活的（因为它是基于 pytest 的），但是推荐使用以下约束：

（1）所有的测试代码都应该写在 tests 文件夹中。

（2）所有的 feature 文件都应定义在 features 文件夹中。

（3）Step Definition 模块应该定义在测试的子目录中并命名为 step_defs。

（4）conftest.py 文件应该存储在与 Step Definition 相同的目录中。

可以使用其他名称和层次结构。例如，大型测试套件可以把具有相同功能的 feature 和 step def 定义在子目录中，如图 11-3 所示。pytest 能够发现测试目录下的所有测试。

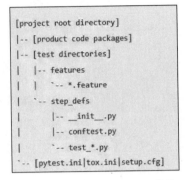

图 11-3 pytest-bdd 的项目结构

11.5.3 BDD 的标准语法

BDD 提供了一套标准的满足需求及用户场景的表达语法，一般为 Feature（功能）、Scenario（场景）、Given（假设，预置条件）、When（操作步骤）、Then（验证及清理）。Gherkin 语言是一种语法定义良好的计算机软件设计交流语言。有的业务人员不懂技术，但 Gherkin 使得业务、开发、测试及其他利益相关人员减少对需求产生歧义和误解。

```
Feature(功能需求)：登录
Scenario(场景)：非中国用户不能登录商旅系统
Given(假如)：笔者持有一个未在系统中注册过的账号 linda
When (当)：笔者输入用户名和密码
Then(那么)：笔者将看到一个提示用户名或密码无效的页面
```

11.5.4 BDD 实现步骤

1. 前提条件

pytest 的最低要求版本是 4.3。

PyCharm 中配置执行工具为 pytest，在 setting→Python Integrated Tools→Default Test Runner to pytest 中进行配置。

2. 实现添加博客功能

（1）在 features 下建立 publish_article.feature 文件。

这里对具体功能进行规范，每个功能文件仅允许使用一个功能。下面的 Feature 实现的功能是"博客",Scenario 场景实现的功能是正确"发布文章",后面编写的代码实现的是前提和步骤,Given 一般是前提条件需要的数据,And 是并列关系,可以写多个并列关系,When 一般是步骤,Then 是期望的结果。

```
Feature: 博客
    一个可以在其中发布文章的网站

    Scenario: 发布文章
        Given 笔者是作者用户
        And 笔者有一篇文章

        When 转到文章页面时
        And 笔者按了发布按钮

        Then 笔者应该不会看到错误信息
        And 应该发布该文章          #注意:将查询数据库
```

(2)在 step_defs 创建 test_publish_article.py 文件。

在这个文件的装饰器中所实现的功能与上面文件所实现的功能一致,例如:上面 Scenario 实现的功能是发布文章,下面文件@scenario 中实现的功能也是发布文章,并且在@scenario 参数中调用上面的文件 publish_article.feature。其他细节请仔细阅读代码。

代码如下:

```
#File: test_publish_article.py
from pytest_bdd import scenario,given,when,then

@scenario('../features/publish_article.feature','发布文章')
def test_publish():
    pass


@given("笔者是一位作者")
def author_user(): 类似一个 pytest 的 fixture 方法
    author = "linda"
    return author


@given("笔者有一篇文章",target_fixture = "article")
def article():
    article_name = "测试 BDD"
    return article_name


@when("转到文章页面时")
```

```python
def go_to_article(article,author_user):
    print("作者: ",author_user)
    print("文章标题: " + article,"文章内容：略过一万字")


@when("笔者按了发布按钮")
def publish_article():
    print("提交")


@then("笔者应该不会看到错误信息")
def no_error_message():
    print("没有错误信息")


@then("应该发布该文章")
def article_is_published(article):
    print("刷新网页")
    print("断言发布成功")
    assert article == "测试BDD"
```

右击后 pytest 运行结果如图 11-4 所示。

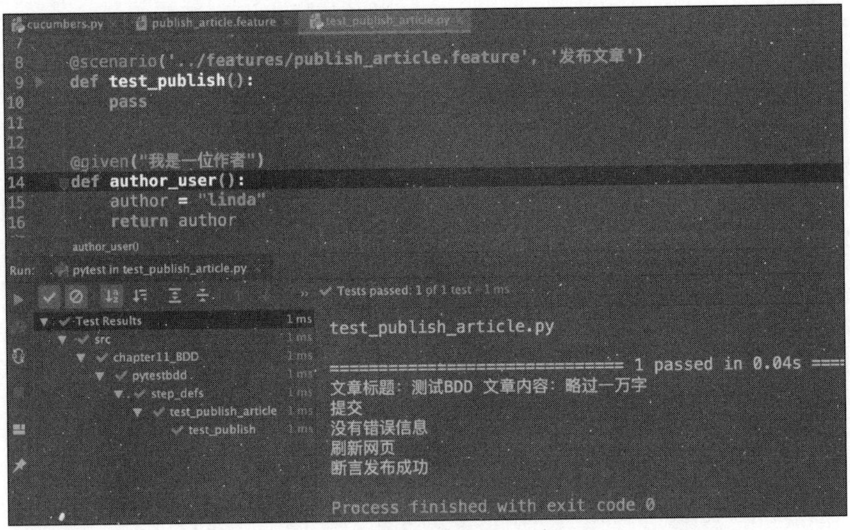

图 11-4　发布文章执行结果

11.5.5　BDD 单元测试实践：添加功能（单一数据）

（1）开发人员开发一个用篮子装黄瓜的功能。

在初始化的方法中传入初值和最大值，默认最大值为 10，添加初值和最大值限制，这两个值不能小于0,建立一个添加和删除的方法。

代码如下：

```python
# File: cucumbers.py
class CucumberBasket:

    def __init__(self, initial_count = 0, max_count = 10):
        if initial_count < 0:
            raise ValueError("未放物品,空篮子,不能是负值")
        if max_count < 0:
            raise ValueError(f"最多放{max_count}个,不能是负值")

        self._count = initial_count
        self._max_count = max_count

    @property
    def count(self):
        return self._count

    @property
    def full(self):
        return self.count == self.max_count

    @property
    def empty(self):
        return self.count == 0

    @property
    def max_count(self):
        return self._max_count

    def add(self, count = 1):
        new_count = self.count + count
        if new_count > self.max_count:
            raise ValueError("超过篮子的最大值")
        self._count = new_count

    def remove(self, count = 1):
        new_count = self.count - count
        if new_count < 0:
            raise ValueError("没有那么多的黄瓜")
        self._count = new_count
```

（2）创建 cucumbers.feature 文件。

本书使用 BDD 方式进行单元测试,在 features 包下创建 cucumbers.feature 文件,对用篮子装黄瓜这一功能进行描述,这是最简单的添加功能的方法。

```
Feature: 用篮子装黄瓜
  As 笔者是菜农
  I 想要将摘到的黄瓜放到篮子中
  So 笔者不会把它们都扔了

  Scenario: 将摘到的黄瓜放到篮子中
     Given 篮子原来有 2 个黄瓜
     When 将 4 个黄瓜添加到篮子中
     Then 篮子中有 6 个黄瓜
```

(3) 在 step_defs 包下创建测试文件 test_cucumbers.py。

代码如下：

```python
#File: test_cucumbers.py
from pytest_bdd import scenario,given,when,then

from cucumbers import CucumberBasket


@scenario('../features/cucumbers.feature','将摘到的黄瓜放到篮子中')
def test_add():
    pass


@given("篮子原来有 2 个黄瓜")
def basket():
    #初始化要测试的篮子方法,传入初始化数据,起始数量为 2
    return CucumberBasket(initial_count = 2)


@when("将 4 个黄瓜添加到篮子中")
def add_cucumbers(basket):
    #basket 参数类似 pytest.fixture,basket()函数的返回值
    #是 CucumberBasket 类的实例
    #执行添加功能
    basket.add(4)


@then("篮子中有 6 个黄瓜")
def basket_has_total(basket):
    #获得 CucumberBasket 的属性并断言添加是否成功
    assert basket.count == 6
```

执行过程大致如下：
(1) 执行 test_add()函数。
(2) 先执行@scenario 装饰器中的内容。
(3) 找到 cucumbers.feature 中的 scenario"将摘到的黄瓜放到篮子中"的场景。

（4）执行"将摘到的黄瓜放到篮子中"的场景。
（5）按"将摘到的黄瓜放到篮子中"场景的步骤执行 Given。
（6）按"将摘到的黄瓜放到篮子中"场景的步骤执行 when。
（7）按"将摘到的黄瓜放到篮子中"场景的步骤执行 then。

@given 类似于@pytest.fixture，basket 函数的调用通过函数名并以参数方式调用，数据传递过程：

（1）在 basket()函数中初始化 CucumberBasket(initial_count=2)，并且将初始值设置为 2。
（2）add_cucumbers(basket)函数中 basket 是上面函数的返回值。
（3）basket.add(4)是添加 4 根黄瓜，并返回合计值。
（4）basket_has_total(basket)函数中 basket 是上面函数的返回值，这个返回类似于篮子的实例。
（5）assert basket.count==6 用于断言数量是否正确。

11.5.6　BDD 单元测试实践：添加和删除功能（数据通过参数传递）

在 features 包下创建 cucumbers_variable.feature 文件，对添加和删除功能进行描述，要测试的数据通过传递参数的形式传递。此外，还要在传递的数据上加" "。

代码如下：

```
Feature: 用篮子装黄瓜
  As 笔者是菜农
  I 想要将摘到的黄瓜放到篮子中
  So 笔者不会把它们都扔了

  Scenario: 将摘到的黄瓜放到篮子中
    Given 篮子原来有 "2" 根黄瓜
    When 将 "4" 根黄瓜添加到篮子中
    Then 篮子中有 "6" 根黄瓜

  Scenario: 从篮子中拿出黄瓜
    Given 篮子原来有 "8" 根黄瓜
    When 从篮子中拿出 "3" 根黄瓜
    Then 篮子中剩余 "5" 根黄瓜
```

在 step_defs 包下创建测试文件 test_cucumbers_varible.py，在 feature 中的" "中的数字通过"{参数名:传递值}"传递参数。

代码如下：

```
#File: test_cucumbers_varible.py

from pytest_bdd import scenarios,parsers,given,when,then
from cucumbers import CucumberBasket
```

```python
scenarios('../features/cucumbers_varible.feature')

EXTRA_TYPES = {
    'Number': int,
}


# 通过外部传递参数的方式,initial 对应参数,Number 对应要传递的数据
@given(parsers.cfparse('篮子原来有 "{initial:Number}" 根黄瓜', extra_types = EXTRA_TYPES))
def basket(initial):
    return CucumberBasket(initial_count = initial)


@when(parsers.cfparse('将 "{some:Number}" 根黄瓜添加到篮子中', extra_types = EXTRA_TYPES))
def add_cucumbers(basket, some):
    basket.add(some)


@when(parsers.cfparse('从篮子中拿出 "{some:Number}" 根黄瓜', extra_types = EXTRA_TYPES))
def remove_cucumbers(basket, some):
    basket.remove(some)


@then(parsers.cfparse('篮子中剩余 "{total:Number}" 根黄瓜', extra_types = EXTRA_TYPES))
def basket_has_total(basket, total):
    print(basket.count)
    assert basket.count == total
```

执行结果如图 11-5 所示。

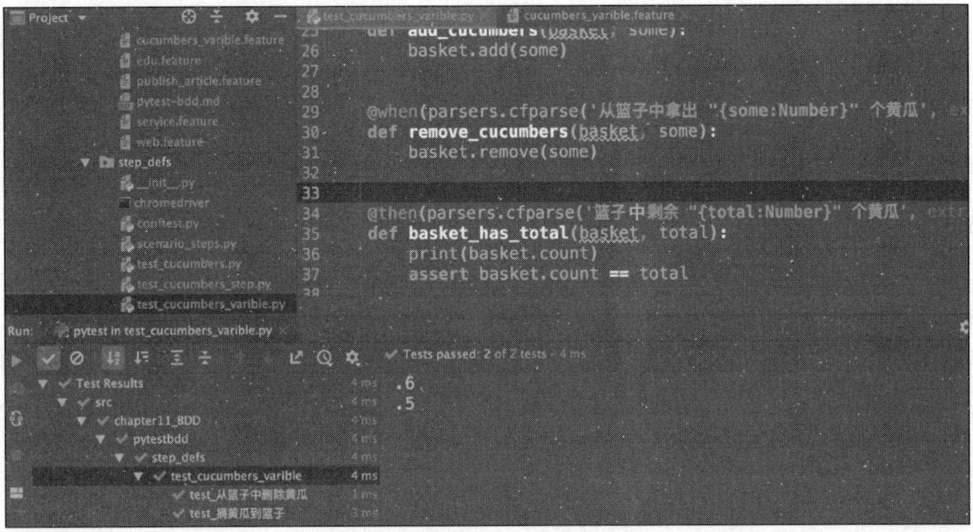

图 11-5　basket 添加功能执行结果

11.5.7 BDD 单元测试实践：数据参数化

在 features 包下创建 cucumbers_param.feature 文件，对添加和删除功能进行描述。添加功能通过变量传递数据，通过下面的多组数据（用例）测试结果是否正确。

将 Scenario 修改为 Scenario Outline，在 Examples 中加入变量及数据，以 | 竖线间隔对齐。

代码如下：

```
Feature: 用篮子装黄瓜
  As 笔者是菜农
  I 想要将摘到的黄瓜放到篮子中
  So 笔者不会把它们都扔了

  @add
  Scenario Outline: 将摘到的黄瓜放到篮子中
    Given 篮子中原来有 "<initial>" 根黄瓜
    When 将 "<some>" 根黄瓜添加到篮子中
    Then 篮子中剩余 "<total>" 根黄瓜


    Examples: Amounts
      | initial | some | total |
      | 2       | 4    | 6     |
      | 0       | 3    | 3     |
      | 5       | 5    | 10    |


  @remove
  Scenario: 从篮子中拿出黄瓜
    Given 篮子原来有 "8" 根黄瓜
    When 从篮子中拿出 "3" 根黄瓜
    Then 篮子中剩余 "5" 根黄瓜
```

在 step_defs 包下创建测试文件 test_cucumbers_param.py，在每个装饰器中添加调用，例如：@given('篮子原来有 "<initial>" 根黄瓜')，并添加变量的类型，在 scenarios 调用中增加 example_converters=CONVERTERS。

代码如下：

```
#File: test_cucumbers_param.py

from pytest_bdd import scenarios, parsers, given, when, then

from cucumbers import CucumberBasket
```

```python
EXTRA_TYPES = {
    'Number': int,
}

CONVERTERS = {
    'initial': int,
    'some': int,
    'total': int,
}
scenarios('../features/cucumbers_param.feature', example_converters=CONVERTERS)


@given(parsers.cfparse('篮子原来有 "{initial:Number}" 根黄瓜', extra_types=EXTRA_TYPES))
@given('篮子原来有 "<initial>" 根黄瓜')
def basket(initial):
    return CucumberBasket(initial_count=initial)


@when(parsers.cfparse('将 "{some:Number}" 根黄瓜添加到篮子中', extra_types=EXTRA_TYPES))
@when('将 "<some>" 根黄瓜添加到篮子中')
def add_cucumbers(basket, some):
    basket.add(some)


@when(parsers.cfparse('从篮子中拿出 "{some:Number}" 根黄瓜', extra_types=EXTRA_TYPES))
@when('从篮子中拿出 "<some>" 根黄瓜')
def remove_cucumbers(basket, some):
    basket.remove(some)


@then(parsers.cfparse('篮子中剩余 "{total:Number}" 根黄瓜', extra_types=EXTRA_TYPES))
@then('篮子中剩余 "<total>" 根黄瓜')
def basket_has_total(basket, total):
    print(basket.count)
    assert basket.count == total
```

11.5.8 BDD 接口测试实践：requests 和 pytest-bdd 实现 bing 的搜索接口

使用 BDD 简单参数化测试 bing 的搜索接口，在 features 目录下新建 interface_api.feature 文件，并添加不同的关键字。

代码如下：

```
@interface_api @lindafang
Feature: lindafang测试必应搜索
```

```
As 一个测试人员
I 想要通过对 REST API 的调用进行测试
So 能获得正确的返回


Scenario Outline: 测试必应搜索
    Given 查询关键字 "<phrase>"
    Then 响应状态码是 "200"
    And 结果包含 "<phrase>"

    Examples: 关键字
      | phrase   |
      | python   |
      | java     |
      | selenium |
```

在 step_defs 包下创建测试文件 test_interface_api.py。
代码如下:

```
#File: test_interface_api.py

import requests

from pytest_bdd import scenarios,given,then,parsers

#Shared Variables
HRM_API = 'https://cn.bing.com/search'

#Scenarios

scenarios('../features/interface_api.feature',example_converters=dict(phrase=str))


#Given Steps

@given('查询关键字 "<phrase>"')
def ddg_response(phrase):
    params = {'q': phrase,'cvid':'214DA874AFE04224A1125033896419D2'}
    headers = {
        'user-agent':'Mozilla/5.0 (Macintosh; Intel Mac OS X 10_15_7) AppleWebKit/537.36 (KHTML,like Gecko) Chrome/86.0.4240.111 Safari/537.36'
    }
    response = requests.get(HRM_API,params=params,headers=headers)
    return response
```

```
# Then Steps

@then('结果包含 "<phrase>"')
def ddg_response_contents(ddg_response,phrase):
    assert phrase in ddg_response.text


@then(parsers.parse('响应状态码是 "{code:d}"'))
def ddg_response_code(ddg_response,code):
    assert ddg_response.status_code == code
```

11.5.9　BDD UI 自动化测试实践：selenium 和 pytest-bdd 实现搜索功能

在 features 目录下新建 web_jianshu.feature 文件，并约定搜索功能的规范。
代码如下：

```
@web @lindafang
Feature: lindafang 浏览自己分享的信息
    As 在互联网上
    I 发现一条自己分享的信息
    So 笔者在网站看到的文章


    Background:
        Given LindaFang 简书页


    Scenario: 搜索 LindaFang 的文章
        When 搜索"接口测试"的文章
        Then 显示"接口测试"的文章
```

在 step_defs 目录下新建 test_web_jianshu.py 文件，chomedriver 的文件应存储在同一工程路径中。
代码如下：

```
# File: test_web_jianshu.py

import pytest

from pytest_bdd import scenarios,given,when,then,parsers
from selenium import webdriver
from selenium.webdriver.common.keys import Keys
import time
```

```python
# Constants
LindaFang_HOME = 'https://www.jianshu.com/u/030c07091ea4'


# Scenarios
scenarios('../features/web_jianshu.feature')

@pytest.fixture(scope='module')
def browser():
    # 打开浏览器
    driver = webdriver.Chrome(executable_path='Chromedriver')
    driver.implicitly_wait(10)
    yield driver
    driver.quit()


# Shared Given Steps
@given('LindaFang 简书页')
def ddg_home(browser):
    # 进入 LindaFang 简书页
    browser.get(LindaFang_HOME)
    assert '测试星云' in browser.page_source

# When Steps
@when(parsers.parse('搜索"{phrase}"的文章'))
def search_phrase(browser, phrase):
    search_input = browser.find_element_by_name('q')
    search_input.send_keys(phrase + Keys.RETURN)


# Then Steps
@then(parsers.parse('显示"{phrase}"的文章'))
def results_have_one(browser, phrase):
    time.sleep(2)
    results = browser.page_source
    assert phrase in results
```

11.5.10　BDD UI 自动化测试实践：selenium 和 pytest-bdd 实现搜索功能参数化

在 step_defs 目录下创建 conftest.py 文件，把公共的方法复制到这里。代码如下：

```python
# File: conftest.py
import pytest
from pytest_bdd import given
from selenium import webdriver

# Constants

LindaFang_HOME = 'https://www.jianshu.com/u/030c07091ea4'

# Fixtures

@pytest.fixture(scope = 'module')
def browser():
    # 打开浏览器
    driver = webdriver.Chrome(executable_path = 'Chromedriver')
    driver.implicitly_wait(10)
    yield driver
    driver.quit()


# Shared Given Steps

@given('LindaFang 简书页')
def ddg_home(browser):
    # 进入 LindaFang 简书的页
    browser.get(LindaFang_HOME)
    assert '测试星云' in browser.page_source
```

在 web_jianshu.feature 中加入参数搜索的场景。

代码如下：

```
Scenario: 搜索任何 LindaFang 的文章
    When 搜索任何一个"<phrase>"的文章
    Then 显示这个"<phrase>"所包含的文章

    Examples: 单词
        | phrase    |
        | test      |
        | interface |
        | App       |
```

在 test_web_jianshu.py 文件的 search_phrase 函数中添加@when(parsers.parse('搜索任何一个"<phrase>"的文章')),再添加一个函数,用于测试参数化数据后的结果,修改整个文件。

代码如下:

```python
# File: test_web_jianshu.py

import pytest

from pytest_bdd import scenarios, given, when, then, parsers
from selenium import webdriver
from selenium.webdriver.common.keys import Keys
import time

# Constants

LindaFang_HOME = 'https://www.jianshu.com/u/030c07091ea4'


# Scenarios

scenarios('../features/web_jianshu.feature')

# When Steps

@when(parsers.parse('搜索"{phrase}"的文章'))
@when(parsers.parse('搜索任何一个"<phrase>"的文章'))
def search_phrase(browser, phrase):
    search_input = browser.find_element_by_name('q')
    search_input.send_keys(phrase + Keys.RETURN)


# Then Steps

@then(parsers.parse('显示"{phrase}"的文章'))
def results_have_one(browser, phrase):
    time.sleep(2)
    results = browser.page_source
    assert phrase in results


@then(parsers.parse('显示这个"<phrase>"所包含的文章'))
def search_results(browser, phrase):
    time.sleep(2)
    results = browser.page_source
    assert phrase in results
    # Check search phrase
    search_input = browser.find_element_by_name('q')
    assert search_input.get_attribute('value') == phrase
```

最终运行的结果如图 11-6 所示。

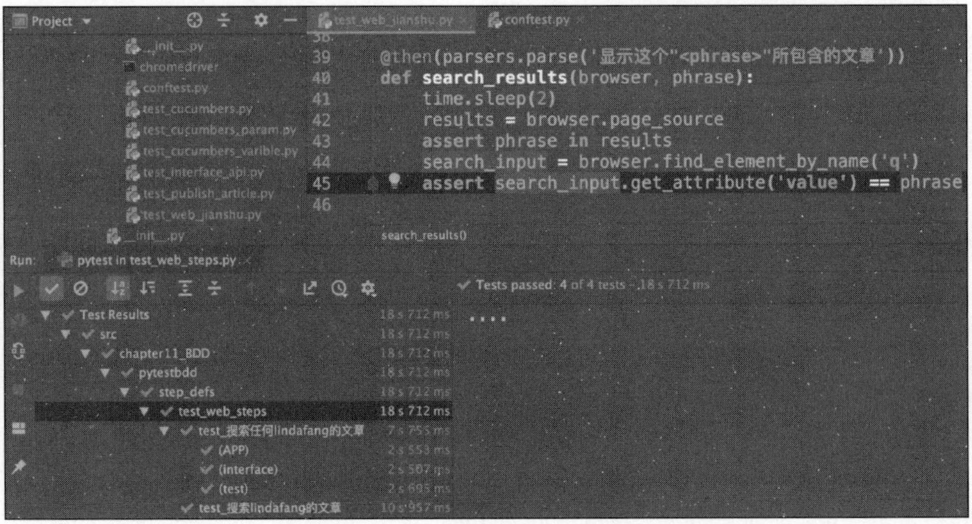

图 11-6　UI 搜索功能执行结果

11.6　本章小结

这一章主要讲解如何使用 BDD 行为驱动开发实现自动化测试，这是一种将客户及老板在开发测试中可能出现的沟通误解通过自然语言来解决，即通过技术解决沟通成本和问题：

（1）BDD 中 pytest-bdd 的介绍安装。

（2）pytest-bdd 在单元测试中的应用。

（3）pytest-bdd 在接口测试中的应用。

（4）pytest-bdd 在 Web-UI 测试中的应用。

第 12 章 pytest.ini 配置及其他配置

本书使用 pytest 运行各种范围的测试用例,跳过和标记失败用例,运行中断错误管理及查看运行结果和设置报告,可通过 pytest.ini 和其他配置改变其中的设置。

12.1 pytest 中的各种配置

pytest.ini 配置文件能够改变 pytest 框架代码的运行规则。例如修改 pytest 收集用例的规则,添加命令行参数等。下面来一一讲解常用的一些配置项。

通过命令 pytest --help 查看配置文件中可以添加的一些参数及选项,这些选项可以添加到 pytest 的配置文件。

代码如下:

```
[pytest] ini-options in the first pytest.ini|tox.ini|setup.cfg file found:

  markers (linelist):   markers for test functions
  empty_parameter_set_mark (string):
                        default marker for empty parametersets
  norecursedirs (args): directory patterns to avoid for recursion
  testpaths (args):     directories to search for tests when no files or
                        directories are given in the command line.
  usefixtures (args):   list of default fixtures to be used with this project
  python_files (args):  glob-style file patterns for Python test module
                        discovery
  python_classes (args):
                        prefixes or glob names for Python test class discovery
  python_functions (args):
                        prefixes or glob names for Python test function and
                        method discovery
  disable_test_id_escaping_and_forfeit_all_rights_to_community_support (bool):
                        disable string escape non-ascii characters, might
                        cause unwanted side effects(use at your own risk)
```

```
console_output_style (string):
                    console output: "classic", or with additional progress
                    information ("progress" (percentage) | "count").
xfail_strict (bool): default for the strict parameter of xfail markers when
                    not given explicitly (default: False)
enable_assertion_pass_hook (bool):
                    Enables the pytest_assertion_pass hook. Make sure to
                    delete any previously generated pyc cache files.
junit_suite_name (string):
                    Test suite name for JUnit report
junit_logging (string):
                    Write captured log messages to JUnit report: one of
                    no|log|system-out|system-err|out-err|all
junit_log_passing_tests (bool):
                    Capture log information for passing tests to JUnit
                    report:
junit_duration_report (string):
                    Duration time to report: one of total|call
junit_family (string):
                    Emit XML for schema: one of legacy|xunit1|xunit2
doctest_optionflags (args):
                    option flags for doctests
doctest_encoding (string):
                    encoding used for doctest files
cache_dir (string): cache directory path.
filterwarnings (linelist):
                    Each line specifies a pattern for
                    warnings.filterwarnings. Processed after
                    -W/--pythonwarnings.
log_level (string): default value for --log-level
log_format (string): default value for --log-format
log_date_format (string):
                    default value for --log-date-format
log_cli (bool):     enable log display during test run (also known as
                    "live logging").
log_cli_level (string):
                    default value for --log-cli-level
log_cli_format (string):
                    default value for --log-cli-format
log_cli_date_format (string):
                    default value for --log-cli-date-format
log_file (string):  default value for --log-file
log_file_level (string):
                    default value for --log-file-level
log_file_format (string):
                    default value for --log-file-format
```

```
log_file_date_format (string):
                        default value for --log-file-date-format
log_auto_indent (string):
                        default value for --log-auto-indent
faulthandler_timeout (string):
                        Dump the traceback of all threads if a test takes more
                        than TIMEOUT seconds to finish.
addopts (args):         extra command line options
minversion (string):    minimally required pytest version
required_plugins (args):
                        plugins that must be present for pytest to run
pep8ignore (linelist):
                        each line specifies a glob pattern and whitespace
                        separated PEP8 errors or warnings which will be
                        ignored, example: *.py W293
pep8maxlinelength (string):
                        max. line length (default: 79)
yaml_loader (string):   Which loader to use when parsing yaml
flakes-ignore (linelist):
                        each line specifies a glob pattern and whitespace
                        separated pyflakes errors which will be ignored,
                        example: *.py UnusedImport
base_URL (string):      base URL for the application under test.
render_collapsed (bool):
                        Open the report with all rows collapsed. Useful for
                        very large reports
rsyncdirs (pathlist):   list of (relative) paths to be rsynced for remote
                        distributed testing.
rsyncignore (pathlist):
                        list of (relative) glob-style paths to be ignored for
                        rsyncing.
looponfailroots (pathlist):
                        directories to check for changes
selenium_capture_debug (string):
                        when debug is captured ('never', 'failure', 'always')
selenium_exclude_debug (string):
                        debug to exclude from capture
saucelabs_job_auth (string):
                        Authorization options for the Sauce Labs job: ('none',
                        'token', 'hour', 'day')
sensitive_URL (string):
                        regular expression for identifying sensitive URLs.
bdd_features_base_dir (string):
                        Base features directory.
bdd_strict_gherkin (bool):
                        Parse features to be strict gherkin.
```

下面按顺序解释一下常用的参数的使用。由于在使用时报的警告比较多,所以单独列出进行讲解。

12.1.1 @pytest.marker 标记用例

在 pytest.ini 中设置 markers,这样在代码中使用时报的警告才会消失。记得把 testpaths 禁用,否则只会在 testpaths 路径中搜索。

实践的代码可参考 10.3.6 节和 10.3.7 节。

```
;testpaths = Testcases

markers =
    web
    app
    p0
```

创建 test_markers.py 文件,代码如下:

```
#File: test_markers.py
import pytest

@pytest.mark.web
def test_pass():
    print("web")
    assert 1

@pytest.mark.app
def test_new_pass():
    print("app")
    assert 1

@pytest.mark.p0
def test_new_mark():
    print("p0")
    assert 1
```

大家可以试试在 cmd 中执行,命令如下:

```
pytest -s -m "p0"
pytest -s -m "app"
pytest -s -m "web"
```

最后一条命令执行结果如图 12-1 所示。

第12章 pytest.ini配置及其他配置

```
collected 16 items / 15 deselected / 1 selected

test_markers.py web
.
========================= 1 passed, 15 deselected in 0.06s ==
lindafang@cpe-172-115-250-122 chapter12 % pytest -s -m "web"
```

图 12-1 pytest.ini 标记后警告消失

12.1.2 添加测试用例路径

添加测试用例路径,代码如下:

```
[pytest]
testpaths = TestCases
```

通过这一项设置,可以把用例所在的目录添加到配置文件中,这样在运行用例的时候,pytest 会直接在配置文件所在的目录中搜索用例。在目录 chapter12 下输入 pytest,正常应该搜索到所有用例,但测试目录设置后只执行测试目录下符合要求的文件,如图 12-2 所示。

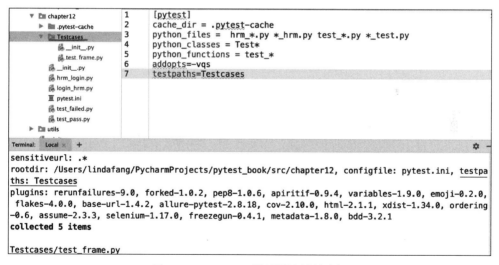

图 12-2 pytest.ini 设置测试用例路径

12.1.3 指定 pytest 忽略哪些搜索目录

代码如下:

```
[pytest]
norecursedirs = .* venv
```

标记部分为系统默认不会搜索的路径,前面是用户自定义的路径,注意,当自定义时最好把系统默认的搜索路径添加到后面。

12.1.4　usefixtures 的默认配置

在 pytest.ini 中配置 usefixtures 的值,如果在使用时与配置值不同,则可以通过参数传入配置值。

```
usefixtures =
    driver
    login
```

如果 usefixtures 后面没有跟具体的 fixture,则读取配置文件中默认的值,这个默认值必须存在,否则会报错。

代码如下:

```python
#File: test_usefixtures.py
import pytest

@pytest.fixture()
def driver():
    driver = 8
    print(driver)
@pytest.fixture()
def login():
    print("登录")

@pytest.mark.usefixtures
def test_driver():
    print(driver)
```

执行结果如图 12-3 所示。

12.1.5　修改测试用例的搜索匹配规则

在帮助中我们知道 pytest 默认查找用例匹配规则如下:
- python_files (args)匹配 Python 用例文件,如 test_*.py、*_test.py。
- python_classes (args)匹配 class 类名称,如 Test*。
- python_functions (args)匹配函数和 class 里面的方法,如 test_*。

也就是说测试文件以 test_开头或者以_test 结尾。测试类以 Test 开头,并且不能带有 init 方法。测试函数以 test_开头。

我们可以通过修改匹配规则搜索具有特定名的文件、类和函数。

在 pytest.ini 文件添加一项 python_files,代码如下:

第12章 pytest.ini配置及其他配置

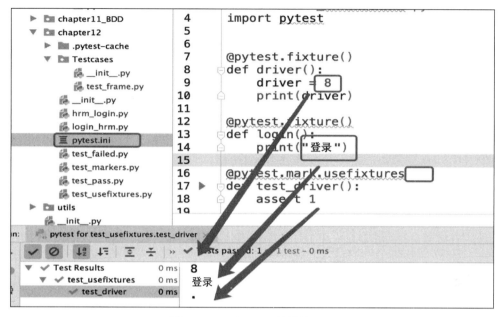

图 12-3 usefixtures 的默认配置

```
[pytest]

python_files = xxx_*.py
```

例如：在项目根目录的 pytest.ini 中添加如下信息：

```
[pytest]

python_files =  hrm_*.py *_hrm.py
python_classes = Test*
python_functions = test_*
```

在路径中有两个文件，一个是 hrm_login.py，另一个是 login_hrm.py，每个文件下有两种方法。在终端输入命令 pytest -s -v 执行测试，查看结果。
代码如下：

```
pytest -s -v
========================= test session starts =========================
platform darwin -- Python 3.6.8, pytest-6.0.1, py-1.9.0, pluggy-0.13.1 -- /Library/Frameworks/Python.framework/Versions/3.6/bin/python3.6
cachedir: .pytest-cache
...
```

```
configfile: pytest.ini
plugins: ......
collected 4 items

hrm_login.py::test_login hello,login
PASSED
hrm_login.py::test_hrm world,hrm
PASSED
login_hrm.py::test_1 hello
PASSED
login_hrm.py::test_2 world
PASSED

====================== 4 passed in 0.04s ======================
```

结果显示 hrm_login.py 和 login_hrm.py 文件也执行测试了。这样我们就不用重命名所有测试文件以 test_ 开头了。

12.1.6 ids 中解决中文显示乱码问题

参考 4.3.4 节，在 pytest.ini 中将这个值设置为 True 后，ids 参数中的中文可正常显示。

```
[pytest]
disable_test_id_escaping_and_forfeit_all_rights_to_community_support = True
```

12.1.7 console_output_style 输出样式配置

在运行测试时设置控制台输出样式：
- classic：经典 pytest 输出。
- progress：类似于经典的 pytest 输出，但带有进度指示器。
- count：与进度类似，但进度显示为已完成的测试数而不是百分比。

默认值为 progress，但可以退回到 classic 模式，代码如下：

```
# content of pytest.ini
[pytest]
console_output_style = classic
```

在 classic 模式下执行时没有显示进度的百分比，如图 12-4 所示。
将输出样式设置为 count，输出如图 12-5 所示。

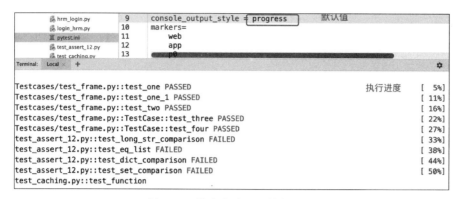

图 12-4　输出格式配置执行进度

图 12-5　输出格式配置执行进度为数字

12.1.8　xfail_strict 不是预期的失败显示 FAILED

使用的代码可参考 2.7.1 节。pytest.ini 配置如下：

```
[pytest]
xfail_strict = true
```

如果用例的失败不是由所期望的异常导致的，则 pytest 会把测试结果标记为 FAILED。

12.1.9　cache_dir 缓存目录设置

参考 2.5.2 节缓存对于失败运行的实践。下面再举例说明，在大数据处理时如果可以使用缓存数据进行计算处理，则速度会更快。也就是可以将数据库中要计算的数据先保存到缓存目录下，计算是先读取这个目录下的值，如果没有，则可以先设置这个值，再进行计算。这样下次就可以直接使用了。

在执行过程中，如果第一次没有缓存，则应在缓存目录的相对路径中对相应文件设置缓存数据，然后进行计算，当再一次运行时，可以直接从缓存中读取数据进行运行。

代码如下:

```
import pytest

def expensive_computation():
    print("运行一个大数据量的计算...")


@pytest.fixture
def mydata(request):
    #从缓存中读取数据
    val = request.config.cache.get("bigdata/value", None)
    if val is None:
        val = 40
        #如果缓存中没有数据,则先设置缓存中的值
        request.config.cache.set("bigdata/value", val)
    expensive_computation()
    return val


def test_function(mydata):
    assert mydata == 18
```

第 1 次执行,如果缓存中没有数据,则应将缓存中的值设置为 40,断言 40!＝18 失败,如图 12-6 所示。

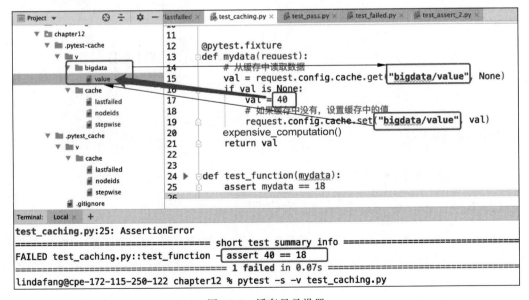

图 12-6　缓存目录设置

第 2 次执行时,把 val＝40 改成 val＝18,执行时,未重新设置 val 的值,如图 12-7 所示。

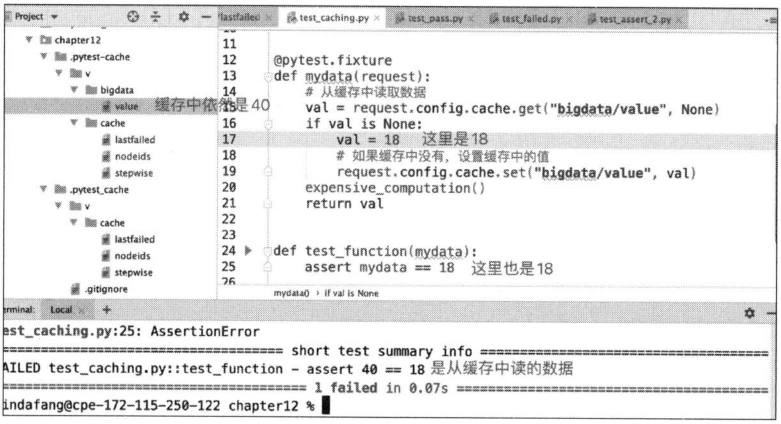

图 12-7　缓存目录设置，先从缓存中读数据

12.1.10　filterwarnings 警告过滤

设置应对匹配警告所采取的筛选器和操作的列表。默认情况下，测试会话期间发出的所有警告都将显示在测试会话结束时的摘要中。

```
# content of pytest.ini
[pytest]
filterwarnings =
    error
    ignore::DeprecationWarning
```

这样 pytest 会忽略（ignore）"拒绝警告"（DeprecationWarning）并将所有其他警告变为错误。详细例子可参考 12.2 节。

12.1.11　log 相关配置

参考 10.3.12 节 App 自动化测试代码中 log 的使用。

pytest.ini 中相关 log 的说明分为三部分：日志、运行中日志、日志文件配置说明。

```
log_level(string):默认值 -- log level
log_format(string):默认值 -- log format
log_date_format (string):日志日期格式的默认值
log_cli(bool):在测试运行期间启用日志显示,为 True 或 1 表示启动日志显示
log_cli_level: -- log cli level 的默认值,例如 INFO、ERROR...
log_cli_format(string): -- log cli format 的默认值
log_cli_date_format(string): -- log cli 日期格式的默认值
log_file(string):默认值 -- log file,标记文件所有地址
log_file_level (string):默认值 -- log file level
log_file_format (string):日志文件格式的默认值
log_file_date_format (string):日志文件日期格式的默认值
log_auto_indent (string):默认值 -- log auto indent
```

在pytest.ini中设置日志，log_cli＝1是启动实时日志，log_cli_level＝INFO是日志INFO，以上的信息全部显示。下面列举3种与log相关配置的具体格式：log_cli_format＝％(时间)s［％(日志级别)8个位置］％(信息)s（％(文件名)s：％(行号)s）、log_cli_date_format＝％年-％月-％日 ％时：％分：％秒、log_file＝logs/pytest-logs.txt，日志文件保存位置。

代码如下：

```
[pytest]

log_cli = 1
log_cli_level = INFO
log_cli_format = %(asctime)s [%(levelname)8s] %(message)s (%(filename)s:%(lineno)s)
log_cli_date_format = %Y-%m-%d %H:%M:%S
log_file = logs/pytest-logs.txt
```

在任何一个文件中加入一句上述日志便可进行测试。

在设置的路径中生成文件，在控制台中打印输出，如图12-8所示。

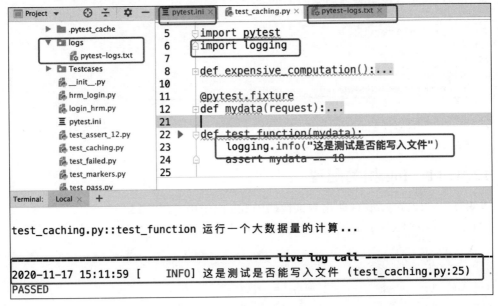

图12-8 log配置及执行结果

因为写入文件的格式未定义，所以写入格式与控制台输出不同，如图12-9所示。

12.1.12 添加pytest执行默认参数选项

pytest可以在cmd中使用命令行运行脚本，通常采用命令pytest -vqs运行脚本，通过这个命令来运行一下第1章中的脚本。

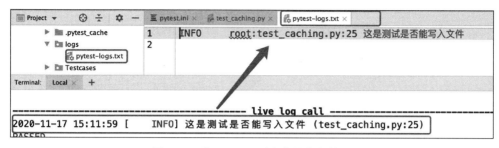

图 12-9　将 log 配置到文件的执行结果

代码如下：

```
pytest -vqs test_frame.py
=============== test session starts ========================================
platform darwin -- Python 3.6.8, pytest-6.0.1, py-1.9.0, pluggy-0.13.1
sensitiveURL: . *
rootdir: /Users/PyCharmProjects/pytest_book
plugins: rerunfailures-9.0, forked-1.0.2, pep8-1.0.6, apiritif-0.9.4, variables-1.9.0,
emoji-0.2.0, flakes-4.0.0, base-URL-1.4.2, allure-pytest-2.8.18, cov-2.10.0, html-
2.1.1, xdist-1.34.0, ordering-0.6, assume-2.3.3, selenium-1.17.0, freezegun-0.4.1,
metadata-1.8.0, bdd-3.2.1
collected 5 items

test_frame.py
setup_module:整个 test_module.py 模块只执行一次
setup_function:每个非类函数测试用例开始前都会执行
正在执行测试模块 ---- test_one
.teardown_function:每个非类函数测试用例结束后都会执行
setup_function:每个非类函数测试用例开始前都会执行
正在执行测试模块 ---- test_one_1
.teardown_function:每个非类函数测试用例结束后都会执行
setup_function:每个非类函数测试用例开始前都会执行
正在执行测试模块 ---- test_two
.teardown_function:每个非类函数测试用例结束后都会执行


setup_class:在类中所有测试用例执行之前
setup_method:每个类中测试方法用例开始前执行
setup:每个类中测试方法用例开始前都会执行
正在执行测试类 ---- test_three
.teardown:每个类中测试方法用例结束后都会执行
teardown_method:每个类中测试方法用例结束后执行
setup_method:每个类中测试方法用例开始前执行
setup:每个类中测试方法用例开始前都会执行
正在执行测试类 ---- test_four
.teardown:每个类中测试方法用例结束后都会执行
```

```
teardown_method：每个类中测试方法用例结束后执行
teardown_class：在类中所有测试用例执行之后
teardown_module：整个 test_module.py 模块只执行一次
```

```
========================= 5 passed in 0.06s =============================
```

在项目根目录下新建 pytest.ini 文件并输入下面的选项：

```
[pytest]
addopts = -vqs
```

当再次运行脚本时仅通过命令 pytest test_frame.py 即可运行，这时候会发现输出信息和上面的输出信息相同，这就是配置文件的作用。

```
addopts = -v --rerun 1 --alluredir=./reports
```

12.1.13　minversion 的设置及限制

当在 pytest.ini 中设置 minversion=7.0 时，执行测试，由于当前版本是 6.0.1，所以无法执行这个测试。也就是当最低版本不符合要求时，无法执行。报错信息如图 12-10 所示。

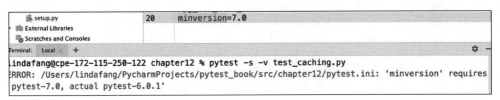

图 12-10　minversion 的设置及限制

12.1.14　required_plugins 需要的插件

插件必须存在，pytest 才能运行此插件，可以采用空格的形式分隔列表。插件可以直接在其名称后面列出，也可以不带版本说明符。不同版本说明符之间不允许有空格。如果没有找到任何一个插件，则发出一个错误。

```
[pytest]
required_plugins = pytest-Django>=3.0.0,<4.2.0 pytest-html pytest-xdist>=1.0.0
```

插件未满足条件时，执行的效果如图 12-11 所示。
满足条件后，执行结果如图 12-12 所示。

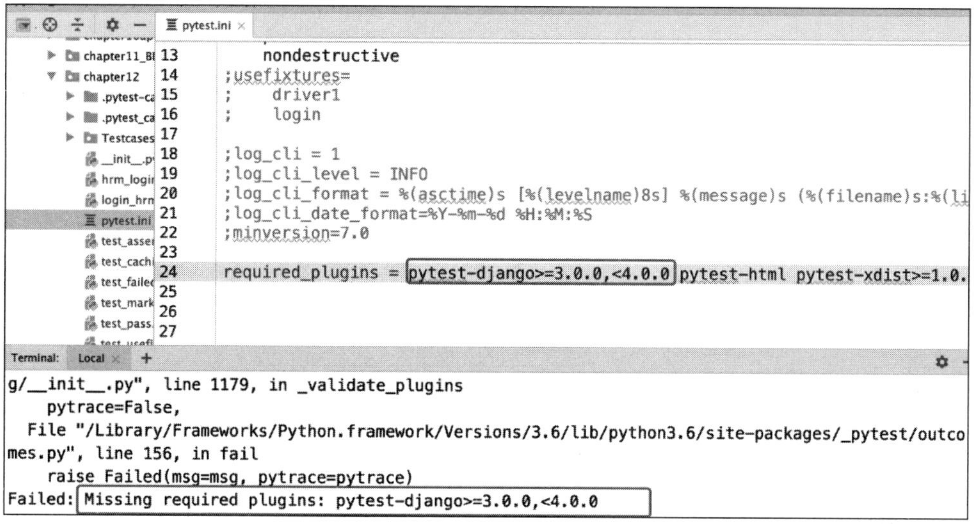

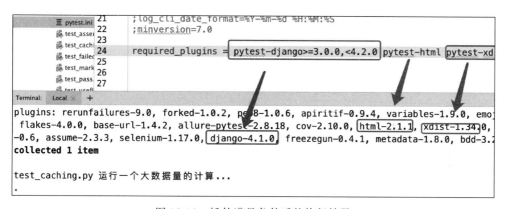

图 12-11 插件未满足条件的执行结果

图 12-12 插件满足条件后的执行结果

12.2 警告相关配置

12.2.1 警告信息的默认捕获行为

pytest 可以自动捕获测试中产生的警告信息,并在执行结束后进行展示。下面的这个例子,在测试中人为地产生一条警告。

代码如下:

```
import warnings
import pytest
```

```
def api_v1():
    warnings.warn(UserWarning('请使用新版本的API.'))
    return 1

def test_one():
    assert api_v1() == 1
```

也可以通过-W arg 命令行选项来自定义警告的捕获行为。

```
arg 参数的格式为:action:message:category:module:lineno;
action 只能在"error" "ignore" "always(all)" "default" "module" "once"中取值,默认取值为 default;
category 必须是 Warning 的子类,默认取值为 Warning 类,表示所有的警告;
module 必须为字符串,表示特定模块(文件)产生的警告信息;
```

下面是一些常见的使用场景：忽略某一种类型的警告信息，也就是当出现这类警告信息时当作没有警告。例如，忽略 UserWarning 类型的警告(-W ignore::UserWarning)。

如果不加参数，则执行的结果如图 12-13 所示。

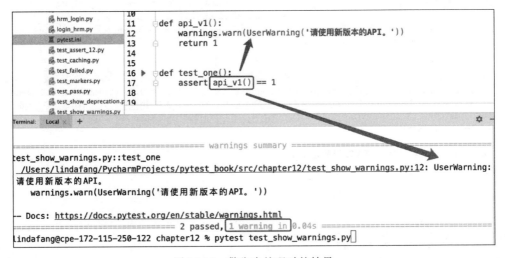

图 12-13　警告未处理时的结果

12.2.2　命令行配置警告是否捕获

如果加上-W ignore::UserWarning 这个参数，则执行后就没有警告了。
代码如下：

```
pytest -W ignore::UserWarning test_show_warnings.py
========================= test session starts =========================
platform win32 -- Python 3.7.3, pytest-5.1.3, py-1.8.0, pluggy-0.13.0
```

```
rootdir: D:\Personal Files\Projects\pytest-chinese-doc
collected 1 item

test_show_warnings.py .                                    [100%]

========================= 1 passed in 0.02s =========================
```

12.2.3 将警告转换成异常失败

将某一种类型的警告转换为异常来处理,也就是显示为不正确的测试。将 UserWarning 警告转换为异常处理(-W error::UserWarning),如图 12-14 所示。

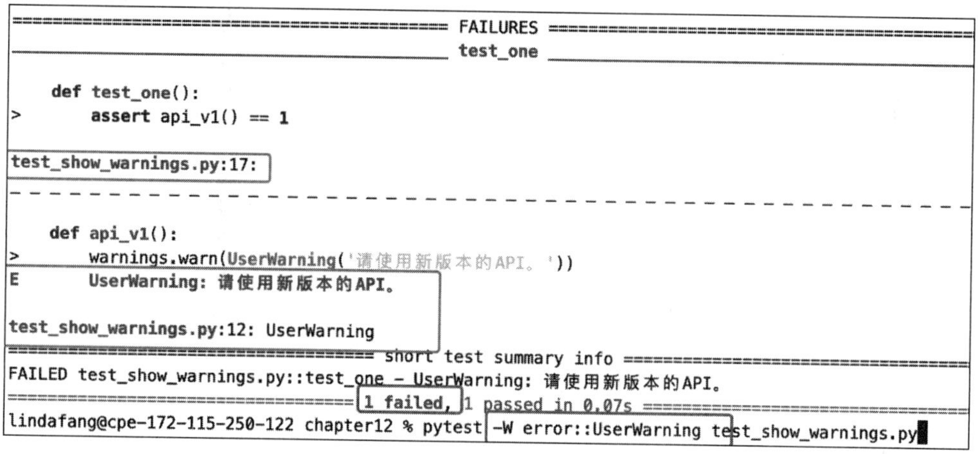

```
========================= FAILURES =========================
_____ test_one _____

    def test_one():
>       assert api_v1() == 1

test_show_warnings.py:17:
_ _ _ _ _ _ _ _ _ _ _ _ _ _ _ _ _ _ _ _ _ _ _ _ _ _ _ _ _ _

    def api_v1():
>       warnings.warn(UserWarning('请使用新版本的API。'))
E       UserWarning: 请使用新版本的API。

test_show_warnings.py:12: UserWarning
===================== short test summary info =====================
FAILED test_show_warnings.py::test_one - UserWarning: 请使用新版本的API。
=================== 1 failed, 1 passed in 0.07s ===================
lindafang@cpe-172-115-250-122 chapter12 % pytest -W error::UserWarning test_show_warnings.py
```

图 12-14　警告转换为异常时的结果

12.2.4 通过 pytest.ini 设置 filterwarnings 实现

也可以通过在 pytest.ini 文件中配置 filterwarnings 项,实现同样的效果。
修改代码如下:

```python
import warnings
import pytest


def api_v1():
    warnings.warn(UserWarning('请使用新版本的 API.'))
    return 1
def api_v2():
    warnings.warn(UnicodeWarning('未转换成汉语'))
    return 2
```

```python
def test_one():
    assert api_v1() == 1

def test_two():
    assert api_v1() == 1
    assert api_v2() == 2
```

在 pytest.ini 文件中将 filterwarnings 配置为：error 表示所有警告都为异常，ignore 表示忽略其中一种警告，也就是依然是警告，但其他警告被执行时显示为错误 F。

代码如下：

```
[pytest]
filterwarnings =
    error
    ignore::UserWarning
```

不带 -W 选项，然后执行，代码如下：

```
pytest test_show_warnings.py
=============== test session starts ========================================
platform darwin -- Python 3.6.8, pytest-6.0.1, py-1.9.0, pluggy-0.13.1
sensitiveURL: .*
rootdir: /Users/lindafang/PyCharmProjects/pytest_book/src/chapter12, configfile: pytest.ini
plugins: rerunfailures-9.0, forked-1.0.2, pep8-1.0.6, apiritif-0.9.4, variables-1.9.
0, emoji-0.2.0, flakes-4.0.0, base-URL-1.4.2, allure-pytest-2.8.18, cov-2.10.0,
html-2.1.1, xdist-1.34.0, ordering-0.6, assume-2.3.3, selenium-1.17.0, Django-4.1.0,
freezegun-0.4.1, metadata-1.8.0, bdd-3.2.1
collected 2 items

test_show_warnings.py::test_one PASSED
[ 50%]
test_show_warnings.py::test_two FAILED
[100%]

=================== FAILURES ========================================
_____ test_two _____

    def test_two():
        assert api_v1() == 1
>       assert api_v2() == 2

test_show_warnings.py:26:
_____

    def api_v2():
```

```
>           warnings.warn(UnicodeWarning('未转换成汉语'))
E           UnicodeWarning: 未转换成汉语

test_show_warnings.py:15: UnicodeWarning
============= short test summary info =====================================
FAILED test_show_warnings.py::test_two - UnicodeWarning: 未转换成汉语
================ 1 failed, 1 passed in 0.07s ==============================
```

-W 其实是 Python 本身自带的命令行选项，可以通过访问官方文档来了解更多关于此选项的说明，网址为 https://docs.python.org/3.7/library/warnings.html#warning-filter。

12.2.5　使用@pytest.mark.filterwarnings 装饰器实现警告忽略

上述操作是在命令行中实现的，如果想要在用例、类、甚至模块级别上自定义警告的捕获行为，上面的方法就不是很便利了，这里，可以通过为测试项添加警告过滤器实现这种需求。

在 test_show_warnings 文件中的 test_two 设置禁止捕获由它所触发的用户警告，代码如下：

```
@pytest.mark.filterwarnings('ignore::UserWarning')
def test_two():
    assert api_v1() == 1
```

执行这个用例，pytest.ini 中未设置：

```
pytest -k "test_two"
```

执行的结果如图 12-15 所示。

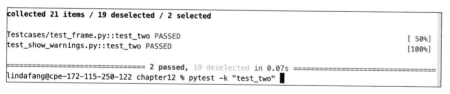

图 12-15　使用 mark 标记警告的结果

没有捕获任何警告信息，这说明通过@pytest.mark.filterwarnings 添加的过滤器优先级要高于命令行或 pytest.ini 添加的过滤器。也可以通过执行 test_one 用例来对比它们之间的不同。

可以通过将@pytest.mark.filterwarnings 应用于测试类来为这个类中所有的用例添加警告过滤器，例如：@pytest.mark.filterwarnings('ignore::UnicodeWarning')。

加上此过滤器就忽略警告，不加此过滤器就不忽略警告。如图 12-16 所示，注释后显示警告。

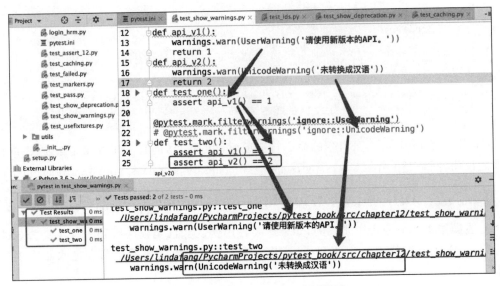

图 12-16　警告过滤和未过滤的结果

12.2.6　设置 pytestmark 变量实现添加警告

也可以通过设置 pytestmark 变量为整个测试模块中所有的用例添加警告过滤器，例如，将模块中所有的警告转换为异常处理。

设置 pytestmark＝pytest.mark.filterwarnings("error")，执行结果如图 12-17 所示。

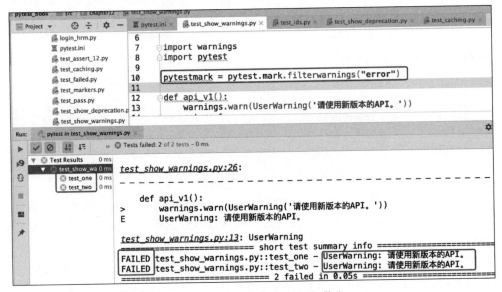

图 12-17　pytestmark 设置添加警告

12.2.7 命令行选项参数去掉警告信息

可以通过--disable-warnings 命令行选项来禁止警告信息的展示,在测试输出中不展示 test_one 用例所产生的警告信息:

```
pytest -k "test_one" --disable-warnings test_show_warnings.py
====================== test session starts =============================
platform darwin -- Python 3.6.8, pytest-6.0.1, py-1.9.0, pluggy-0.13.1
sensitiveURL: .*
rootdir: /Users/lindafang/PyCharmProjects/pytest_book/src/chapter12, configfile: pytest.ini
plugins: rerunfailures-9.0, forked-1.0.2, pep8-1.0.6, apiritif-0.9.4, variables-
1.9.0, emoji-0.2.0, flakes-4.0.0, base-URL-1.4.2, allure-pytest-2.8.18, cov-2.10.0,
html-2.1.1, xdist-1.34.0, ordering-0.6, assume-2.3.3, selenium-1.17.0, Django-4.1.0,
freezegun-0.4.1, metadata-1.8.0, bdd-3.2.1
collected 2 items / 1 deselected / 1 selected

test_show_warnings.py::test_one PASSED
[100%]


============= 1 passed, 1 deselected, 1 warning in 0.03s =================
```

12.2.8 通过触发警告自定义失败时的提示消息

当使用一段代码并且期望其触发警告时,可以通过以下方法,自定义失败时的提示消息,增加其可读性。

```
import warnings
import pytest


def api_call_v1():
    # warnings.warn('v1 版本已废弃,请使用 v2 版本的 API; ', DeprecationWarning)
    return 200


def test_deprecation():
    assert pytest.deprecated_call(api_call_v1) == 200

def test_deprecation1():
    with pytest.warns(Warning) as records:
        rsp = api_call_v1()
        if not records:
            pytest.fail('期望 api_call_v1 触发一个警告,实际上没有; ')
        assert rsp == 200
```

如果 api_call_v1 中的警告不注释，也就是应该有的警告确实有，则执行的效果是都通过。

```
test_show_deprecation.py::test_deprecation PASSED
[ 50%]
test_show_deprecation.py::test_deprecation1 PASSED
[100%]

========================= 2 passed in 0.04s =========================
```

如果 api_call_v1 没有触发任何警告，也就是注释一下 api_call_v1 中的警告，pytest 就会显示 fail 提示信息。test_deprecation1 通过上下文管理器编写一个自定义的提示信息。代码及执行结果如图 12-18 所示。

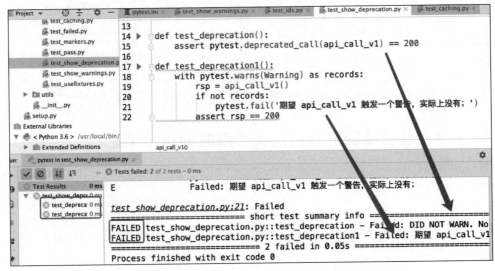

图 12-18　通过 with 自定义提示信息

12.3　内置 fixture 之临时目录

pytest 预打包的内置 fixture 可以帮助用户在测试中轻松地完成一些非常有用的事情，例如，处理临时文件、访问命令行选项、在测试会话之间通信、验证输出流、修改环境变量和询问警告等。内置的 fixture 是 pytest 核心功能的扩展。

12.3.1　tmp_path

tmp_path 是一个用例级别的 fixture，其作用是返回一个唯一的临时目录对象（pathlib.Path）。

代码如下：

```
CONTENT = "在文件中编写内容"


def test_create_file(tmp_path):
    d = tmp_path / "sub"
    d.mkdir()    # 创建一个子目录
    p = d / "hello.txt"
    p.write_text(CONTENT)
    assert p.read_text() == CONTENT
    assert len(list(tmp_path.iterdir())) == 1   # iterdir() 迭代目录,返回迭代器
    assert 0   # 为了展示,强制置为失败
```

执行结果会显示临时路径,如图 12-19 所示。

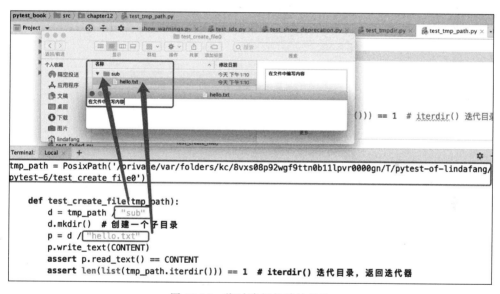

图 12-19　临时路径的默认地址

tmp_path 在不同的操作系统中,返回的是不同类型的 pathlib.Path 对象,这里在 Windows 系统下返回的是 WindowsPath 对象,它是 Path 的子类对象。

默认情况下,临时目录被创建为系统临时目录的子目录。基本名称为 pytest-NUM,其中 NUM 在每个测试运行时其值递增。此外,早于 3 个临时目录的条目将被删除。

可以像这样覆盖默认的临时目录进行设置：

```
pytest -- basetemp = mydir
```

注意：mydir 将被完全删除,因此需确保仅将目录用于该目的。

12.3.2　tmp_path_factory

tmp_path_factory 是一个会话（Session）级别的 fixture，其作用是在其他 fixture 或者用例中创建任意的临时目录。

代码如下：

```python
# _pytest.tmpdir

@pytest.fixture
def tmp_path(request, tmp_path_factory):
    """Return a temporary directory path object
    which is unique to each test function invocation,
    created as a sub directory of the base temporary
    directory.  The returned object is a :class:`pathlib.Path`
    object.

    .. note::

        in python < 3.6 this is a pathlib2.Path
    """

    return _mk_tmp(request, tmp_path_factory)

@pytest.fixture(scope = "session")
def tmp_path_factory(request):
    """Return a :class:`_pytest.tmpdir.TempPathFactory` instance for the test session.
    """
    return request.config._tmp_path_factory
```

可以看出，tmp_path 调用了 tmp_path_factory，tmp_path_factory 返回一个 _pytest. tmpdir.TempPathFactory 对象，进一步查看 _mk_tmp 的源码：

```python
def _mk_tmp(request, factory):
    name = request.node.name
    name = re.sub(r"[\W]", "_", name)
    MAXVAL = 30
    name = name[:MAXVAL]
    return factory.mktemp(name, numbered = True)
```

可以看出，tmp_path 最终调用了 TempPathFactory.mktemp() 方法，它返回的是一个 pathlib.Path 对象。

12.3.3　tmpdir

tmpdir 和 tmpdir_factory 内置功能用于在测试运行之前创建一个临时文件目录，并在测试完成后删除该目录。在某些项目中，需要一个目录来存储数据库使用的临时数据库文件，但是，由于测试希望使用无法在测试会话后仍存在的临时数据库进行测试，所以可以使用 tmpdir 和 tmpdir_factory 来创建和清理目录。

如果正在测试读取、写入或修改文件的内容，则可以使用 tmpdir 创建单个测试文件或目录，并且当要为多个测试设置目录时，可以使用 tmpdir_factory 进行设置。

tmpdir 功能具有功能作用域，tmpdir_factory 功能具有会话作用域。任何需要临时目录或文件的单独测试都可以使用 tmpdir。

代码如下：

```python
def test_tmpdir(tmpdir):
    # 在临时目录上创建 something.txt 文件
    a_file = tmpdir.join('something.txt')
    # 创建目录
    a_sub_dir = tmpdir.mkdir('anything')
    # 在目录下创建文件
    another_file = a_sub_dir.join('something_else.txt')
    # 在'something.txt'中编写
    a_file.write('这个内容为测试！')
    # 在'anything/something_else.txt'中编写
    another_file.write('这个内容有些不同！')
    # 可以读出来，并且可以断言
    assert a_file.read() == '这个内容为测试！'
    assert another_file.read() == '这个内容有些不同！'
```

--basetemp=/Users/lindafang/PyCharmProjects/pytest_book/src/chapter12/mydir 可以设置临时目录，执行下面命令的效果如图 12-20 所示。

tmpdir 是一个装饰器，代码如下：

```python
# _pytest.tmpdir

@pytest.fixture
def tmpdir(tmp_path):
    """Return a temporary directory path object
    which is unique to each test function invocation,
    created as a sub directory of the base temporary
    directory.  The returned object is a `py.path.local`_
    path object.
```

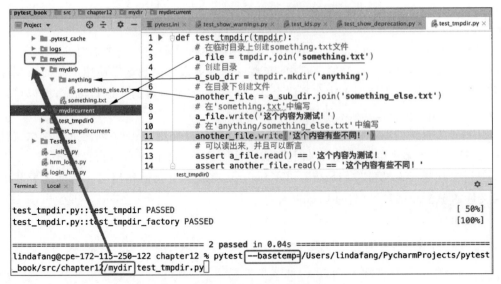

图 12-20 设置临时路径

```
.. _`py.path.local`: https://py.readthedocs.io/en/latest/path.html
"""
return py.path.local(tmp_path)
```

12.3.4 tmpdir_factory

tmpdir_factory 是一个会话级别的 fixture,其作用是在其他 fixture 或者用例中创建任意的临时目录。

代码如下：

```
def test_tmpdir_factory(tmpdir_factory):

    a_dir = tmpdir_factory.mktemp('mydir')
    base_temp = tmpdir_factory.getbasetemp()
    print('base:', base_temp)

    a_file = a_dir.join('something_factory.txt')
    a_sub_dir = a_dir.mkdir('anything')
    another_file = a_sub_dir.join('something_else_factory.txt')

    a_file.write('这个是工厂设置内容!')
    another_file.write('这个是与工厂设置不同的内容!!')

    assert a_file.read() == '这个是工厂设置内容!'
    assert another_file.read() == '这个是与工厂设置不同的内容!'
```

执行结果如下，临时路径被打印出来了/private/var/folders/kc/8vxs08p92wgf9ttn0b11lpvr0000gn/T/pytest-of-lindafang/pytest-2。

代码如下：

```
pytest -s -v test_tmpdir.py::test_tmpdir_factory
==================== test session starts ============================
platform darwin -- Python 3.6.8, pytest-6.0.1, py-1.9.0, pluggy-0.13.1 -- /Library/Frameworks/Python.framework/Versions/3.6/bin/python3.6
cachedir: .pytest_cache
sensitiveURL: .*
...
collected 1 item

test_tmpdir.py::test_tmpdir_factory base: /private/var/folders/kc/8vxs08p92wgf9ttn0b11lpvr0000gn/T/pytest-of-lindafang/pytest-2
PASSED

========================== 1 passed in 0.04s =======================
```

假设，一个测试会话需要使用一个很大的并由程序生成的图像文件，相比于每个测试用例生成一次文件，更好的做法是每个会话只生成一次，两个测试方法只使用一张图片。

代码如下：

```
import pytest


@pytest.fixture(scope="session")
def image_file(tmpdir_factory):
    img = compute_expensive_image()
    fn = tmpdir_factory.mktemp("data").join("img.png")
    img.save(str(fn))
    return fn


def test_histogram(image_file):
    print(image_file)

def test_histogram1(image_file):
    print(image_file)
```

执行结果如图 12-21 所示。

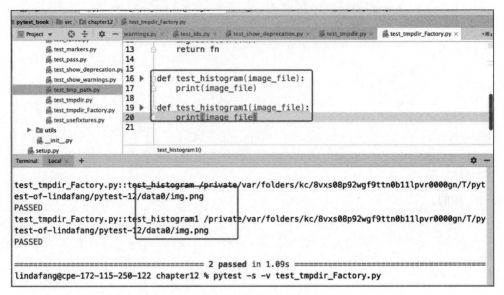

图 12-21　临时目录的设置

12.4　输出及捕获级别配置

12.4.1　标准输出/标准错误输出/标准输入的默认捕获行为

在测试执行期间，任何标准输出和标准错误输出都将被捕获。如果测试失败或者发生异常，异常信息的堆栈也将一同显示，可以通过--show-capture 命令行选项来自定义这些行为。

--show-capture 的配置项可以为：no、stdout、stderr、log、all，默认为 all。

另外，标准输入被设置为一个"null"对象，这是因为在自动化测试中，很少需要使用交互输入的场景。

实际上，当测试想要使用标准输入时，会得到一个错误：OSError：reading from stdin while output is captured。

通常情况下，捕获行为通过拦截对低级别文件描述符的写入操作实现。这就使得程序可以捕获简单的 print() 语句及测试中子程序的输出行为。

12.4.2　修改和去掉捕获行为

pytest 有两种捕获行为，可以通过--capture 命令行选项来指定。

1．文件描述符级别的捕获行为（默认）

所有向操作系统的文件描述符 1（标准输入）和 2（标准错误输入）进行写入操作的行为都会被捕获，这是 pytest 的默认捕获行为，也可以通过--capture=fd 来指定。

文件描述符是与当前进程打开的文件相对应的整数。例如，标准输入的文件描述符通常是 0，标准输出的文件描述符通常是 1，标准错误的文件描述符通常是 2，之后被进程打开的文件的描述符依次被指定为 3、4、5 等。

2．sys 级别的捕获行为

只有向 Python 的 sys.stdout 和 sys.stderr 进行写入操作的行为才会被捕获，不执行对文件描述符进行写入的行为，通过 --capture=sys 来指定。

3．去掉捕获行为

通过 --capture=no 可以去掉 pytest 的捕获行为。

也可以通过 -s 命令行选项实现相同的效果，它只是 --capture=no 的一个快捷方式，本质上是一样的。

12.4.3　在测试用例中访问捕获到的信息

可以通过 capsys、capsysbinary、capfd 和 capfdbinary fixtures 访问测试执行过程中产生的输出信息。

下面这个例子用于检查测试中的输出信息，代码如下：

```python
# test_output.py

import sys


def test_output(capsys):
    print('hello')
    print('world', file = sys.stderr, end = '&')    # 标准错误输出，修改结束符
    captured = capsys.readouterr()
    assert captured.out == 'hello\n'                # print() 默认的结束符是换行符
    assert captured.err == 'world&'
    print('next')
    captured = capsys.readouterr()
    assert captured.out == 'next\n'
```

readouterr() 方法会返回一个命名元组（包含 out 和 err 属性），表示到目前为止所有的标准输出和标准错误输出，然后重置缓存区。

如果想访问文件描述符级别的测试输出，可以使用 capfd fixture，它提供了完全相同的接口。

如果想访问的是非文本型的数据，可以使用 capsysbinary fixture，它的 readouterr() 方法返回的是字节流，可参考下面的例子。

代码如下：

```
# test_output.py

def test_binary_output(capsysbinary):
    print('hello')
    captured = capsysbinary.readouterr()
    assert captured.out == b'hello\n'
```

如果想临时去掉捕获行为,可以使用 capsys.disabled() 方法,它作为一个上下文管理器来使用,可以禁止 with 作用域中的捕获行为,可参考下面的例子。

代码如下:

```
def test_disabling_capturing(capsys):
    print("hello")
    with capsys.disabled():
        print("world")
    captured = capsys.readouterr()
    assert captured.out == "hello\n"
```

执行结果如图 12-22 所示。

图 12-22 在测试用例中访问捕获到的信息

12.5 Mock

工作场景:开发应用和服务常常需要依赖后端或者服务的接口或第三方接口,有时还没有开发完毕但需要返回数据进行前端调试,有时第三方接口不方便被调用,但还是希望知道调用情况。

举例来讲,开发移动应用 App,可能后端接口还在开发中,这时 App 的开发因为无法调用后端,调试很不方便。程序会依赖第三方的接口,例如微信支付功能在本地开发时不能直接调用。

如何解决依赖产生的痛点呢？通常有 3 种方法：第 1 种方法,使用代理工具实现 mock 服务器。第 2 种方法,在测试代码中编写 mock,编写 mock 服务器。第 3 种方法,真实编写临时接口返回数据。第 2 种方法就是本节要讲解的方法。

Mock 在 Python 3.x 中作为一个模块被内嵌到 unittest 标准库(现在基本被 pytest 代替)。简单地说,mock 就是制造假数据(对象)的模块,以此来模拟多种代码运行的情景,而无须真地发生这种情景。

测试人员通常使用 Mock 这个模块制造一些假的数据,完成测试。

12.5.1 使用 Mock 对象模拟测试情景

测试单元的代码实现,代码如下：

```
#filename: client.py

import requests

#该函数不属于测试范畴
#而是需要被模拟的 Python 对象
def send_request(URL):
    r = requests.get(URL)
    return r.status_code

#待测试的单元
#其功能是访问 URL
#存在两种结果
#访问成功:200
#访问失败:404
def visit_baidu():
    return send_request('http://www.baidu.com')
```

单元测试用例,代码如下：

```
#File: test_client.py
#filename: test_client.py

import mock
import sys
#sys.path.append('/Users/lindafang/PyCharmProjects/pytest_book/src/chapter12/Mock')
```

```
import client


class TestClient():

    def test_success_request(self):
        #测试访问生成的情况
        #实例化一个 Mock 对象,用于替换 client.send_request 函数
        #这个 Mock 对象会返回 HTTP Code 200
        success_send = mock.Mock(return_value = '200')
        client.send_request = success_send
        assert client.visit_baidu() == '200'

    def test_fail_request(self):
        #测试访问失败的情况
        #实例化一个 Mock 对象,用于替换 client.send_request 函数
        #这个 Mock 对象会返回 HTTP Code 404
        fail_send = mock.Mock(return_value = '404')
        client.send_request = fail_send
        assert client.visit_baidu() == '404'
```

在单元测试用例中通过构建模拟对象(Class Mock 的实例化)来模拟待测试代码中指定的 Python 对象的属性和行为,通过这种方式在单元测试用例中模拟出代码运行可能会发生的各种情况。

上述示例中将 client.visit_baidu() 作为测试单元,使用 Mock 对象 success_send/fail_send 模拟了 client.send_request() 的成功及失败并将结果返回。在测试单元中被模拟的 Python 对象,往往是这种"会发生变化的对象"或"通过外部接口获取的对象"。使用 mock 模块大致上可以总结出这样的流程:

确定测试单元中要模拟的 Python 对象,可以是一个类、一个函数、一个实例对象或者一个输入。

在编写单元测试用例的过程中,实例化 Mock 对象并设置其属性、行为与被替代的 Python 对象的属性、行为能够符合预期。例如:被调用时会返回预期的值,被访问实例属性时会返回预期的值等。

12.5.2　Mock 类的原型

```
class Mock(spec = None, side_effect = None, return_value = DEFAULT, wraps = None, name = None, spec_set = None, **kwargs)
```

name:命名一个 Mock 实例化对象,起到标识作用。

side_effect:指定一个可调用对象,一般为函数。当 Mock 实例化对象被调用时,如果该可调用对象返回的不是 DEFAULT,则以该可调用对象的返回值作为 Mock 实例化对象

调用的返回值。

return_value：显示指定返回一个值（或对象），当 Mock 实例化对象被调用时，如果 side_effect 指定的可调用对象返回的是 DEFAULT，则以 return_value 作为 Mock 实例化对象调用的返回值。

注意：更详细的信息建议查看官方文档。

12.5.3　MonkeyPatching 返回的对象：临时修改全局配置

有时候，测试用例需要调用某些依赖于全局配置的功能，或者这些功能本身又调用了某些不容易被测试的代码（例如：网络接入）。fixture monkeypatch 可以帮助用户安全地设置/删除一个属性、字典项或者环境变量，甚至改变导入模块时的 sys.path 路径。

monkeypatch 提供了以下方法：

```
monkeypatch.setattr(obj, name, value, raising = True)
monkeypatch.delattr(obj, name, raising = True)
monkeypatch.setitem(mapping, name, value)
monkeypatch.delitem(obj, name, raising = True)
monkeypatch.setenv(name, value, prepend = False)
monkeypatch.delenv(name, raising = True)
monkeypatch.syspath_prepend(path)
monkeypatch.chdir(path)
```

所有的修改将在测试用例或者 fixture 执行完成后被撤销。raising 参数表明：当设置/删除操作的目标不存在时，是否上报 KeyError 和 AttributeError 异常。

12.5.4　修改 MonkeyPatching 函数功能或者类属性

使用 monkeypatch.setattr() 可以将函数或者属性修改为所希望的结果，使用 monkeypatch.delattr() 可以删除测试用例所使用的函数或者属性。

代码如下：

```
#test_module.py
from pathlib import Path


def getssh():
    return Path.home() / ".ssh"


def test_getssh(monkeypatch):
    def mockreturn():
        return Path("/abc")
```

```
# 替换 Path.home
# 需要在真正调用之前执行
monkeypatch.setattr(Path, "home", mockreturn)

# 使用 mockreturn 代替 Path.home
x = getssh()
assert x == Path("/abc/.ssh")
```

在这个例子中,使用 monkeypatch.setattr() 修改 Path.home 方法,在测试运行期间,它一直返回的是固定的 Path("/abc"),这样就移除了它在不同平台上的依赖。测试运行完成后,对 Path.home 的修改会被撤销。使用 monkeypatch.setattr() 结合类,模拟函数的返回对象。

假设有一个简单的功能,访问一个 URL 并返回网页内容。

代码如下:

```python
# app.py
import requests


def get_json(URL):
    # 返回响应 json 格式
    r = requests.get(URL)
    return r.json()
```

代码如下:

```python
# File: test_app.py
import requests

import app

# 这是模拟类,返回响应
class MockResponse:

    # 模拟返回方法及返回值
    @staticmethod
    def json():
        return {"mock_key": "mock_response"}

def test_get_json(monkeypatch):

    # 任何参数都会返回这个模拟类
    def mock_get(*args, **kwargs):
```

```
        return MockResponse()

    # 应用 monkeypatch 替换 requests.get 为模拟 mock_get
    monkeypatch.setattr(requests, "get", mock_get)

    # app.get_json, 包括 requests.get, 应用 monkeypatch
    result = app.get_json("https://172.10.1.125")
    assert result["mock_key"] == "mock_response"
```

monkeypatch 将模拟应用于 requests.get 与 mock_get 功能。这个 mock_get 函数返回 MockResponse 类,其中有一个 json() 方法被定义为返回已知的测试字典,不需要任何外部 API 连接。

可以建立 MockResponse 为正在测试的场景使用适当的复杂性来初始化。例如,它可以包括 ok 始终返回的属性 True 或返回不同的值,json() 基于输入字符串的模拟方法。

fixture 可以跨用例共享 mock_response,代码如下:

```
# File: test_app_fixture.py
import pytest
import requests

# app.py 有 get_json()方法
import app

class MockResponse:
    @staticmethod
    def json():
        return {"mock_key": "mock_response"}

@pytest.fixture
def mock_response(monkeypatch):

    def mock_get(*args, **kwargs):
        return MockResponse()

    monkeypatch.setattr(requests, "get", mock_get)


def test_get_json(mock_response):
    result = app.get_json("https://172.10.1.125")
    assert result["mock_key"] == "mock_response"
```

注意:测试用例使用的 fixture 由原先的 mock_response 替换为 monkeypatch。

因为 monkeypatch 是 function 级别的作用域,所以 mock_response 也只能是 function 级别,否则会报 ScopeMismatch: You tried to access the 'function' scoped fixture 'monkeypatch' with a 'module' scoped request object 错误。

如果想让 mock_response 应用于所有的测试用例，可以考虑将它移到 conftest.py 文件中，并标记 autouse=True。

12.5.5　修改 MonkeyPatching 环境变量

如果使用的是环境变量，那么为了测试的目的，通常需要安全地更改这些值或从系统中删除它们。monkeypatch 提供了一种使用 setenv 和 delenv 的方法。

测试的代码如下：

```python
#File: code.py
import os


def get_os_user_lower():

    username = os.getenv("USER")

    if username is None:
        raise OSError("用户环境还没设置。")

    return username.lower()
```

有两条可能的路径。首先，USER 环境变量设置为值。其次，USER 环境变量不存在。使用 monkeypatch 两条路径都可以在不影响运行环境的情况下进行安全测试。

代码如下：

```python
#File: test_code.py

import pytest
from code import get_os_user_lower


def test_upper_to_lower(monkeypatch):

    monkeypatch.setenv("USER", "TestingUser")
    assert get_os_user_lower() == "testinguser"


def test_raise_exception(monkeypatch):

    monkeypatch.delenv("USER", raising=False)

    with pytest.raises(OSError):
        _ = get_os_user_lower()
```

也可以使用 fixture，实现跨用例共享。

代码如下：

```python
import pytest


@pytest.fixture
def mock_env_user(monkeypatch):
    monkeypatch.setenv("USER", "TestingUser")


@pytest.fixture
def mock_env_missing(monkeypatch):
    monkeypatch.delenv("USER", raising=False)


# notice the tests reference the fixtures for mocks
def test_upper_to_lower(mock_env_user):
    assert get_os_user_lower() == "testinguser"


def test_raise_exception(mock_env_missing):
    with pytest.raises(OSError):
        _ = get_os_user_lower()
```

12.5.6　修改 MonkeyPatching 字典

monkeypatch.setitem 可用于在测试期间安全地将字典值设置为特定值。以这个简化的连接字符串为例进行讲解。

在 app.py 中添加如下代码：

```python
DEFAULT_CONFIG = {"user": "user1", "Database": "db1"}


def create_connection_string(config=None):
    config = config or DEFAULT_CONFIG
    return f"User Id = {config['user']}; Location = {config['Database']};"
```

出于测试目的，可以修补 DEFAULT_CONFIG 特定值的字典。

代码如下：

```python
import app
def test_connection(monkeypatch):
```

```python
# 为测试设置默认值
monkeypatch.setitem(app.DEFAULT_CONFIG, "user", "test_user")
monkeypatch.setitem(app.DEFAULT_CONFIG, "Database", "test_db")

# 基于模拟设置预期结果
expected = "User Id = test_user; Location = test_db;"

# 调用,断言
result = app.create_connection_string()
assert result == expected
```

可以使用 monkeypatch.delitem 删除指定的项。

代码如下:

```python
def test_missing_user(monkeypatch):

    # 将user这个默认的配置属性删除
    monkeypatch.delitem(app.DEFAULT_CONFIG, "user", raising = False)
    # Key error expected because a config is not passed, and the
    # default is now missing the 'user' entry.
    # expected = "User Id = test_user; Location = test_db;"
    with pytest.raises(KeyError):
        _ = app.create_connection_string()
    # result = app.create_connection_string()
    # assert result == expected
```

fixture 共享,在 test_app.py 中添加下面的代码:

```python
@pytest.fixture
def mock_test_user(monkeypatch):

    monkeypatch.setitem(app.DEFAULT_CONFIG, "user", "test_user")

@pytest.fixture
def mock_test_Database(monkeypatch):

    monkeypatch.setitem(app.DEFAULT_CONFIG, "Database", "test_db")


@pytest.fixture
def mock_missing_default_user(monkeypatch):

    monkeypatch.delitem(app.DEFAULT_CONFIG, "user", raising = False)
```

```python
def test_connection(mock_test_user, mock_test_Database):

    expected = "User Id = test_user; Location = test_db;"

    result = app.create_connection_string()
    assert result == expected


def test_missing_user(mock_missing_default_user):

    with pytest.raises(KeyError):
        _ = app.create_connection_string()
```

12.6 钩子——Hook 方法的作用

pytest 提供了很多钩子（Hook）方法，如图 12-23 所示，这些方法方便对测试用例框架进行二次开发，可以根据自己的需求进行改造。

```
pytest.hookspec.pytest_pycollect_makemodule (Python 函数, in API引用)
pytest.hookspec.pytest_pyfunc_call (Python 函数, in API引用)
pytest.hookspec.pytest_report_collectionfinish (Python 函数, in API引用)
pytest.hookspec.pytest_report_header (Python 函数, in API引用)
pytest.hookspec.pytest_report_teststatus (Python 函数, in API引用)
pytest.hookspec.pytest_runtest_call (Python 函数, in API引用)
pytest.hookspec.pytest_runtest_logfinish (Python 函数, in API引用)
pytest.hookspec.pytest_runtest_logreport (Python 函数, in API引用)
pytest.hookspec.pytest_runtest_logstart (Python 函数, in API引用)
pytest.hookspec.pytest_runtest_makereport (Python 函数, in API引用)
pytest.hookspec.pytest_runtest_protocol (Python 函数, in API引用)
pytest.hookspec.pytest_runtest_setup (Python 函数, in API引用)
pytest.hookspec.pytest_runtest_teardown (Python 函数, in API引用)
pytest.hookspec.pytest_runtestloop (Python 函数, in API引用)
pytest.hookspec.pytest_sessionfinish (Python 函数, in API引用)
pytest.hookspec.pytest_sessionstart (Python 函数, in API引用)
pytest.hookspec.pytest_terminal_summary (Python 函数, in API引用)
pytest.hookspec.pytest_unconfigure (Python 函数, in API引用)
pytest.hookspec.pytest_warning_captured (Python 函数, in API引用)
pytest.hookspec.pytest_warning_recorded (Python 函数, in API引用)
```

图 12-23 部分 Hook 方法截图

12.6.1 pytest_runtest_makereport 修改测试报告内容

首先查看一下相关的源码，在_pytest/runner.py 文件下，可以在导入之后，进去查看。

```python
from _pytest import runner

# 对应源码
def pytest_runtest_makereport(item, call):
    """ return a :py:class:`_pytest.runner.TestReport` object
    for the given :py:class:`pytest.Item` and
    :py:class:`_pytest.runner.CallInfo`.
    """
```

这里 item 是测试用例，call 是测试步骤，具体执行过程如下：

先执行 when='setup'，返回 setup 的执行结果，然后执行 when='call'，返回 call 的执行结果，最后执行 when='teardown'，返回 teardown 的执行结果。

运行案例，代码如下：

```python
# conftest.py 写 pytest_runtest_makereport 内容，打印运行过程和运行结果

# conftest.py
import pytest


@pytest.hookimpl(hookwrapper=True, tryfirst=True)
def pytest_runtest_makereport(item, call):
    print('-----------------------------------------')

    # 获取钩子方法的调用结果
    out = yield
    print('用例执行结果', out)

    # 从钩子方法的调用结果中获取测试报告
    report = out.get_result()

    print('测试报告:%s' % report)
    print('步骤:%s' % report.when)
    print('nodeid:%s' % report.nodeid)
    print('description:%s' % str(item.function.__doc__))
    print(('运行结果: %s' % report.outcome))

# test_a.py 写一个简单的用例

def test_a():
    '''用例描述:test_a'''
    print("linda")
```

运行结果如下:

```
======================= test session starts ========================
platform darwin -- Python 3.6.8, pytest-6.0.1, py-1.9.0, pluggy-0.13.1
sensitiveURL: . *
rootdir: /Users/lindafang/PyCharmProjects/pytest_book/src/chapter12, configfile: pytest.ini
plugins: ...
collected 1 item

test_a.py
------------------------------------
用例执行结果 < pluggy.callers._Result object at 0x10effcdd8 >
测试报告:< TestReport 'hook/test_a.py::test_a' when = 'teardown' outcome = 'passed'>
步骤:teardown
nodeid:hook/test_a.py::test_a
description:用例描述:test_a
运行结果: passed

                                                           [100%]

===================== 1 passed in 0.04s ==========================
------------------------------------
用例执行结果 < pluggy.callers._Result object at 0x10effcdd8 >
测试报告:< TestReport 'hook/test_a.py::test_a' when = 'setup' outcome = 'passed'>
步骤:setup
nodeid:hook/test_a.py::test_a
description:用例描述:test_a
运行结果: passed
------------------------------------
用例执行结果 < pluggy.callers._Result object at 0x10effcc88 >
测试报告:< TestReport 'hook/test_a.py::test_a' when = 'call' outcome = 'passed'>
步骤:call
nodeid:hook/test_a.py::test_a
description:用例描述:test_a
运行结果: passed
.linda

Process finished with exit code 0
```

从运行结果可以看出,运行用例的过程会经历 3 个阶段:setup→call→teardown,每个阶段都会返回 Result 对象和 TestReport 对象,以及对象属性。

由于 setup 和 teardown 上面的用例默认都没有,所以其结果都是 passed。

使用 fixture 添加初始化操作。

给用例写个 fixture,以此增加用例的前置和后置操作,conftest.py 文件的代码如下:

```python
import pytest


@pytest.hookimpl(hookwrapper = True, tryfirst = True)
def pytest_runtest_makereport(item, call):
    print('------------------------------------')

    #获取钩子方法的调用结果
    out = yield
    print('用例执行结果', out)

    #从钩子方法的调用结果中获取测试报告
    report = out.get_result()

    print('测试报告:%s' % report)
    print('步骤:%s' % report.when)
    print('nodeid:%s' % report.nodeid)
    print('description:%s' % str(item.function.__doc__))
    print(('运行结果: %s' % report.outcome))


@pytest.fixture(scope = "session", autouse = True)
def fix_a():
    print("setup 前置操作")
    yield
    print("teardown 后置操作")
```

setup 失败情况：当 setup 执行失败时，setup 的执行结果为 failed，后面的 call 用例和 teardown 都不会被执行了，此时用例的状态是：error，也就是用例（call）还没开始执行，便异常了。

call 失败情况：虽然 setup 正常执行，但是测试用例 call 失败了。

代码如下：

```python
@pytest.fixture(scope = "session", autouse = True)
def fix_a():
    print("setup 前置操作")
    yield
    print("teardown 后置操作")
#test_a.py用例

def test_a():
    '''用例描述:test_a'''
    print("linda")
    assert 1 == 0
```

此时运行的结果为 failed，如图 12-24 所示。

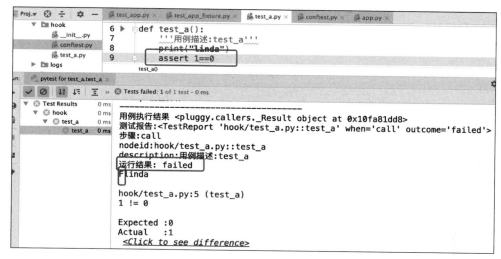

图 12-24　call 步骤失败的执行结果

teardown 失败情况：虽然 setup 正常执行，并且测试用例 call 正常执行，但 teardown 失败了。

代码如下：

```
@pytest.fixture(scope = "session", autouse = True)
def fix_a():
    print("setup 前置操作")
    yield
    print("teardown 后置操作")
    raise Exception("teardown 失败了")
# teat_a.py 用例

def test_a():
    '''用例描述:test_a'''
    print("linda")
```

只获取 call 的结果。

在写用例的时候，如果保证 setup 和 teardown 不报错，并且只关注测试用例本身的运行结果，则前面的 pytest_runtest_makereport 钩子方法执行了 3 次。

可以加个判断语句：if report.when=="call"。

代码如下：

```
import pytest
from _pytest import runner
'''
```

```python
# 对应源码
def pytest_runtest_makereport(item, call):
    """ return a :py:class:`_pytest.runner.TestReport` object
    for the given :py:class:`pytest.Item` and
    :py:class:`_pytest.runner.CallInfo`.
    """
'''

@pytest.hookimpl(hookwrapper = True, tryfirst = True)
def pytest_runtest_makereport(item, call):
    print('----------------------------------------')

    # 获取钩子方法的调用结果
    out = yield
    # print('用例执行结果', out)

    # 从钩子方法的调用结果中获取测试报告
    report = out.get_result()
    if report.when == "call":
        print('测试报告:%s' % report)
        print('步骤:%s' % report.when)
        print('nodeid: %s' % report.nodeid)
        print('description:%s' % str(item.function.__doc__))
        print(('运行结果: %s' % report.outcome))

@pytest.fixture(scope = "session", autouse = True)
def fix_a():
    print("setup 前置操作")
    yield
    print("teardown 后置操作")
```

12.6.2　pytest_collection_modifyitems 改变用例执行顺序

测试人员一直想改变 pytest 用例的执行顺序，实际上在测试用例设计原则上就不要依赖顺序。如果非要改变顺序，则可以用第三方的插件实现，即通过钩子函数实现。

pytest 默认执行用例先根据项目下的文件夹名称按 ASCII 码去收集，module 里面的用例是从上往下执行的。

pytest_collection_modifyitems 这个钩子函数顾名思义可以改变用例的执行顺序。

pytest_collection_modifyitems 的功能是当测试用例收集完成后，可以改变测试用例集合(items)的顺序。

代码如下：

```python
def pytest_collection_modifyitems(session, config, items):
    '''called after collection is completed.
```

```
you can modify the ``items`` list
:param _pytest.main.Session session: the pytest session object
:param _pytest.config.Config config: pytest config object
:param List[_pytest.nodes.Item] items: list of item objects
'''
```

items 是用例对象的一个列表,只要改变 items 里面用例的顺序就可以改变用例的执行顺序了。

1. pytest 默认执行顺序

先设计一个简单的 pytest 项目,有 a 和 b 两个包,分别在 test_a.py 和 test_b.py 文件中写测试用例。

代码如下:

```
# conftest.py 内容

import pytest

def pytest_collection_modifyitems(session, items):
    print("收集到的测试用例:%s" % items)


# test_a.py 内容

def test_a_1():
    print("测试用例 a_1")

def test_a_2():
    print("测试用例 a_2")
# test_b.py 内容

def test_b_2():
    print("测试用例 b_2")

def test_b_1():
    print("测试用例 b_1")
```

运行完成后可以看到收集到的测试用例会在测试用例开始时执行。

```
pytest test_aaa.py test_bbb.py
========================= test session starts =========================
platform darwin -- Python 3.6.8, pytest-6.0.1, py-1.9.0, pluggy-0.13.1
sensitiveURL: .*
rootdir: /Users/lindafang/PyCharmProjects/pytest_book/src/chapter12, configfile: pytest.ini
```

```
plugins: ...
collecting ... 收集到的测试用例:[<Function test_a_1>,<Function test_a_2>,<Function test_
b_2>,<Fuest_b_1>]
collected 4 items

test_aaa.py ..
[ 50%]
test_bbb.py ..
[100%]

================================== 4 passed in 0.04s =================
```

从结果可以看出运行的时候先按模块名称的 ASCII 码去收集,单个 py 文件里面的用例按从上到下的顺序收集。

2. items 用例排序

如果笔者想改变上面用例的执行顺序,则应以用例名称的 ASCII 码进行排序。先获取用例的名称,以用例名称进行排序就可以了。

代码如下:

```
def pytest_collection_modifyitems(session, items):
    print(type(items))
    print("收集到的测试用例:%s" % items)
    # sort 排序,根据用例名称 item.name 排序
    items.sort(key = lambda x: x.name)
    print("排序后的用例:%s" % items)
    for item in items:
        print("用例名:%s" % item.name)
```

重新执行后的结果:

```
pytest test_aaa.py test_bbb.py
========================= test session starts =========================
platform darwin -- Python 3.6.8, pytest-6.0.1, py-1.9.0, pluggy-0.13.1
sensitiveURL: .*
rootdir: /Users/lindafang/PyCharmProjects/pytest_book/src/chapter12, configfile: pytest.ini
plugins: ...
collecting ... <class 'list'>
收集到的测试用例:[<Function test_a_1>,<Function test_a_2>,<Function test_b_2>,
<Function test_b_1>]
排序后的用例:[<Function test_a_1>,<Function test_a_2>,<Function test_b_1>,<Function
test_b_2>]
用例名:test_a_1
用例名:test_a_2
用例名:test_b_1
```

```
用例名:test_b_2
collected 4 items

test_aaa.py ..
[ 50%]
test_bbb.py ..
[100%]

======================= 4 passed in 0.05s ================
```

重新排序后就可以按用例的名称顺序执行了。

12.6.3 pytest_terminal_summary

用例执行完成后,希望能获取执行的结果,这样方便快速统计用例的执行情况。

也可以把获取的结果当成总结报告,发邮件的时候可以先统计测试结果,再加上 html 的报告。

关于 TerminalReporter 类可以在_pytest.terminal 中查看最后的结果汇总,可以获得所有的执行结果。

```
from _pytest import terminal

pytest_terminal_summary(terminalreporter, exitstatus, config)
#参数:
# - terminalreporter (_pytest.terminal.TerminalReporter) - 内部使用的终端测试报告对象
# - exitstatus (int) - 返回操作系统的返回码
# - config(_pytest.config.Config) - pytest config 对象
```

TerminalReporter 部分代码如下:

```
class TerminalReporter(object):
    def __init__(self, config, file = None):
        import _pytest.config

        self.config = config
        self._numcollected = 0
        self._session = None
        self._showfspath = None

        self.stats = {}
        self.startdir = config.invocation_dir


    def report_collect(self, final = False):
```

```python
        if self.config.option.verbose < 0:
            return

        if not final:
            # Only write "collecting" report every 0.5s.
            t = time.time()
            if (
                self._collect_report_last_write is not None
                and self._collect_report_last_write > t - REPORT_COLLECTING_RESOLUTION
            ):
                return
            self._collect_report_last_write = t

        errors = len(self.stats.get("error", []))
        skipped = len(self.stats.get("skipped", []))
        deselected = len(self.stats.get("deselected", []))
        selected = self._numcollected - errors - skipped - deselected
        if final:
            line = "collected "
        else:
            line = "collecting "
        line += (
            str(self._numcollected) + " item" + ("" if self._numcollected == 1 else "s")
        )
        if errors:
            line += " / %d errors" % errors
        if deselected:
            line += " / %d deselected" % deselected
        if skipped:
            line += " / %d skipped" % skipped
        if self._numcollected > selected > 0:
            line += " / %d selected" % selected
        if self.isatty:
            self.rewrite(line, bold=True, erase=True)
            if final:
                self.write("\n")
        else:
            self.write_line(line)
```

在 conftest.py 文件中写一个 pytest_terminal_summary 函数，用于收集测试结果：

```python
import time
from _pytest import terminal


def pytest_terminal_summary(terminalreporter, exitstatus, config):
```

```
    '''收集测试结果'''
    print(terminalreporter.stats)
    print("total:", terminalreporter._numcollected)
    print('passed:', len(terminalreporter.stats.get('passed', [])))
    print('failed:', len(terminalreporter.stats.get('failed', [])))
    print('error:', len(terminalreporter.stats.get('error', [])))
    print('skipped:', len(terminalreporter.stats.get('skipped', [])))
    #terminalreporter._sessionstarttime 会话开始时间
    duration = time.time() - terminalreporter._sessionstarttime
    print('total times:', duration, 'seconds')
```

运行结果如下：

```
pytest test_aaa.py test_bbb.py
========================= test session starts =========================
platform darwin -- Python 3.6.8, pytest-6.0.1, py-1.9.0, pluggy-0.13.1
sensitiveURL: .*
rootdir: /Users/lindafang/PyCharmProjects/pytest_book/src/chapter12, configfile: pytest.ini
plugins: rerunfailures-9.0, forked-1.0.2, pep8-1.0.6, apiritif-0.9.4, variables-
1.9.0, emoji-0.2.0, flakes-4.0.0, base-URL-1.4.2, allure-pytest-2.8.18, cov-2.10.0,
html-2.1.1, xdist-1.34.0, ordering-0.6, assume-2.3.3, selenium-1.17.0, Django-4.1.0,
freezegun-0.4.1, metadata-1.8.0, bdd-3.2.1
collected 10 items

test_bbb.py ...F
[100%]
test_aaa.py ...s.F
[ 60%]
============================ FAILURES =========================
_____ test_4 _____
    def test_4():
        print("测试用例 44444444")
>       assert 1 == 2
E       assert 1 == 2

test_aaa.py:29: AssertionError
------------------------ Captured stdout call ------------------------
测试用例 44444444
_____ test_6 _____
    def test_6():
        print("测试用例 66666666")
        time.sleep(3)
>       assert 1 == 2
E       assert 1 == 2
```

```
test_bbb.py:21: AssertionError
------------------------------ Captured stdout call ------------------------------
测试用例 66666666
{'': [<TestReport 'hook/test_aaa.py::test_a_1' when='setup' outcome='passed'>, <TestReport
'hook/test_aaa.py::test_a_1' when='teardown' outcome='passed'>, <TestReport 'hook/test_aaa.
py::test_a_2' when='setup' outcome='passed'>, <TestReport 'hook/test_aaa.py::test_a_2' when=
'teardown' outcome='passed'>, <TestReport 'hook/test_aaa.py::test_1' when='setup' outcome=
'passed'>, <TestReport 'hook/test_aaa.py::test_1' when='teardown' outcome='passed'>,
<TestReport 'hook/test_aaa.py::test_2' when='teardown' outcome='passed'>, <TestReport 'hook/
test_aaa.py::test_3' when='setup' outcome='passed'>, <TestReport 'hook/test_aaa.py::test_3'
when='teardown' outcome='passed'>, <TestReport 'hook/test_aaa.py::test_4' when='setup'
outcome='passed'>, <TestReport 'hook/test_aaa.py::test_4' when='teardown' outcome='passed'>,
<TestReport 'hook/test_bbb.py::test_b_2' when='setup' outcome='passed'>, <TestReport 'hook/test_
bbb.py::test_b_2' when='teardown' outcome='passed'>, <TestReport 'hook/test_bbb.py::test_b_1'
when='setup' outcome='passed'>, <TestReport 'hook/test_bbb.py::test_b_1' when='teardown'
outcome='passed'>, <TestReport 'hook/test_bbb.py::test_5' when='setup' outcome='passed'>,
<TestReport 'hook/test_bbb.py::test_5' when='teardown' outcome='passed'>, <TestReport 'hook/
test_bbb.py::test_6' when='setup' outcome='passed'>, <TestReport 'hook/test_bbb.py::test_6'
when='teardown' outcome='passed'>], 'passed': [<TestReport 'hook/test_aaa.py::test_a_1' when=
'call' outcome='passed'>, <TestReport 'hook/test_aaa.py::test_a_2' when='call' outcome=
'passed'>, <TestReport 'hook/test_aaa.py::test_1' when='call' outcome='passed'>, <TestReport
'hook/test_aaa.py::test_3' when='call' outcome='passed'>, <TestReport 'hook/test_bbb.py::
test_b_2' when='call' outcome='passed'>, <TestReport 'hook/test_bbb.py::test_b_1' when='call
' outcome='passed'>, <TestReport 'hook/test_bbb.py::test_5' when='call' outcome='passed'>],
'skipped': [<TestReport 'hook/test_aaa.py::test_2' when='setup' outcome='skipped'>], 'failed':
[<TestReport 'hook/test_aaa.py::test_4' when='call' outcome='failed'>, <TestReport 'hook/
test_bbb.py::test_6' when='call' outcome='failed'>]}
total: 10
passed: 7
failed: 2
error: 0
skipped: 1
total times: 6.123329162597656 seconds
======================== short test summary info ==========================
FAILED test_aaa.py::test_4 - assert 1 == 2
FAILED test_bbb.py::test_6 - assert 1 == 2
==================== 2 failed, 7 passed, 1 skipped in 6.12s ==================
```

setup 和 teardown 异常情况：如果 setup 出现异常，则可以对 test_bbb.py 文件的代码进行修改，代码如下：

```
import time
import pytest
```

```python
@pytest.fixture(scope = "function")
def setup_demo():
    raise TypeError("ERROR!")


def test_5(setup_demo):
    print("测试用例 55555555")
    time.sleep(3)


def test_6():
    print("测试用例 66666666")
    time.sleep(3)
    assert 1 == 2
```

重新运行用例,结果如下:

```
total: 10
passed: 7
failed: 2
error: 0
skipped: 1
total times: 6.123329162597656 seconds
```

此时统计结果没什么问题,接下来查看 teardown 的异常情况。

代码如下:

```python
import time
import pytest


@pytest.fixture(scope = "function")
def setup_demo():
    yield
    raise TypeError("ERROR!")
```

运行结果如下:

```
total: 10
passed: 7
failed: 2
error: 0
skipped: 1
total times: 6.1097190380096436 seconds
```

12.7　本章小结

本章主要讲解了 pytest.ini 的各种配置及与警告相关配置、测试临时目录的配置，以及 Mock 和 Hook，具体内容如下：

（1）pytest.ini 的各种配置。

（2）与警告相关配置的设置。

（3）临时路径和目录的配置。

（4）输出与捕获级别配置。

（5）Mock 的配置使用。

（6）Hook 方法的使用例子。

图 书 推 荐

书 名	作 者
鸿蒙应用程序开发	董昱
鸿蒙操作系统开发入门经典	徐礼文
鸿蒙操作系统应用开发实践	陈美汝、郑森文、武延军、吴敬征
华为方舟编译器之美——基于开源代码的架构分析与实现	史宁宁
鲲鹏架构入门与实战	张磊
华为 HCIA 路由与交换技术实战	江礼教
Flutter 组件精讲与实战	赵龙
Flutter 实战指南	李楠
Dart 语言实战——基于 Flutter 框架的程序开发(第 2 版)	亢少军
Dart 语言实战——基于 Angular 框架的 Web 开发	刘仕文
IntelliJ IDEA 软件开发与应用	乔国辉
Vue+Spring Boot 前后端分离开发实战	贾志杰
Vue.js 企业开发实战	千锋教育高教产品研发部
Python 人工智能——原理、实践及应用	杨博雄 主编,于营、肖衡、潘玉霞、高华玲、梁志勇 副主编
Python 深度学习	王志立
Python 异步编程实战——基于 AIO 的全栈开发技术	陈少佳
物联网——嵌入式开发实战	连志安
智慧建造——物联网在建筑设计与管理中的实践	[美]周晨光(Timothy Chou)著;段晨东、柯吉译
TensorFlow 计算机视觉原理与实战	欧阳鹏程、任浩然
分布式机器学习实战	陈敬雷
计算机视觉——基于 OpenCV 与 TensorFlow 的深度学习方法	余海林、翟中华
深度学习——理论、方法与 PyTorch 实践	翟中华、孟翔宇
深度学习原理与 PyTorch 实战	张伟振
ARKit 原生开发入门精粹——RealityKit + Swift + SwiftUI	汪祥春
Altium Designer 20 PCB 设计实战(视频微课版)	白军杰
Cadence 高速 PCB 设计——基于手机高阶板的案例分析与实现	李卫国、张彬、林超文
SolidWorks 2020 快速入门与深入实战	邵为龙
UG NX 1926 快速入门与深入实战	邵为龙
西门子 S7-200 SMART PLC 编程及应用(视频微课版)	徐宁、赵丽君
三菱 FX3U PLC 编程及应用(视频微课版)	吴文灵
全栈 UI 自动化测试实战	胡胜强、单镜石、李睿
软件测试与面试通识	于晶、张丹
深入理解微电子电路设计——电子元器件原理及应用(原书第 5 版)	[美]理查德·C. 耶格(Richard C. Jaeger)、[美]特拉维斯·N. 布莱洛克(Travis N. Blalock)著;宋廷强 译
深入理解微电子电路设计——数字电子技术及应用(原书第 5 版)	[美]理查德·C. 耶格(Richard C. Jaeger)、[美]特拉维斯·N. 布莱洛克(Travis N. Blalock)著;宋廷强 译
深入理解微电子电路设计——模拟电子技术及应用(原书第 5 版)	[美]理查德·C. 耶格(Richard C. Jaeger)、[美]特拉维斯·N. 布莱洛克(Travis N. Blalock)著;宋廷强 译